Advances in Experimental Medicine and Biology

Volume 1204

Advances in Experimental Medicine and Biology provides a platform for scientific contributions in the main disciplines of the biomedicine and the life sciences. This series publishes thematic volumes on contemporary research in the areas of microbiology, immunology, neurosciences, biochemistry, biomedical engineering, genetics, physiology, and cancer research. Covering emerging topics and techniques in basic and clinical science, it brings together clinicians and researchers from various fields.

Advances in Experimental Medicine and Biology has been publishing exceptional works in the field for over 40 years, and is indexed in SCOPUS, Medline (PubMed), Journal Citation Reports/Science Edition, Science Citation Index Expanded (SciSearch, Web of Science), EMBASE, BIOSIS, Reaxys, EMBiology, the Chemical Abstracts Service (CAS), and Pathway Studio.

2018 Impact Factor: 2.126.

More information about this series at http://www.springer.com/series/5584

Shie-Liang Hsieh
Editor

Lectin in Host Defense Against Microbial Infections

Springer

Editor
Shie-Liang Hsieh
Genomics Research Center
Academia Sinica
Taipei, Taiwan

ISSN 0065-2598 ISSN 2214-8019 (electronic)
Advances in Experimental Medicine and Biology
ISBN 978-981-15-1579-8 ISBN 978-981-15-1580-4 (eBook)
https://doi.org/10.1007/978-981-15-1580-4

This Springer imprint is published by the registered company Springer Nature Singapore Pte Ltd.
The registered company address is: 152 Beach Road, #21-01/04 Gateway East, Singapore 189721, Singapore

Foreword

Lectin proteins recognise carbohydrate structures and those derived from plant sources such as ricin, wheat germ agglutinin and Concanavilin A have been of interest to immunologists for some time. More recently, lectin-like molecules in mammalian species have become widely studied as mediators of cellular recognition and responses such as adhesion, phagocytosis and cytotoxicity. They contribute to host–pathogen interactions, both invasion and evasion strategies, providing clues to ongoing evolution. Their ability to discriminate self from non-and modified self has been of perennial interest to immunologists. In combination with other membrane molecules, they contribute to homeostasis, inflammation and repair, with special reference to innate, acquired and autoimmunity; lectin receptors and potential ligands are expressed by all haematopoietic cells on the plasma membrane for extracellular recognition, and in cytoplasm during membrane glycoprotein and glycolipid biosynthesis and processing. The discipline of Glycobiology has grown enormously with technical advances in glycomic and structural analysis, facilitated by immunochemical tools such as monoclonal antibodies and contemporary genomics, cell biology and in vivo gene manipulation.

The present volume edited by Shie-Liang Edmond Hsieh from the Academia Sinica in the Republic of China (Taiwan) fulfils an unmet need in focussing on an important subgroup of receptors with canonical C-Type Lectin (CLEC) domains. The field of glycobiology research has flourished in Taiwan, with the return to Chang Gung University of Albert M Wu, a collaborator of Elvin Kabat at Columbia University in New York, and Chi-Huey Wong who was the founder director of the Genomics Research Center and President of Academia Sinica, which is the most prestigious research institute in Taiwan, from the development of a major research centre in Life Sciences at the Academia Sinica in Taipei. Shie-Liang Hsieh received his medical education at the National Yang-Ming University, and a doctorate in Biochemistry in Porter's laboratory at Oxford. After postdoctoral studies with Hugh McDevitt at Stanford, he became a professor at National Yang-Ming University (from 1993–2003) and a Distinguished Research Fellow (from 2013 to the present) at the Academia Sinica, with broad research interests in Glycoimmunology, host–pathogen interactions and cancer immunology. He has published extensively on the role of glycan recognition in Dengue and other viral infections, by CLEC5A, and on a Decoy receptor 3 with pleiotropic roles in immunomodulation and immunopathology. In addition to several Taiwanese colleagues, notably Fu-Tong Liu, a galectin specialist, has established close ties with scientists in China, Japan, the UK and USA, as is evident from the list of contributors to the present volume.

Highlights of Individual Chapters (Corresponding Authors) Chapter 1 C-type lectin receptors (CLRs) in antifungal immunity, GD Brown, Aberdeen University

After a broad introductory summary of CLRs in innate and adaptive immunity, 4 fungal infections—Candida albicans, Aspergillus fumigatus, Cryptococcus neoformans and Pneumocystis jiravecii are discussed, dealing with fungal wall composition, receptor–ligand interactions and signalling pathways. Innate receptors implicated in microbial recognition include CLRs, Toll-, RIGI- and NOD-like Receptors, as well as Collectins. The authors describe a novel melanin sensing receptor, MelLec. They review CLR structure, phylogeny and expression on myeloid cells. Signalling, direct or via the FcR gamma chain, induces a range of effector responses. Activatory or Inhibitory ITAM or ITIM motifs initiate kinase and phosphatase activation, respectively.

Chapter 2 Immune recognition of pathogen-derived glycolipids through Mincle, Y Miyake, Saga University, Japan

Ligands for Mincle include pathogen glycolipids and trehalose dimycolate (TDM) induced activation or inhibition depends on the setting. The Mincle receptor gene cluster includes Dectin-2, MCL, a Mincle partner and DCAR. The role of Mincle in Mycobacterial, Klebsiella and Strep pneumonia infections is considered in detail. Leishmania major releases soluble Mincle ligand, a possible evasion mechanism. Sterile ligands include damaged cells and cholesterol crystals. Mycobacterial adjuvant may enhance vaccine efficacy, especially in early life.

Chapter 3 CLEC5A: A promiscuous pattern recognition receptor to microbes and beyond, S-L Hsieh, Academia Sinica, Taiwan

CLEC5A is a promiscuous, low-affinity CLR; multivalent heterocomplexes with DC-SIGN and mannose receptor to activate Syk. This CLR, together with TLR2, has been implicated in the cytokine storm and neutrophil extracellular trap formation associated with the Dengue shock syndrome; Clec5A can participate in inflammasome activation, neurologic, vasculitis and osteoclast complications of Dengue viral infection. Microbial ligands include other viral glycans and Listeria monocytogenes coactivate CLEC5A/TLR2 complex to induce the differentiation of gamma/delta Th17 cells in intestine.

Chapter 4 Collectins: Innate immune pattern recognition molecules, U Kishore, Brunel University, UK

These soluble lectins include Mannose-binding lectin (MBL), bovine conglutinins and Surfactant proteins A and D, produced in lung and other tissues, e.g. placenta (CL-P1). Collectins mediate microbial agglutination, neutralisation, opsonisation and complement activation; they modulate inflammatory responses and link innate and adaptive immunity. Collagenous polypeptides and multimerisation play an important role in these activities. MBL is known to associate with serum proteases (Masps). An extensive description includes collectin structures, sources and induction in lung and beyond. Collectin ligands are broad, mannose-like or galactose-like, or even non glycans; they are present in various bacteria, viruses, fungi, parasites and helminths. Binding may occur to a C1Q Receptor protein and/or calreticulin. Collectins can play a role as anti-opsonins as well as in opsonisation. Human deficiency and possible functional consequences are described.

Chapter 5 Insect C-type lectins in microbial infections, G Cheng, Tsinghua University, China

Insect lectins are involved in Innate immunity. They are pleiotropic and contribute to recognition, agglutination, encapsulation, melanisation, phagocytosis by hemocytes and propenol oxidation. Lectin–microbiome interactions may contribute to homeostasis. Insects provide genetically tractable models to study infections by bacteria, viruses, parasites and fungi. Apart from host defence, it is possible that lectins facilitate the transmission of infection.

Chapter 6 Galectins in host defense against microbial infections, F-T Liu Academia Sinica, Taiwan

Secretion of galectins occurs by unidentified mechanisms. Various galectins contribute to pathogen recognition, innate and adaptive immunity. An intriguing function to promote autophagy has been described for Intracellular galectins. Gal8 binding has been demonstrated to luminal host glycans of ruptured endo-lysosomes. Ligands of galectins include bacteria, viruses such as HIV and influenza, fungi, parasites and helminths. Galectins may play a role in the budding and transmission of viruses. The authors caution against using galectins exogenously added to cells to study the functions of endogenous galectins, and remind us that galectins are mainly located inside the cells and may bind to intracellular proteins independent of glycans.

Chapter 7 Galectins in host–pathogen interactions: Structural, functional and evolutionary aspects, GR Vasta, University of Maryland, USA

Pathogens use galectins to gain entry into host cells and also to evade effector mechanisms, with implications for evolutionary selection. The author argues for involvement of galectins in self/non-self-recognition and a role in the infection of Dendritic Cells by HIV. A Zebra fish model has been used to study galectin function in vivo.

Chapter 8 Siglecs at the host–pathogen interface, YC Chang, National Taiwan University, Taiwan

This chapter is written in collaboration with V Nizet, an expert on PMN. There are 2 phylogenetic families of Siglecs, named after the combination of Immunoglobulin superfamily and lectin domains (Crocker et al, Glycobiology, 1998, v8): CD33- and Siglec-1-related, there are 14 in human and 9 in the mouse, with a single ligand-binding Vset domain. Siglecs have no functional intracellular domain; ITIM domains generate inhibitory signals through SHP phosphatase, activating ITAM domain is contributed by the adaptor DAP 12. An example given of pathogen interaction is Siglec 9 which downregulates the respiratory burst in human PMN in response to Group B Streptococcal infection, the sialylated residues of capsular polysaccharide conferring resistance to killing. Soluble Siglec 9 provides a decoy signal which renders PMN more resistant to infection. This may set the stage for an evolutionary dynamic, to counteract inhibitory receptor expression. Whereas medically important bacteria and viruses can take advantage of Sialic acid–Siglec interactions. Highly restricted Siglec-1 (CD169) expression by metallophilic macrophages in mouse spleen can serve as the reservoir for systemic Streptococcal pneumonia infection. A loss of function truncation variant of the Siglec-1 gene is thought to be relatively common in human populations. Narrow cell tropism of Siglec-1 may account for striking species specificity of PPRRSV, a devastating porcine reproductive and respiratory syndrome virus infection.

Chapter 9 Siglecs that associate with DAP12, T Angata, Academia Sinica, Taiwan

Polymorphisms of receptor pairs may contribute to bacterial susceptibility or resistance. The author, a former postdoc with A Varki, discusses the concept that immune cells may sense sialic acid residues on their own cells, perhaps relevant to auto immunity. Factor H, a regulator of complement, can bind Siglec-1. Another important molecule that associates with DAP-12 is TREM, implicated in a Neurologic genetic disorder and Alzheimer's disease.

PERSPECTIVE

The focus of this volume on a variety of CTL and a range of host-pathogens provides a valuable resource of the existing literature, and a perspective on common themes as well as the specificity or redundancy of lectin–glycan interactions. Lectins constitute only a subset of potential membrane, cytoplasmic or soluble pathogen receptors, which include opsonic (Fc and complement), Toll-like and other non-opsonic sensors, such as various Scavenger Receptors (SR) A and B, NOD-Like (NLR), RIG-like (RLR) and DNA-recognition molecules, among others. The lectin-like receptors can be relatively selective for terminal sialyl, mannosyl and galactosyl structures, but tend to be of low affinity and promiscuous. They can signal directly e.g. Dectin-1 which activates a well-defined syk and CARD 9 pathway through an ITAM domain, or depend on a partner transmembrane protein such as DAP12 or the FcR common gamma chain to signal; Inhibitory receptors contain an ITIM motif. Structural diversity endows myeloid cell populations, for example, with functional plasticity.

Bacteria and fungi provide a great deal of scope for recognition of foreign, often complex structures, distinct from self, in contrast with viral glycoproteins which incorporate host glycans, and tumour cells which are subtly modified self, at best. The role of the highly polymorphic MHC complex in CLR biology has not been sufficiently studied, by contrast with the lectin-dependent activation pathway of the complement cascade. Nevertheless, the CLRs have been amply demonstrated to play a vital role in viral entry and transmission, not least by the studies of Shie-Liang Hsieh.

Another theme of this collection of articles is the role of multimerisation and complex formation of CLR polypeptides with other host molecules during cell adhesion, phagocytosis and receptor-mediated endocytosis. The cellular biology of these molecular interactions is now amenable to high-resolution study. The use of genetic depletion of defined CLRs will be of considerable value in uncovering pathogen invasion and evasion strategies.

The variation in CLR genes and their expression in different species is suggestive of ongoing evolution, e.g. among primates and other mammals, resulting from gene duplication, mutation and selection pressure by infection. While it is essential to study this in all species, including invertebrates and microorganisms themselves, it also means that rodent and insect models may not reflect human CLR functions. Natural variants in human populations deserve further study and experimental methods are now available to delete selected genes in human immune cells in vitro, for example, single cell mRNA analysis of human tissues and immunohistochemical and FACS analysis of CLR expression in situ and ex vivo will provide much-needed further evidence of tissue specificity of CLR expression. Such information will yield new tools for diagnosis and hopefully therapy.

The restricted focus of the present volume was a wise choice, which confined much of the content to the immune system and pathogens. There are tantalising hints that CLR biology extends beyond immunity, to all aspects of development, physiology and pathogenesis. When sufficient information accrues, it may be worthwhile to undertake a follow-up companion volume along these lines. In the meantime, all the contributors have performed a valuable service in summarising what is known, in order to catalyse a deeper understanding of structures and functions of the role of CLRs in immune homeostasis and possible therapy.

Siamon Gordon
Chang Gung University, Taoyuan
City Taiwan
University of Oxford, Oxford UK

Contents

Contributors

Takashi Angata Institute of Biological Chemistry, Academia Sinica, Taipei, Taiwan

Emerson S. Bernardes Nuclear and Energy Research Institute - IPEN, São Paulo, Brazil

Gordon D. Brown Medical Research Council Centre for Medical Mycology, University of Exeter, Exeter, UK;
Previous address: MRC Centre for Medical Mycology, University of Aberdeen, Aberdeen, UK

Wei-Chiao Chang School of Pharmacy, Taipei Medical University, Taipei, Taiwan

Yung-Chi Chang Graduate Institute of Microbiology, College of Medicine, National Taiwan University, Taipei, Taiwan

Gong Cheng Tsinghua-Peking Joint Center for Life Sciences, Beijing Advanced Innovation Center for Structural Biology, School of Medicine, Tsinghua University, Beijing, China;
Institute of Pathogenic Organisms, Shenzhen Center for Disease Control and Prevention, Shenzhen, Guangdong, China

Anthony G. Tsolaki College of Health and Life Sciences, Brunel University London, London, UK

Shie-Liang Hsieh School of Medicine, Genomics Research Center, Academia Sinica, Taipei, Taiwan

Uday Kishore College of Health and Life Sciences, Brunel University London, London, UK

Fang-Yen Li Institute of Biomedical Sciences, Academia Sinica, Taipei, Taiwan

Fu-Tong Liu Institute of Biomedical Sciences, Academia Sinica, Taipei, Taiwan

Yasunobu Miyake Division of Molecular and Cellular Immunoscience, Department of Biomolecular Sciences, Faculty of Medicine, Saga University, Saga, Japan

Valarmathy Murugaiah College of Health and Life Sciences, Brunel University London, London, UK

Christina Nikolakopoulou Medical Research Council Centre for Medical Mycology, University of Exeter, Exeter, UK;
Previous address: MRC Centre for Medical Mycology, University of Aberdeen, Aberdeen, UK

Victor Nizet Division of Host-Microbe Systems and Therapeutics, Department of Pediatrics, and Skaggs School of Pharmacy and Pharmaceutical Sciences, UC San Diego, La Jolla, CA, USA

Pei-Shan Sung Genomics Research Center, Academia Sinica, Taipei, Taiwan

Gerardo R. Vasta Department of Microbiology and Immunology, University of Maryland School of Medicine, UMB, Baltimore, MD, USA;
Institute of Marine and Environmental Technology, Columbus Center, University of Maryland, Baltimore, MD, USA

Sheng-Fan Wang Department of Medical Laboratory Science and Biotechnology, Kaohsiung Medical University, Kaohsiung, Taiwan

Janet A. Willment Medical Research Council Centre for Medical Mycology, University of Exeter, Exeter, UK;
Previous address: MRC Centre for Medical Mycology, University of Aberdeen, Aberdeen, UK

Sho Yamasaki Department of Molecular Immunology, Research Institute for Microbial Diseases, Osaka University, Suita, Japan;
Laboratory of Molecular Immunology, Immunology Frontier Research Center, Osaka University, Osaka, Japan

Xi Yu Tsinghua-Peking Joint Center for Life Sciences, Beijing Advanced Innovation Center for Structural Biology, School of Medicine, Tsinghua University, Beijing, China;
School of Life Science, Tsinghua University, Beijing, China

Yibin Zhu Tsinghua-Peking Joint Center for Life Sciences, Beijing Advanced Innovation Center for Structural Biology, School of Medicine, Tsinghua University, Beijing, China;
Institute of Pathogenic Organisms, Shenzhen Center for Disease Control and Prevention, Shenzhen, Guangdong, China;
School of Life Science, Tsinghua University, Beijing, China

Chapter 1
C-Type Lectin Receptors in Antifungal Immunity

Christina Nikolakopoulou, Janet A. Willment and Gordon D. Brown

Abstract Most fungal species are harmless to humans and some exist as commensals on mucocutaneous surfaces. Yet many fungi are opportunistic pathogens, causing life-threatening invasive infections when the immune system becomes compromised. The fungal cell wall contains conserved pathogen-associated molecular patterns (PAMPs), which allow the immune system to distinguish between self (endogenous molecular patterns) and foreign material. Sensing of invasive microbial pathogens is achieved through recognition of PAMPs by pattern recognition receptors (PRRs). One of the predominant fungal-sensing PRRs is the C-type lectin receptor (CLR) family. These receptors bind to structures present on the fungal cell wall, eliciting various innate immune responses as well as shaping adaptive immunity. In this chapter, we specifically focus on the four major human fungal pathogens, *Candida albicans*, *Aspergillus fumigatus*, *Cryptococcus neoformans* and *Pneumocystis jirovecii*, reviewing our current understanding of the CLRs that are involved in their recognition and protection of the host.

Keywords Pathogenic fungi · Antifungal immunity · PRRs · CLRs

1.1 Introduction

Fungi are eukaryotic organisms that include yeasts, moulds and mushrooms. The fungal kingdom is large and distinct from animals and plants, with over one million

C. Nikolakopoulou · J. A. Willment · G. D. Brown (✉)
Medical Research Council Centre for Medical Mycology, University of Exeter, Exeter EX4 4QD, UK
e-mail: Gordon.Brown@exeter.ac.uk

C. Nikolakopoulou
e-mail: cn359@exeter.ac.uk

J. A. Willment
e-mail: j.willment@exeter.ac.uk

Previous address: MRC Centre for Medical Mycology, University of Aberdeen, Aberdeen AB25 2ZD, UK

S.-L. Hsieh (ed.), *Lectin in Host Defense Against Microbial Infections*,
Advances in Experimental Medicine and Biology 1204,
https://doi.org/10.1007/978-981-15-1580-4_1

fungal species mainly found on organic substrates, such as plant debris and soil (Mora et al. 2011). It has been estimated that around 600 fungal species can cause disease in humans (Brown et al. 2012). Most human-associated fungi are found as commensals on mucocutaneous surfaces, such as *Candida* and *Malassezia* species and do not cause any disease in healthy individuals (Havlickova et al. 2008). Fungal pathogens rarely cause disease in immunocompetent people, with the most frequent infections being onychomycosis, keratitis (corneal infection that can lead to blindness) and dermatophytosis (infection of the skin) and in rare occasions invasive infections (Brown et al. 2012). However, commensal fungi and many environmental fungi can become opportunistic pathogens when there is a disruption in physical barriers, when the immune system of the host is compromised, or in times of fungal dysbiosis (Brown 2011; Iliev and Leonardi 2017). The incidence of invasive fungal infections (IFI) has increased significantly over the last 30 years mainly because of the HIV/AIDS pandemic and invasive and immunosuppressive modern medical treatments (Erwig and Gow 2016). Recent estimates indicate more people die globally from fungal infections than malaria and about the same number die from tuberculosis (Brown et al. 2012); yet significantly less research is conducted on fungal infections. Even though in the UK, between 1997 and 2010, £2.6 billion was invested in research on all infectious diseases (Head et al. 2013), only 2% of this was put towards medical mycology research (Head et al. 2014). As a result, many people are unaware of their huge impact and most of the times fungal diseases are undiagnosed. Some significant problems facing the field are the lack of appropriate diagnostics and vaccines. Although antifungal drugs are available, they suffer from toxicity, rising resistance and other complications. About 1.6 million people die each year because of lethal IFI (https://www.gaffi.org/why/fungal-disease-frequency/) (Brown et al. 2012). In the developed world, the mortality rate due to common life-threatening fungal infections, even in non-AIDS patients, often exceeds 50% despite treatment (https://www.gaffi.org/why/fungal-disease-frequency/).

Fungal pathogens are primarily recognised by a family of pattern recognition receptors (PRRs), known as C-type lectin receptors (CLRs). In this chapter, we will focus on the CLRs involved in the recognition of the four fungal pathogens with the highest disease burden; *Candida albicans*, *Aspergillus fumigatus*, *Cryptococcus neoformans* and *Pneumocystis jirovecii*. These cause invasive candidiasis, invasive and chronic pulmonary aspergillosis (CPA), cryptococcal meningitis and pneumocystis pneumonia, respectively (https://www.gaffi.org/why/fungal-disease-frequency/). A brief introduction to the fungal cell wall and the major CLRs recognising these structures will be presented, and then the roles of these receptors in the control of each pathogen will be discussed in detail.

1.2 Structure of the Fungal Cell Wall

The cell wall of fungal pathogens is a perfect target for our immune system to separate "self" from "non-self" as it consists of proteins and polysaccharides that are not present in mammalian cells. However, recognition of fungi is complicated by their ability to adopt different morphologies, such as yeasts, hyphae, encapsulated cells and enlarged cells, making it hard for phagocytic cells to detect and kill these pathogens. Fungi become a "moving target" for the immune system in response to metabolic adaptation during infection. For example, *C. albicans* changes its cell wall components upon exposure to the host metabolite lactate, preventing immune recognition (Ballou et al. 2016). As the fungal cell wall has been extensively reviewed elsewhere (Erwig and Gow 2016), the following paragraphs will provide a brief context for our review.

Fungi have an inner and an outer cell wall that are made of different components and the proportion of these components varies in the different morphologies and species (see following sections for more details). Approximately 90% of the cell wall is comprised of carbohydrates and the remainder are proteins. The inner cell wall of almost all fungi is made up of β-glucans and chitin (Fig. 1.1). In the case of

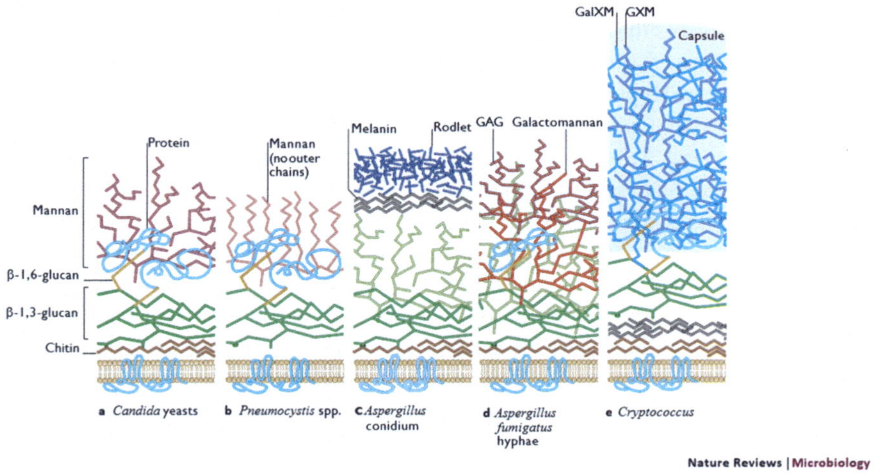

Fig. 1.1 Fungal cell wall composition and recognition by phagocytes. Figure and legend from Erwig and Gow (2016) with permission. The cell wall structures of selected fungal cells are depicted schematically showing the various layers. The core inner skeletal layer is comprised of the highly pro-inflammatory β-1,3-glucan, β-1,6-glucan and chitin. **a** In the cell walls of *Candida* species, the outer layer comprises mainly pro-inflammatory O-linked mannans, N-linked mannans and phosphorylated mannans. **b** Recent genomics analysis suggests that *Pneumocystis* spp. lack the enzymes that are required to synthesize chitin and the outer chain N-mannans, but have an outer wall with core N-mannan and O-mannan proteins. **c**, **d** The conidial spore wall of *Aspergillus* spp. consists of an immunologically inert outer hydrophobin rodlet layer and includes an inner melanin layer (part **c**), whereas the hyphae of *Aspergillus fumigatus* have a typical inner cell wall composition but have α-1,3-glucan, galactomannan and galactosaminoglycan (GAG) in the outer cell wall, along with a reduced amount of glycoprotein compared with the walls of *Candida* species (part **d**). **e** The cell wall of *Cryptococcus* is surrounded by a thick capsule of glucuronoxylomannan (GXM) and galactoxylomannan (GalXM), which is sloughed off in large quantities and can inhibit phagocyte function, and the inner wall has a layer of melanin

P. jirovecii, though, it is still debatable whether chitin is present in the inner cell wall (Skalski et al. 2015; Ma et al. 2016).

The outer cell wall layer of *C. albicans* contains mainly O- and N-linked mannans that form glycoproteins through their covalent association with proteins (Erwig and Gow 2016). The inner cell wall is more dense than the outer cell wall, primarily due to the covalent attachment of chitin, galactomannan and other proteins to β-glucans (Van De Veerdonk et al. 2017) (Fig. 1.1a). *C. albicans* can transit from the yeast to hyphal form, modifying the composition of the cell wall with changes in protein content, increased chitin, modified glucan structures (closed β-glucan linked structures) and glycosylation (Chaffin et al. 1998; Heilmann et al. 2011; Lowman et al. 2014).

During its life cycle, *A. fumigatus* germinates from conidia to form hyphae, which changes the composition of the cell wall. As a result, different PAMPs are exposed (Fig. 1.1c, d). Specifically, the conidia cell wall contains α-1,3-glucan, 1,8-dihydroxynaphthalene-(DHN) melanin (synthesised from the pentaketide pathway) and a rodlet layer made of hydrophobins (highly organised hydrophobic proteins). This rodlet layer coats the conidial outer cell wall (Fig. 1.1c) (Van De Veerdonk et al. 2017). Upon hyphae formation, the rodlet layer is removed and the melanin layers are lost completely. Hyphae then form an extracellular matrix (ECM) that contains α-1,3-glucans, galactomannan, galactosaminogalactan (GAG) and proteins that can enhance disease (Fig. 1.1d).

A characteristic feature of *C. neoformans* cell wall that is not found in *C. albicans* or *A. fumigatus* cell wall is a thick gelatinous capsule. This surrounds the outer wall of *C. neoformans* and is composed of polysaccharides including glucuronoxylomannan (GXM) and galactoxylomannan (GalXM) (Fig. 1.1e). The capsule masks cell wall β-glucan from immune recognition (see below), inhibiting the production of pro-inflammatory cytokines and protects the pathogen from phagocytosis, due to its negative charge, and from antibody and complement depletion (Buchanan and Murphy 1998). The inner wall of *C. neoformans* contains 3,4-dihydroxyphenylalanine (DOPA) melanin, which unlike *A. fumigatus* 1,8-DHN melanin, is formed by laccase and protects *C. neoformans* against the host immune response through its antioxidant ability (Casadevall et al. 2000). In the cell wall of *C. neoformans*, melanin is accumulated in melanin granules that consist of fungal melanosomes. These granules are associated with proteins that contribute to the generation of melanin as well as its structure. Melanisation of *C. neoformans* is thought to be similar to mammalian melanisation; in both systems melanin synthesis occurs inside specialised vesicles, the melanosomes, that provide protection to the host cell from toxic reaction intermediates (Camacho et al. 2019). *C. neoformans* can form titan cells during infection that have a gigantic size of up to 100 μm, in contrast to the size of typical cells, which are 5–7 μm in cell diameter (Zaragoza et al. 2010). These cells expose different PAMPs including altered chitin content in the cell wall (Mukaremera et al. 2018).

Lastly, the outer cell wall of *P. jirovecii* contains highly mannosylated (N-mannan and O-mannan) glycoproteins that mask β-glucan. Unlike *C. albicans* and *A. fumigatus*, however, the outer cell wall of *P. jirovecii* contains a heavily mannosylated glycoprotein A (gpA or MSG), but not the outer chain of N-mannans (Fig. 1.1b)

(Ma et al. 2016). The life cycle of *Pneumocystis* is uniquely adapted to growth in its host having lost the ability to grow outside the lung, and it is difficult to grow this fungus in vitro. Inside the lung, *Pneumocystis* undergoes three life cycle stages; cyst, precystic form and trophic form, which differ in size and shape. The precyst and trophic form are similar in cell wall composition (Chabé et al. 2011) and lack β-glucan (Nollstadt et al. 1994).

1.3 Pattern Recognition Receptors (PRRs)

The first line of defence against fungal pathogens is provided by the skin, mucosal surfaces and antimicrobial peptides (AMPs). AMPs are secreted by epithelial cells or produced in secretions and are active against many fungi, including *C. albicans*, *A. fumigatus* and *C. neoformans* (Schop 2007; Brown 2011). However, this protection is lost when these physical barriers are breached, and the pathogen invades deeper tissues. This invasion is sensed by the innate immune system which becomes activated, promoting the recognition and killing of the invading organisms. Innate recognition occurs through germline-encoded pattern recognition receptors (PRRs), which are expressed in nearly all cell types and recognise pathogen-associated molecular patterns (PAMPs) (Janeway 1992). As mentioned above, the cell wall contains many fungal PAMPs, which are recognised by PRRs via two mechanisms. The first is indirect recognition, where the fungal pathogens are opsonised by soluble PRRs, e.g. the complement receptor (CR3), enabling recognition through opsonic receptors. The second mechanism is direct recognition, where cell surface PRRs bind directly to the pathogen (Brown 2011) (Fig. 1.2).

All cells express PRRs. Epithelial and endothelial cell-expressed PRRs play an important role in antimicrobial immunity by stimulating the release of AMPs and inflammatory cytokines. For example, oral epithelial cells contribute to the host defence against oropharyngeal candidiasis (OPC) as they can sense *C. albicans* and upon invasion, they secrete AMPs and pro-inflammatory cytokines (Swidergall and Filler 2017). Another example is the expression of the newly identified CLR, Melanin sensing C-type lectin receptor (MelLec, which is discussed further below), which is expressed predominantly by endothelial cells in mice, and recognises *A. fumigatus* 1,8-DHN melanin (Stappers et al. 2018). Neutrophils, macrophages and other phagocytic cells use PRRs to take up and kill pathogens, as well as stimulating pro-inflammatory responses (Fig. 1.2). Following uptake and killing, antigen-presenting cells (APCs), such as dendritic cells (DCs) process antigenic fragments on major histocompatibility complex (MHC) class I or II molecules for presentation and activation of naïve T cells, and ultimately induce an adaptive immune response. The control of fungal infections is primarily reliant on Th1 and Th17 adaptive immunity (Brown 2011).

There are four families of PRRs that recognise microbial pathogens; Toll-like receptor (TLR), Nod-like receptor (NLR), RIG-I-like receptor (RLR) and C-type

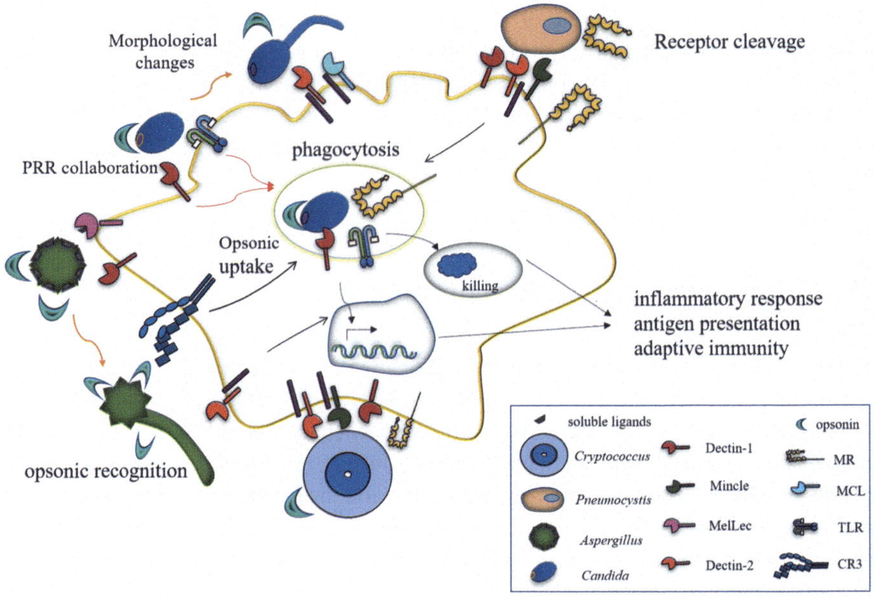

Fig. 1.2 Recognition of fungal pathogens by pattern recognition receptors (PRRs). Recognition of fungi takes place via numerous mechanisms, both direct and indirect, by cells of the innate immune system such as macrophages and neutrophils, and non-immune cells such as epithelial and endothelial cells. Indirect mechanisms of recognition are by Fc Receptors or complement receptors, such as CR3 (shown here as a heterodimer of CD11b and CD18), of fungi coated by serum opsonins (green crescent), such as antibodies and complement, respectively. Soluble PPRs, such as collectins, surfactant proteins-A and D and mannose-binding lectins can act as opsonins. Direct mechanisms of recognition by membrane bound PPRs, such as mannose receptor (MR), Dectin-1, Dectin-2, macrophage C-type lectin (MCL), Melanin sensing C-type lectin receptor (MelLec) and Toll-like receptors (TLRs) interact with fungi on the cell surface and some are recruited to the phagosome. Intracellular recognition of fungi occurs, both in the cytoplasm and vacuoles by Nod-like receptors (NLRs) and TLRs, respectively (not shown). Collaboration between various PRRs, such as Dectin-1 and the TLRs (red arrows) or between CLRs can alter the antifungal response. Depending on the cell type, recognition and uptake induces the production of inflammatory mediators and release of reactive oxygen species and antifungal peptides (black arrows). Fungal antigen presentation and cytokine production by antigen presenting cells, such as dendritic cells, shapes the adaptive immune response. Fungi undergo morphological changes (orange arrows) which alter their PAMP exposure (e.g. *Candida* masks its β-glucan exposure) and consequently the range of PRRs that recognise them differs. Fungi, such as *Pneumocystis*, have been shown to induce cleavage of PRRs, e.g. MR which binds to the fungus and alters the immune response. Fungal derived cell wall components, such as 1,8-dihydroxynaphthalene (black trapezium), which forms the melanin layer in *Aspergillus* conidia, can be recognised by the endothelial membrane bound receptor MelLec, but is shed once swelling and germination occurs

lectin receptor (CLR) families. We will focus here on the CLRs, which are the main PRRs involved in antifungal immunity.

1.4 C-Type Lectin Receptors (CLRs)

The CLRs are a diverse family of receptors that contain one or more C-type lectin-like domain (CTLD). These proteins can be soluble or transmembrane bound and recognise most human fungal pathogens through conserved motifs found in the CTLD. These motifs include QPD (Gln–Pro–Asp) and EPN (Glu–Pro–Asn), which specifically recognise galactose-type and mannose-type carbohydrates, respectively (Weis et al. 1998; Zelensky and Gready 2005). Soluble CLRs including the collectins, surfactant protein A (SP-A), SP-D, mannose-binding lectin (MBP) and collectin kidney protein 1 (CL-K1; also known as collectin 11), can act as opsonins and antimicrobial proteins (Brown et al. 2018). Here, we will focus only on transmembrane receptors. Initially, it was thought that CLRs bound only to carbohydrates in a Ca^{2+}-dependent manner, hence the name of this class of receptors. However, it is now known that many CLRs not only bind non-carbohydrates (e.g. lipids and proteins) via CTLDs but can also bind ligands in a Ca^{2+}-independent fashion (e.g. β-glucan). The CLRs have been separated into 17 groups based on their structure and phylogeny (Zelensky and Gready 2005). See http://www.imperial.ac.uk/research/animallectins/ctld/mammals/humanvmousedata.html for the full list of CTLD receptor groups.

CLR members of the group II, V and VI receptors (Fig. 1.3) are predominantly expressed on myeloid cells and are involved in antifungal immunity and will be discussed here. The group VI CLRs are type I transmembrane receptors with

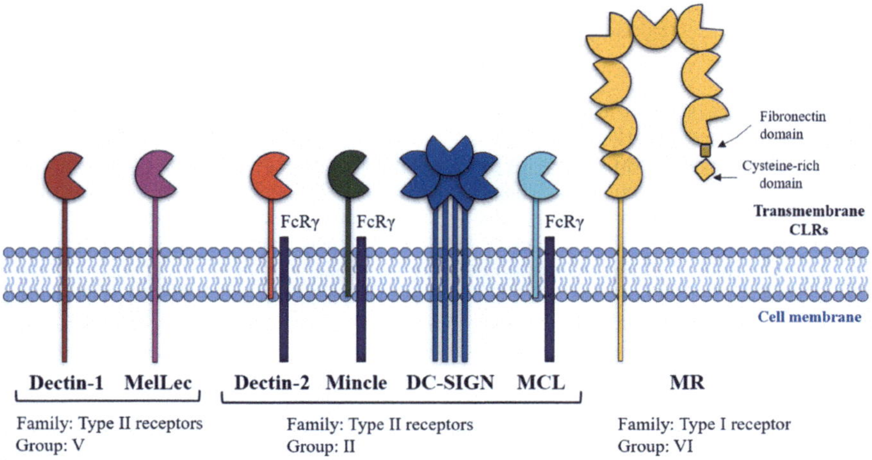

Fig. 1.3 Schematic representation of C-type lectin receptors. The structure of the CLRs described in the main text and the groups they belong to are depicted here

multiple CTLDs, and group II and V CLRs are type II transmembrane receptors with single extracellular CTLDs (Fig. 1.3). Some of these transmembrane receptors can trigger intracellular signalling directly or through association with other signalling molecules (for e.g. FcRγ) (Fig. 1.4), forming signalling heterodimeric and sometimes heterotrimeric complexes. An example of heterotrimeric complex is the recently described one formed between macrophage-inducible C-type lectin (Mincle)/macrophage C-type lectin (MCL)/FcεRI-γ (Lobato-Pascual et al. 2013). According to the signalling pathways they utilise, CLRs can be further divided into activation or inhibitory receptors.

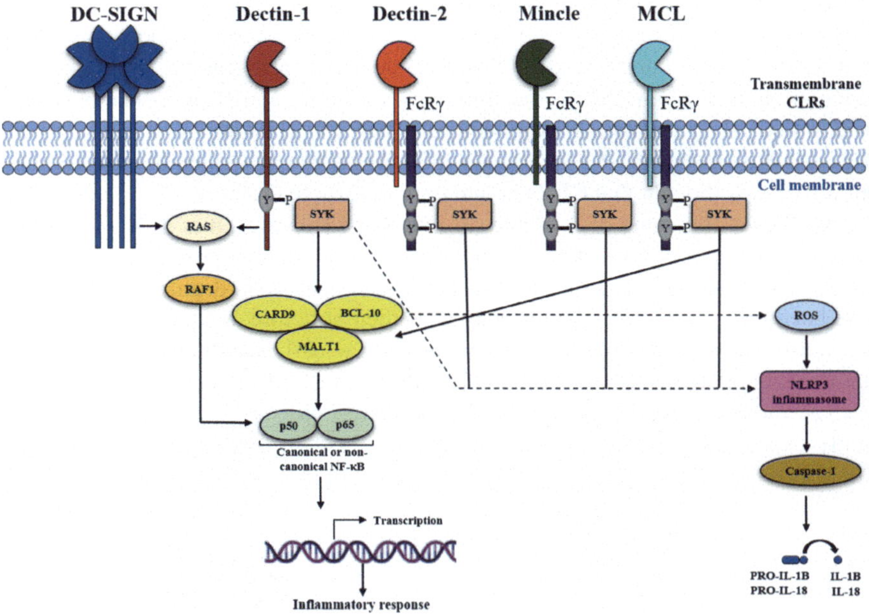

Fig. 1.4 Signalling pathways in innate recognition of fungi. The signalling pathways through which CLRs act are depicted here. Upon binding of activation CLRs (DC-SIGN, Dectin-1, Dectin-2, Mincle and MCL) to fungal PAMPs, they trigger intracellular signalling through an immunoreceptor tyrosine-based activation motif (ITAM) domain(s) located inside their cytoplasmic tails or via association with ITAM-bearing Fc receptor γ-chain (FcRγ) adaptor molecules. Phosphorylated ITAMs promote the recruitment of spleen tyrosine kinase (Syk) kinase, which is then phosphorylated resulting in the binding of caspase recruitment domain-containing protein 9 (CARD9), B-cell lymphoma/leukaemia 10 (BCL-10) and mucosa-associated lymphoid tissue 1 (MALT1). The resulting complex stimulates the canonical (c-Rel and p65) or the NF-κB-inducing kinase (NIK)-dependent non-canonical (RelB) NF-κB subunit and the NOD-, LRR- and pyrin domain-containing 3 (NLRP3) inflammasome. This leads to the production of reactive oxygen species (ROS) and the proteolytic activation of the pro-inflammatory cytokines IL-1β and IL-18 by caspase 1. Ligand recognition can also drive, in a cell and receptor specific manner, phagocytosis and the respiratory burst, which do not require transcription and in some cases are Syk-independent. Dectin-1 and DC-SIGN can also recruit the GTPase Ras proteins, which then activate the serine/threonine-protein kinase RAF1. RAF1 mediates the phosphorylation of the p65 subunit in a Syk-independent manner

Activation CLRs trigger intracellular signalling through immunoreceptor tyrosine-based activation motifs (ITAM; consensus sequence YxxL/I) located inside their cytoplasmic tails or via association with ITAM-bearing FcRγ adaptor molecules. Phosphorylated ITAMs promote the recruitment of spleen tyrosine (Syk) kinase (Gringhuis et al. 2009; Kerrigan and Brown 2011), resulting in a phosphorylation cascade leading to the activation of caspase recruitment domain-containing protein 9 (CARD9), B-cell lymphoma/leukaemia 10 (BCL-10) and mucosa-associated lymphoid tissue 1 (MALT1). The resulting complex stimulates transcription factors, including NF-κB, and results in chemokine and cytokine production. Unexpectedly, the phosphatase SHP-2 (previously associated with inhibitory receptors) has been implicated in stabilising Syk signalling necessary to control anti-*C. albicans* immunity (Deng et al. 2015). Activation of NF-κB can also be mediated by other signalling pathways that are Syk-independent. For example, activation of some CLRs, including Dectin-1 and dendritic cell-specific ICAM-3-grabbing non-integrin (DC-SIGN), can recruit the GTPase Ras proteins, which then activate the serine/threonine-protein kinase RAF1. RAF1 mediates the phosphorylation of the p65 subunit of the NF-κB in a Syk-independent manner (Geijtenbeek and Gringhuis 2009). Ligand recognition can also drive, in a cell type and receptor-specific manner, phagocytosis, the respiratory burst and other cellular responses that do not require transcription and in some cases are Syk-independent (Drummond et al. 2011; Strasser et al. 2012) (Fig. 1.4).

Inhibitory CLRs signal via an immunoreceptor tyrosine-based inhibition motif (ITIM; consensus sequence I/V/L/SxYxxI/L/V) within their cytoplasmic tails. When these receptors are activated by their ligands, tyrosine and inositol phosphatases including SHP-1, SHP-2 and SHIP, are recruited that inhibit activation pathways, leading to suppression of cellular responses. Some CLRs like MelLec lack ITAM or ITIM or other domains with known function and their mechanisms of intracellular signalling are unknown. This was also thought to be true for the mannose receptor (MR) but its association with FcRγ chain was recently described (Rajaram et al. 2017).

1.4.1 CLRs in Antifungal Immunity

The essential nature of CLRs in antifungal immunity is highlighted by the fact that humans (and mice) lacking functional CARD9, the signalling molecule through which most of the CLRs signal (Fig. 1.4), are extremely susceptible to fungal infections (Drummond et al. 2011). Moreover, as we will discuss below, polymorphisms in CLRs are also associated with alterations in human susceptibility to fungal infection. The major CLRs that play a role in fungal recognition and induction of protective immunity are Dectin-1 (CLEC7A), Dectin-2 (CLEC6A), MR (CD206), Mincle

(CLEC4E), DC-SIGN (CD209), MCL/Dectin-3 (CLEC4D) and the recently discovered MelLec (CLEC1A). We will now provide a brief introductory description of each receptor.

Dectin-1

Dectin-1 belongs to the group V CLRs (Fig. 1.3) and is expressed as several isoforms generated by alternative splicing, but primarily as the full-length receptor and a receptor lacking the stalk region (Brown 2006; Heinsbroek et al. 2006). Dectin-1 is predominantly expressed on myeloid cells, including DCs, macrophages and neutrophils and on specific lymphocyte subsets and on B-cells in humans (Brown 2006). Dectin-1 recognises β-glucans of fungal cell walls in a Ca^{2+}-independent manner (Brown 2006) but also ligands on mycobacteria (Wagener et al. 2018), *Leishmania* (Lima-Junior et al. 2017) and invertebrate tropomyosin (Gour et al. 2018). In addition, this receptor recognises several endogenous ligands, including vimentin and galectin-9 (Thiagarajan et al. 2013; Daley et al. 2017). Dectin-1 utilises an ITAM-like motif in its cytoplasmic tail to trigger intracellular signalling through the Syk/CARD9 pathway, but is also able to trigger several other pathways, including RAF1 (Fig. 1.4). Dectin-1 signalling requires clustering into a phagocytic synapse, where regulatory tyrosine phosphatases CD45 and CD148 are excluded. Several other regulators have also been described, including SHIP-1 (Blanco-Menéndez et al. 2015), and internalisation of ligand baring particles is thought to cause cessation of signalling (Hernanz-Falcón et al. 2009).

Dectin-1 is involved in inducing or regulating several cellular and immunological responses. For example, Dectin-1 can induce phagocytosis, the respiratory burst leading to production of reactive oxygen species (ROS) and regulate NETosis. It can also induce the production of cytokines and chemokines, and other inflammatory mediators. Finally, it can regulate the development of Th1 and Th17 adaptive immune responses (Goodridge et al. 2009; Kerrigan and Brown 2009).

MelLec

MelLec has structure similar to that of Dectin-1 and is part of the group V CLRs (Fig. 1.3). Unlike other fungal sensing CLRs, murine MelLec is not expressed on myeloid cells ($CD45^-$) but it is expressed on endothelial ($CD31^+$) cells. Interestingly, MelLec is also expressed on a novel subpopulation of $CD31^+$ cells that co-express the epithelial marker ($EpCAM^+$); these double-positive cells ($CD31^+EpCAM^+$) were only found in the mouse lung and liver (Stappers et al. 2018). In humans, this receptor is expressed on endothelial and myeloid cells, including DCs (Colonna et al. 2000; Sobanov et al. 2001; Sattler et al. 2012). MelLec recognises the naphthalene-diol unit of 1,8-DHN melanin found in several fungal species, including *A. fumigatus* and *Fonsecaea pedrosoi*. This recognition is specific, as MelLec does not recognise other melanin biosynthetic pathways, such as L-DOPA melanin synthesised in *Candida* or *Cryptococcus*. Despite its structural similarity with Dectin-1, MelLec lacks conserved motifs in its cytoplasmic tail and its mechanisms of intracellular signalling are not yet known.

Dectin-2

Dectin-2 is a member of the group II of CLRs (Fig. 1.3), and is expressed on tissue macrophages, DCs, neutrophils, Langerhans cells (LCs) and Kupffer cells (Shiokawa et al. 2017). Dectin-2 recognises mannose-based structures in a Ca^{2+}-dependent manner but has greatest affinity for high-mannose structures (e.g. $Man_9GlcNAc_2$) (McGreal et al. 2006). Dectin-2 forms a heterodimer with MCL (see later), which allows optimal ligand recognition (Zhu et al. 2013). Dectin-2 recognises a wide variety of ligands including numerous fungal species, as well as bacteria (e.g. *Mtb*), schistosome egg antigen and house dust mite allergens (HDM) (Ritter et al. 2010; Parsons et al. 2014; Shiokawa et al. 2017). Dectin-2 lacks a cytoplasmic tail with signalling domains, but like other receptors in this group, dimerizes with the ITAM-containing FcRγ chain for surface expression and intracellular signalling (Sato et al. 2006).

Cellular and immunological functions of Dectin-2 include phagocytosis, production of cytokines and ROS. Dectin-2 has also been shown to mediate extracellular trap formation in plasmacytoid DCs (pDCs) in response to *Aspergillus* hyphae (Loures et al. 2015). Dectin-2 has been implicated in allergic airway inflammation, through stimulation of cysteinyl leukotrienes (Barrett et al. 2009). Indeed, mice lacking Dectin-2 have decreased levels of HDM-mediated Th17 and Th2 cell differentiation as well as reduced inflammation of the airways (Plantinga et al. 2013; Norimoto et al. 2014). This receptor is also involved in mounting protective Th17 and Th1 responses to fungi (Wang et al. 2017).

Mincle

Mincle belongs to the Group II CLRs and is present on macrophages, DCs and neutrophils (Williams 2017) (Fig. 1.3). The CRD binds endogenous ligands (SAP130, sterols and β-glucosylceramides), mycobacterial ligands (such as cord factor, thermoduric bacteria) and fungal ligands (e.g. mannosyloxystearyl mannitol, galectin and β-gentiobiosyl diglycerides). Fungal pathogens recognised by Mincle include *C. albicans*, *P. jirovecii*, *Fonsecaea* and *Malassezia*. Mincle expression is enhanced following stimulation by fungal and microbial components through TLRs signalling (Patin et al. 2017). Mincle forms a heterotrimeric complex with MCL and FcRγ, which is required for surface expression, enhances phagocytosis, and stimulates signalling via the NF-κB pathway (Lobato-Pascual et al. 2013) (Fig. 1.4).

DC-SIGN

DC-SIGN, another member of group II CLRs (Fig. 1.3), has eight orthologues in mice mDC-SIGN, L-SIGN, DC-SIGNR1,2-4 and DC-SIGNR6-8 which differ in structure and expression patterns (Park et al. 2001; Koppel et al. 2005; Powlesland et al. 2006). Humans have two orthologues of this receptor: DC-SIGN and liver/lymph node-specific ICAM-3-grabbing integrin (L-SIGN) (Soilleux et al. 2000; Bashirova et al. 2001). DC-SIGN is mainly present on the surface of immature DCs in peripheral tissue, mature DCs in lymphoid tissue, and some subsets of macrophages (Koppel et al. 2005; Krutzik et al. 2005). DC-SIGN recognises mannose structures, such as galactomannans, in a Ca^{2+}-dependent fashion and recognises a range of endogenous ligands and pathogens, including viruses (e.g. HIV, Hepatitis C, Ebola), bacteria (*Mtb*), *Leishmania* and fungi (e.g. *C. albicans*, *Chrysosporium tropicum*, *A. fumigatus*) (Garcia-Vallejo and van Kooyk 2013). DC-SIGN recognises allergens from peanut, HDM (e.g. Derp1), dog and the Bermuda grass pollen allergen (Salazar et al. 2013). DC-SIGN possesses several internalisation motifs and signals via RAF1 (Gringhuis et al. 2007) (Fig. 1.4).

Most studies on DC-SIGN have been performed using human cells in vitro. Through these types of analyses, DC-SIGN has been associated with allergy as DC-SIGN-deficient DCs biased towards a Th2 cell polarisation in autologous DC-T cell cocultures (Emara et al. 2012). In addition, DC-SIGN can be cleaved by the protease allergen Derp1, which could contribute to driving allergic responses (Furmonaviciene et al. 2007). The cellular and immunological responses of DC-SIGN orthologues in mouse models are still to be examined.

Mannose Receptor (MR)

The MR is part of the group VI CLRs (Fig. 1.3) and can be proteolytically cleaved to form a soluble version. Interestingly the cleavage of the MR can be induced following Dectin-1 signalling (Gazi et al. 2011). This receptor is primarily expressed on tissue macrophages and some DC subtypes. The CTLDs bind to branched N-linked mannans in a Ca^{2+}-dependent manner, while the CR and FN-II domains have both exogenous and endogenous ligands (sulphated carbohydrates found in lymphoid and kidney tissues and collagen I–IV, respectively) (Martinez-Pomares 2012). The MR has also been reported to bind to α-1,3-glucan and chitin (Cambi et al. 2008; Romani 2011), and *P. jirovecii* glycoprotein A (O'Riordan et al. 1995). Other exogenous ligands of the MR include viruses (HIV, Dengue virus, HBV), bacteria (*Mtb*, *Klebsiella*, *S. pneumonia*), allergens (e.g. HDM), parasites (helminths, *Trichuris muris*) and fungi (e.g. *C. neoformans*, *P. jirovecii*) (Martinez-Pomares 2012). The short intracellular cytoplasmic tail of the MR has no known signalling motifs, although it was recently shown to associate with the FcRγ receptor (Rajaram et al. 2017).

The MR has been implicated in endocytosis, phagosomal trafficking, cross presentation, and promotion of antifungal Th17 responses (van de Veerdonk et al. 2009; Romani 2011). This receptor is also involved in allergy and asthma as it is required for allergen-induced Th2 cell polarisation, it is highly expressed in the DCs of allergic patients (Deslée et al. 2002; Royer et al. 2010), and the MR C-type 1 (*MRC1*) gene has been identified as a candidate for allergen (Li et al. 2007; Hattori et al. 2009).

1.4.2 CLRs in Immunity to Candida albicans

C. albicans is normally found in the microflora of 30–70% of healthy individuals, without causing any disease, including the oral cavity, vagina, intestinal mucosa and skin (Rizzetto et al. 2015). However, *C. albicans* is also an opportunistic human fungal pathogen when the immune system of the host is compromised. This fungus can exist as yeast, hyphae and pseudohyphae (filamentous forms), and morphogenetic switching is thought to be a virulence factor (Saville et al. 2003; Wächtler et al. 2011). *C. albicans* can infect oral and vaginal mucosae, skin and nail beds causing chronic mucocutaneous candidiasis (CMC), which arises due to dysfunction of T cells. *C. albicans* can also cause invasive candidiasis upon rupture of the host's physical barriers (e.g. skin and gut mucosal barriers) and/or modulation of host immunity. This pathogen is primarily thought to cross the GI tract allowing dissemination of the fungus into the bloodstream and deep tissues (Gow et al. 2012). Following *Staphylococcus epidermidis* and *Staphylococcus aureus*, invasive candidiasis is the fourth most common cause of bloodstream hospital infections. The mortality rate of invasive candidiasis is 30–40% (Perlroth et al. 2007).

1.4.2.1 *Candida* Infections and Dectin-2

Dectin-2 has been implicated in anti-*Candida* immunity. It recognises yeast and hyphae of *C. albicans* and plays an important role in driving the stimulation of cytokines upon *C. albicans* infection (Sato et al. 2006; Bi et al. 2010; Saijo et al. 2010). In response to infection, Dectin-2 forms heterodimers with MCL, and this heterodimeric complex binds to α-mannans more efficiently, leading to the induction of pro-inflammatory responses (Zhu et al. 2013). Dectin-2 has been found to contribute slightly to the phagocytosis of *C. albicans* (Bi et al. 2010) and neutrophils lacking Dectin-2 produced less ROS, which is critical for the killing of *Candida* (Ifrim et al. 2014).

1.4.2.2 *Candida* Infections and MCL

MCL was found to be a critical CLR for recognising and protecting against *C. albicans* as $MCL^{-/-}$ mice were susceptible to systemic candidiasis (Zhu et al. 2013). Nothing more is known regarding the role of MCL during *Candida* infections in humans.

1.4.2.3 *Candida* Infections and MR

C. albicans is also sensed by the mannose receptor. Despite the lack of phenotype in MR KO mice (Lee et al. 2003), the MR was initially recognised as an opsonin-independent phagocytic receptor able to bind and phagocytose un-opsonised *C. albicans* yeast particles (Porcaro et al. 2003) and zymosan particles (Harris et al. 1992). In contrast, other studies have suggested that the MR is implicated in live *C. albicans* uptake, it is recruited during phagosome maturation and is involved in cytokine release, e.g. monocyte chemoattractant protein 1 (MCP-1), TNF and IL-17 (Netea et al. 2006; Heinsbroek et al. 2008). No MR SNPs associated with increased susceptibility to *Candida* infections have been identified in humans. Thus, more studies are required to decipher the role of MR in protection against *C. albicans.*

1.4.2.4 *Candida* Infections and Dectin-1

The most well studied CLR against *C. albicans* infections is Dectin-1. It is involved in recognition and uptake of *C. albicans*, and it is able to stimulate ROS and nitric oxide (NO) production, phagocytosis and cytokine release (e.g. TNF, IL-6) in response to *C. albicans* infection in vitro (Drummond and Brown 2011; Pinke et al. 2016).

Dectin-1 has been found to be critical for protecting against systemic *C. albicans* infections in the GI tract. This receptor is crucial for activating $CD4^+$-T cell, but not $CD8^+$-T cell, responses in these tissues (Drummond et al. 2016). Dectin-1 is also required for the killing of *C. albicans* and the proper production of cytokines upon infection in the GI tract (Galès et al. 2010; Vautier et al. 2012). Dectin-1 also plays a critical role in protecting against candidiasis during inflammatory bowel disease (IBD) and colitis (Iliev et al. 2012). In humans, a single nucleotide polymorphism in CLEC7A (rs2078178) is significantly associated with ulcerative colitis (Iliev et al. 2012).

However, the role of Dectin-1 in protection against *Candida* infections in vivo depends on the site of infection, and strain of mouse and fungus. Dectin-1 has been reported to be essential for controlling vaginal and gastrointestinal candidiasis in C57BL/6 but not BALB/c mice, phenotypes that may be related to the isoforms of Dectin-1 (full-length Dectin-1A or the stalkless Dectin-1B) predominant in these mice (Heinsbroek et al. 2006; Carvalho et al. 2012). There are also conflicting reports describing the importance of Dectin-1 during systemic infection, however this has been linked to differential adaptations of different fungal strains (Saijo et al. 2007; Taylor et al. 2007; Netea et al. 2010; Marakalala et al. 2013).

Dectin-1 also appears to be associated with protection against chronic mucocutaneous candidiasis (CMC), which is caused by defective production of IL-17 and IL-22 as well as impaired Th17 cell differentiation. Using a skin mouse infection model, it was discovered that different morphological forms of *C. albicans* facilitate the development of different T helper responses that are protective for either systemic or cutaneous infection in a Dectin-1-dependent or independent manner. Specifically, *C. albicans* yeasts mediated a protective Th17 response against cutaneous infection via Dectin-1 (Kashem et al. 2015). Dectin-1 is crucial for defence against mucosal *C. albicans* infection in humans. The Tyr238X Dectin-1 polymorphism (completely lacking Dectin-1 surface expression) has been associated with reduced Th17 responses and increased susceptibility to CMC (Ferwerda et al. 2009). Indeed, stem cell transplant patients carrying this polymorphism had abrogated IL-1β production and increased colonisation with *Candida* (Plantinga et al. 2009).

Dectin-1 has been implicated in the development of liver cirrhosis. Chronic exposure of ethanol in mice led to fungal dysbiosis in the intestine and increased the translocation of β-glucans into the systemic circulation (Yang et al. 2017). This β-glucan induced inflammatory responses in Kupffer cells occurs in a Dectin-1-dependent manner. Associated phenotypes were also observed in human alcoholic patients with cirrhosis (Yang et al. 2017).

C. albicans can enter the skeletal tissue, causing osteomyelitis. This infection is hard to treat and causes significant morbidity (Gamaletsou et al. 2012). It was recently reported that Dectin-1 might contribute to the resolution of *C. albicans* osteomyelitis by inducing Nav1.8-positive-nociceptors, which prevent osteomyelitis and osteoporosis in response to β-glucan (Maruyama et al. 2017).

Other CLRs that we have not described here that have been implicated in host defence against *C. albicans* infections include dendritic cell natural killer lectin group receptor-1 (DNGR-1; CLEC9A). DNGR-1 was recently found to be required for controlling systemic *C. albicans* infection in mice and it reduced tissue damage by controlling the recruitment of neutrophils to *C. albicans* infected tissues (Cano et al. 2018).

1.4.3 CLRs in Immunity to Aspergillus fumigatus

The saprophytic fungus *A. fumigatus* belongs to the genus *Aspergillus* that is comprised of about 200 species (Gugnani 2003). *Aspergillus* species are found in decaying vegetation and soil, and contribute to a great extent to the recycling of carbon and nitrogen (Van De Veerdonk et al. 2017). Only a few *Aspergillus* species cause infections in humans, with *A. fumigatus* being the most common. *A. fumigatus* is very abundant in the environment, releasing spores (conidia) which are inhaled. It is an opportunistic pathogen that causes disease in people whose lung function or immune defences are impaired. About 1–15% of cystic fibrosis (CF) patients and 2% of asthma patients, develop allergic bronchopulmonary aspergillosis (ABPA), a pulmonary disease caused by hypersensitivity of these patients to antigens of *Aspergillus* (especially *A. fumigatus*). It may also contribute to the pathogenesis of CF (Escobar et al. 2016). In immunocompromised individuals, *A. fumigatus* hyphae can penetrate pulmonary tissues causing invasive aspergillosis (IA), which is the most severe disease caused by *Aspergillus* spp. (https://www.gaffi.org/why/fungal-disease-frequency/). Immunocompromised people affected by IA include, hematopoietic stem cell transplant (HSCT) patients, solid-organ (particularly lung) transplantation and people with genetic immunodeficiencies, e.g. chronic granulomatous disease (CGD) (Denning 1998). Emerging populations that are high in risk for IA are patients admitted to the intensive care unit (ICU) (up to 7% incidence) and chronic obstructive pulmonary disease (COPD) patients (Kousha et al. 2011; Schlamm et al. 2013). Despite the availability of antifungal drugs, mortality rates of IA exceed 50% in neutropenic patients and reaches 90% in HSCT recipients (Kousha et al. 2011).

1.4.3.1 *Aspergillus* Infections and Dectin-1

Dectin-1 was found to play an important role on human immature dendritic cells (iDCs) and myeloid DCs (mDCs) during *A. fumigatus* infection, as inhibition of this receptor resulted in decreased production of pro-inflammatory cytokines (Mezger et al. 2008; Hefter et al. 2017). However, the role of Dectin-1 in human pDCs during *A. fumigatus* infection is contradictory; this CLR has been shown to facilitate the uptake and clearance of *A. fumigatus* spores by human pDCs (Maldonado et al. 2015), but it has also been demonstrated that human pDCs recognise *A. fumigatus* hyphae via Dectin-2 but not Dectin-1 (Loures et al. 2015). The different results obtained from the two studies could be attributed to different forms of *A. fumigatus* and/or different strains used.

The requirement of Dectin-1 in protection against IA has been demonstrated by murine studies. For example, Dectin-1$^{-/-}$ mice were susceptible to *A. fumigatus* infection as they had increased fungal burdens in their lung and most of them died within 5 days of infection. There was also a defective recruitment of neutrophils, uncontrolled growth of *A. fumigatus* and reduced release of cytokines (e.g. IL-17) and chemokines in the lungs of Dectin-1$^{-/-}$ mice (Werner et al. 2009). Polymorphisms of Dectin-1, including the Dectin-1 TyrY238X polymorphism, have been linked to susceptibility to IA in transplant patients (Fisher et al. 2017). Indeed, polymorphonuclear cells (PMCs) from individuals carrying the polymorphism released less cytokines such as interferon-γ (IFNγ) (Cunha et al. 2010; Chai et al. 2011).

As mentioned earlier, exposure to *A. fumigatus* increases the severity of asthma in patients. Although Dectin-1 has been explored in asthma, its role is still unclear. Dectin-1 has been found to provide lung protection in mice via IL-22 (a cytokine highly increased in asthmatic patients) against *A. fumigatus* at early innate responses (Gessner et al. 2012). However, following chronic fungal exposure, this CLR might also enhance lung inflammation via release of this cytokine (Lilly et al. 2012). Further investigation is thus needed to identify the role of Dectin-1 in asthma.

Aspergillus species can also cause a disease of the cornea known as fungal keratitis, which can lead to severe visual defects and blindness in immunocompromised individuals. Dectin-1 has been found to play a role in this disease; it is involved in regulating neutrophil recruitment, release of IL-1β and other cytokines (e.g. CXCL1), and *A. fumigatus* killing (Leal et al. 2010; Xu et al. 2015; Yuan et al. 2017).

1.4.3.2 *Aspergillus* Infections and Dectin-2

Recently, Dectin-2, Dectin-1 (in an isoform specific manner) and Mincle were found to be important in the clearance of *A. fumigatus* from the lung of CF patients; cleavage of these three CLRs by neutrophil elastase (NE) (which is upregulated in the lungs of CF patients) and/or *A. fumigatus*-derived proteases, led to impaired antifungal protection, which might be also contributing to disease development in these patients (Griffiths et al. 2018). In a model of *A. fumigatus* keratitis, Dectin-2 enhanced the expression of IL-17RC on murine and human neutrophils (Taylor et al. 2014). Lastly,

Dectin-2 is also expressed in the lungs of patients with IA or ABPA, where it might be collaborating with the MR (Sun et al. 2013).

1.4.3.3 *Aspergillus* Infections and MR

The mannose receptor plays a role in protection against *A. fumigatus* in human corneal epithelial cells (HCECs) as upon stimulation of these cells with *A. fumigatus*, the mRNA and protein expression levels of the MR were upregulated, and this CLR was found to be required for inducing the production of inflammatory cytokines by these cells (Wang et al. 2016). No human polymorphisms in MR have been associated with infection, nor are there published studies examining the impact of MR-deficiency in mouse models.

1.4.3.4 *Aspergillus* Infections and DC-SIGN

A. fumigatus is also recognised by DC-SIGN, and although limited studies have been carried out, this receptor is able to directly mediate the phagocytosis of conidia in IL-4 treated macrophages and monocyte derived (MD)-DCs (Serrano-Gomez et al. 2004). Similarly, DCs overexpressing DC-SIGN phagocytosed *A. fumigatus* more effectively and produced higher levels of Th1 cytokines (Li et al. 2018). Interestingly, this receptor appears to recognise live but not heat-killed *A. fumigatus* conidia (Serrano-Gómez et al. 2005). Some DC-SIGN (rs4804800, rs11465384, rs7248637 and rs7252229) polymorphisms are associated with an elevated risk of developing IA, but this requires validation (Sainz et al. 2012). No studies in mouse models have been conducted so far.

1.4.3.5 *Aspergillus* Infections and MelLec

The novel CLR, MelLec, recognises melanized *A. fumigatus* conidia and provides protection against disseminated *A. fumigatus* infection. Mice lacking this receptor had impaired cytokine production (IL-1β, CCL2 and KC) and elevated fungal loads and were more susceptible to infection. A missense single nucleotide polymorphism (Gly26Ala) in the cytoplasmic tail of human MelLec is associated with reduced IL-8 and IL-1β cytokine release from macrophages, and stem cell transplant patients carrying this polymorphism were more susceptible to disseminated *A. fumigatus* infection (Stappers et al. 2018).

1.4.4 CLRs in Immunity to Cryptococcus neoformans

C. neoformans is an encapsulated, basidiomycete fungus that is typically found in trees, soil and pigeon droppings. It is a human pathogen responsible for causing cryptococcosis and cryptococcal meningitis. This fungus has three serotypes (A, D and AD) and two varieties (*C. neoformans* variety *grubii* (serotype A) and *C. neoformans* variety *neoformans* (serotype D)), which are found in different geographic regions and have different virulence factors to infect humans (Hagen et al. 2015). *C. neoformans* can form titan cells during infection and due to their gigantic size, their clearance by phagocytes is impossible. These cells help the fungus to survive and spread throughout the central nervous system (CNS), and they also exhibit increased resistance to drugs (Zaragoza et al. 2010). Cryptococcosis occurs after inhalation of desiccated yeast cells or spores into the lungs, where they proliferate and disseminate to the CNS, causing meningoencephalitis (Buchanan and Murphy 1998). Spores can also infect the kidneys and liver (Perfect 2005). Cryptococcosis mainly affects immunocompromised persons, such as HIV infected patients (Idnurm et al. 2005; Hagen et al. 2015). Globally about 220,000 million cryptococcosis cases occur among HIV patients worldwide each year, with the majority occurring in sub-Saharan Africa, leading to 181,000 deaths (https://www.cdc.gov/fungal/diseases/cryptococcosis-neoformans/statistics.html). Failure to treat disease of the CNS is fatal in 100% of cases (Buchanan and Murphy 1998).

Very little is known regarding the roles of CLRs in protection against this fungus, and sometimes the results of different studies are contradictory. Thus, more research is required to understand how *Cryptococcus* is recognised by our innate immune system.

1.4.4.1 *Cryptococcus* Infections and Dectin-1

The role of Dectin-1 in protection against cryptococcosis has been debated. This CLR was found to be redundant for providing protection against *C. neoformans*, as mice lacking Dectin-1 were not susceptible to *C. neoformans* infection (Nakamura et al. 2007). Spores of *C. neoformans* expose β-glucan and have been found to be taken up by alveolar macrophages via an interaction with Dectin-1 (Giles et al. 2009). However, it was later reported that Dectin-1 is not necessary for the phagocytosis of *C. neoformans* spores (Walsh et al. 2017). Finally, a recent study showed that Dectin-1 together with Dectin-2 facilitates the uptake of *C. neoformans* (Lim et al. 2018).

1.4.4.2 *Cryptococcus* Infections and Dectin-2

In vitro experiments have indicated a role for Dectin-2 in inflammatory responses to *Cryptococcus*, as cells lacking Dectin-2 released significantly less TNF and IL-12. However, in mice, Dectin-2 appears to play a minor role during infection, as inflammatory responses and susceptibility to infection were unaltered, although there were slight alterations in terms of increased Th2 responses and increased production of mucin (glycoprotein component of the mucous) (Nakamura et al. 2015).

1.4.4.3 *Cryptococcus* Infections and MR

The mannose receptor is required for host defence in cryptococcosis; this CLR is necessary for binding and phagocytosis of *C. neoformans* by DCs (Syme et al. 2002) but it is redundant for the uptake of mannoproteins upon infection (Dan et al. 2008). $MR^{-/-}$ mice exhibited increased susceptibility to pulmonary *C. neoformans* infection and had high fungal burdens in their lung. Furthermore, during infection, the MR is required for the development of mannoprotein-dependent $CD4^{+}$-T cell responses (Dan et al. 2008).

1.4.4.4 *Cryptococcus* Infections and MCL

Contradictory results have been obtained regarding the role of MCL during *C. neoformans* infections. MCL recognises glucuronoxylomannan (GXM), which is the main component of *Cryptococcus* capsule, and mediates the release of pro-inflammatory cytokines by *C. neoformans* infected bone marrow derived macrophages (BMDMs) (Huang et al. 2018). However, $MCL^{-/-}$ macrophages and DCs were able to phagocytose *C. neoformans* in vitro. In addition, no differences were observed in survival rate, cytokine release and leukocyte recruitment between wild-type and $MCL^{-/-}$ mice in response to pulmonary infection with serotype A or serotype D strains of *C. neoformans*, suggesting that MCL is not required for host protection against cryptococcosis (Campuzano et al. 2017). In contrast, a recent report describes increased susceptibility in $MCL^{-/-}$ mice upon infection with serotype AD (hybrid) of *C. neoformans* (Huang et al. 2018). Thus, the impact of MCL deficiency may differ depending on the serotype of *C. neoformans*.

1.4.5 CLRs in Immunity to Pneumocystis jirovecii

P. jirovecii is an opportunistic ascomycetous fungus that causes life-threatening pneumonia. This fungus is not a human commensal. As it can only grow inside the host lung, culture conditions have not been found so far, making it difficult/impossible to grow *P. jirovecii* in vitro (Thomas and Limper 2004). About 20% of healthy individuals get this fungus, which does not cause any disease and it is effectively controlled by the immune system (Medrano et al. 2005). *P. jirovecii* causes disease in immunocompromised patients, especially those with low count of CD4$^+$-T cells (Phair et al. 1990); thus, 40% of immunocompromised people who develop *Pneumocystis* pneumonia are HIV patients. The mortality rate of *Pneumocystis* pneumonia ranges between 30–60% in immunocompromised patients, and 10–20% among HIV patients (Thomas and Limper 2004).

Very little information is available regarding the role of CLRs in *Pneumocystis* infection, e.g. nothing is known about the role of MCL, and some of the CLRs known to recognise this fungus are redundant; more research is necessary to fully understand the innate recognition of this fungus.

1.4.5.1 *Pneumocystis* Infections and MR

One of the first CLRs described to recognise *P. jirovecii* was the MR (Ezekowitz et al. 1991) as it bound to gpA (O'Riordan et al. 1995), and this CLR was shown to be sufficient to induce the phagocytosis of *P. jirovecii* (Ezekowitz et al. 1991). Also, alveolar macrophages (AMs) (first immune cells responsible for the killing of *P. jirovecii* in the lung) from HIV patients were less efficient in binding and killing *P. murina* (a mice-associated species) as the MR was downregulated on these cells (Koziel et al. 1998). A soluble MR immuno-adhesin (sMR-Fc) has been generated which binds to *P. jirovecii*, it participates in phagocytosis via its Fc-portion and augments the phagocytosis of *P. jirovecii* by human polymorphonuclear cells (Stehle et al. 2000). However, *Pneumocystis* promotes the cleavage of the MR from AMs, releasing the sMR; because sMR binds to the organisms in the alveoli, the non-opsonic phagocytosis of this receptor is impeded (Fraser et al. 2000). Notably, one of the mechanisms that *Pneumocystis* has developed to evade immune killing is the release of its gpA, which competes for the MR on the surface of AMs preventing phagocytosis (Lasbury et al. 2004). In vivo studies demonstrated that more phagocytes were recruited into the alveoli of MR$^{-/-}$ mice during infection with *P. murina* but did not develop any signs of disease (Swain et al. 2003).

1.4.5.2 *Pneumocystis* Infections and Dectin-1

Dectin-1 contributes to protection against *P. jirovecii*. In vitro experiments have shown that cysts of *P. jirovecii* are recognised by Dectin-1, facilitating uptake and clearance by AMs. Indeed, overexpression of Dectin-1 in macrophages enhanced binding to this fungus (Steele et al. 2003). In vivo, Dectin-1$^{-/-}$ mice exhibited high susceptibility and displayed impaired generation of ROS in response to *Pneumocystis* lung infection. However, no differences were observed in cytokine release between Dectin-1$^{-/-}$ and wild-type mice (Saijo et al. 2007).

1.4.5.3 *Pneumocystis* Infections and Mincle

Mincle was recently found to be able to recognise *P. murina* (Kottom et al. 2017). Overexpression of Mincle in mouse macrophages enhanced their binding to this fungus and increased the phosphorylation of Syk kinase. Mincle$^{-/-}$ mice had impaired cytokine production in their lungs and decreased killing ability. However, the susceptibility of these mice was not altered during infection. Mincle$^{-/-}$ mice lacking CD4 succumbed to *P. murina* infection faster than wild-type mice lacking CD4. Thus, Mincle is required for the induction of host responses against *Pneumocystis* in immunosuppressed mice. Notably, Mincle may regulate expression of other CLRs including Dectin-1, Dectin-2 and MCL during *Pneumocystis* infection (Kottom et al. 2017).

1.4.5.4 *Pneumocystis* Infections and Dectin-2

Dectin-2 might also be required for host defence against *P. jirovecii* as AMs from mice lacking this receptor could not trigger Syk kinase and they produced significantly lower levels of TNF and IL-6. However, Dectin-2 may not be required for protection against *Pneumocystis* in immunosuppressed mice as Dectin-2$^{-/-}$ mice lacking CD4 were not susceptible to *Pneumocystis* infection (Kottom et al. 2018).

1.5 Conclusion and Future Directions

CLRs are critical for the protection against fungal infection. Although we have learned a great deal about the roles and functions of CLRs in antifungal immunity in general, there is much we still do not understand about their roles during infection with specific pathogens. Moreover, we are also just starting to understand the mechanisms that fungal pathogens have developed to escape recognition by these receptors. For example, fungi can alter their morphology and cell wall composition, as well as mask key components, such as β-glucan. A greater understanding of how CLRs mediate responses to fungal pathogens offers great promise for the development of new diagnostic tools, vaccines and immunotherapies, which are desperately needed to combat these devastating infections.

Acknowledgements We thank the Wellcome Trust (102705) and the Medical Research Council Centre for Medical Mycology and the University of Exeter (MR/N006364/1) for funding.

References

Ballou ER, Avelar GM, Childers DS et al (2016) Lactate signalling regulates fungal β-glucan masking and immune evasion. Nat Microbiol 2:16238. https://doi.org/10.1038/nmicrobiol.2016.238

Barrett NA, Maekawa A, Rahman OM et al (2009) Dectin-2 recognition of house dust mite triggers cysteinyl leukotriene generation by dendritic cells. J Immunol 182:1119–1128. https://doi.org/10.4049/jimmunol.182.2.1119

Bashirova AA, Geijtenbeek TB, van Duijnhoven GC et al (2001) A dendritic cell-specific intercellular adhesion molecule 3-grabbing nonintegrin (DC-SIGN)-related protein is highly expressed on human liver sinusoidal endothelial cells and promotes HIV-1 infection. J Exp Med 193:671–678

Bi L, Gojestani S, Wu W et al (2010) CARD9 mediates Dectin-2-induced IκBα kinase ubiquitination leading to activation of NF-κB in response to stimulation by the hyphal form of *Candida albicans*. J Biol Chem 285:25969–25977. https://doi.org/10.1074/jbc.M110.131300

Blanco-Menéndez N, del Fresno C, Fernandes S et al (2015) SHIP-1 Couples to the Dectin-1 hemITAM and selectively modulates reactive oxygen species production in dendritic cells in response to *Candida albicans*. J Immunol 195:4466–4478. https://doi.org/10.4049/jimmunol.1402874

Brown GD (2006) Dectin-1: a signalling non-TLR pattern-recognition receptor. Nat Rev Immunol 6:33–43. https://doi.org/10.1038/nri1745

Brown GD (2011) Innate antifungal immunity: the key role of phagocytes. Annu Rev Immunol 29:1–21. https://doi.org/10.1146/annurev-immunol-030409-101229

Brown GD, Denning DW, Gow NAR et al (2012) Hidden killers: human fungal infections. Sci Transl Med 4:1–10. https://doi.org/10.1126/scitranslmed.3004404

Brown GD, Willment JA, Whitehead L (2018) C-type lectins in immunity and homeostasis. Nat Rev Immunol 1:1. https://doi.org/10.1038/s41577-018-0004-8

Buchanan KL, Murphy JW (1998) What makes *Cryptococcus neoformans* a pathogen? Emerg Infect Dis 4:71–83. https://doi.org/10.3201/eid0401.980109

Camacho E, Vij R, Chrissian C et al (2019) The structural unit of melanin in the cell wall of the fungal pathogen *Cryptococcus neoformans*. 1–29. https://doi.org/10.1074/jbc.ra119.008684

Cambi A, Netea MG, Mora-Montes HM et al (2008) Dendritic cell interaction with *Candida albicans* critically depends on N-linked Mannan. J Biol Chem 283:20590–20599. https://doi.org/10.1074/jbc.M709334200

Campuzano A, Castro-Lopez N, Wozniak KL et al (2017) Dectin-3 is not required for protection against *Cryptococcus neoformans* infection. PLoS ONE 12:1–14. https://doi.org/10.1371/journal.pone.0169347

Cano E, Planas A, Enamorado M et al (2018) DNGR-1 in dendritic cells limits tissue damage by dampening neutrophil recruitment. Science 362:351–356. https://doi.org/10.1126/science.aan8423

Carvalho A, Giovannini G, De Luca A et al (2012) Dectin-1 isoforms contribute to distinct Th1/Th17 cell activation in mucosal candidiasis. Cell Mol Immunol 9:276–286. https://doi.org/10.1038/cmi.2012.1

Casadevall A, Rosas AL, Nosanchuk JD (2000) Melanin and virulence in *Cryptococcus neoformans*. Curr Opin Microbiol 3:354–358. https://doi.org/10.1016/S1369-5274(00)00103-X

Chabé M, Aliouat-Denis CM, Delhaes L et al (2011) *Pneumocystis*: from a doubtful unique entity to a group of highly diversified fungal species. FEMS Yeast Res 11:2–17. https://doi.org/10.1111/j.1567-1364.2010.00698.x

Chaffin WL, López-Ribot JL, Casanova M et al (1998) Cell wall and secreted proteins of *Candida albicans*: identification, function, and expression. Microbiol Mol Biol Rev 62:130–180

Chai LYA, De Boer MGJ, Van Der Velden WJFM et al (2011) The Y238X stop codon polymorphism in the human β-glucan receptor Dectin-1 and susceptibility to invasive aspergillosis. J Infect Dis 203:736–743. https://doi.org/10.1093/infdis/jiq102

Colonna M, Samaridis J, Angman L (2000) Molecular characterization of two novel C-type lectin-like receptors, one of which is selectively expressed in human dendritic cells. Eur J Immunol 30:697–704. https://doi.org/10.1002/1521-4141(200002)30:2%3c697:AID-IMMU697%3e3.0.CO;2-M

Cunha C, Di Ianni M, Bozza S et al (2010) Dectin-1 Y238X polymorphism associates with susceptibility to invasive aspergillosis in hematopoietic transplantation through impairment of both recipient- and donor-dependent mechanisms of antifungal immunity. Blood 116:5394–5402. https://doi.org/10.1182/blood-2010-04-279307

Daley D, Mani VR, Mohan N et al (2017) Dectin-1 activation on macrophages by galectin 9 promotes pancreatic carcinoma and peritumoral immune tolerance. Nat Med. https://doi.org/10.1038/nm.4314

Dan JM, Kelly RM, Lee CK, Levitz SM (2008) Role of the mannose receptor in a murine model of *Cryptococcus neoformans* infection. Infect Immun 76:2362–2367. https://doi.org/10.1128/IAI.00095-08

Deng Z, Ma S, Zhou H et al (2015) SHP-2 mediates C-type lectin receptors-induced Syk activation and anti-fungal TH17 responses. Nat Immunol 16:642–652. https://doi.org/10.1038/nature13527. Structure

Denning DW (1998) Invasive aspergillosis. Clin Infect Dis 26:781–803

Deslée G, Charbonnier AS, Hammad H et al (2002) Involvement of the mannose receptor in the uptake of Derp1, a major mite allergen, by human dendritic cells. J Allergy Clin Immunol 110:763–770. https://doi.org/10.1067/mai.2002.129121

Drummond RA, Brown GD (2011) The role of Dectin-1 in the host defence against fungal infections. Curr Opin Microbiol 14:392–399. https://doi.org/10.1016/j.mib.2011.07.001

Drummond RA, Saijo S, Iwakura Y, Brown GD (2011) The role of Syk/CARD9 coupled C-type lectins in antifungal immunity. Eur J Immunol 41:276–281. https://doi.org/10.1002/eji.201041252

Drummond RA, Dambuza IM, Vautier S et al (2016) CD4 + T-cell survival in the GI tract requires Dectin-1 during fungal infection. Mucosal Immunol 9:492–502. https://doi.org/10.1038/mi.2015.79

Emara M, Royer PJ, Mahdavi J et al (2012) Retagging identifies dendritic cell-specific intercellular adhesion molecule-3 (ICAM3)-grabbing non-integrin (DC-SIGN) protein as a novel receptor for a major allergen from house dust mite. J Biol Chem 287:5756–5763. https://doi.org/10.1074/jbc.M111.312520

Erwig LP, Gow NAR (2016) Interactions of fungal pathogens with phagocytes. Nat Rev Microbiol 14:163–176. https://doi.org/10.1038/nrmicro.2015.21

Escobar N, Ordonez SR, Wösten HAB et al (2016) Hide, keep quiet, and keep low: properties that make *Aspergillus fumigatus* a successful lung pathogen. Front Microbiol 7:1–13. https://doi.org/10.3389/fmicb.2016.00438

Ezekowitz RAB, Williams DJ, Koziel H et al (1991) Uptake of *Pneumocystis carinii* mediated by the macrophage mannose receptor. Nature 351:155–158. https://doi.org/10.1038/351155a0

Ferwerda B, Ferwerda G, Plantinga TS et al (2009) Human Dectin-1 deficiency and mucocutaneous fungal infections. N Engl J Med 361:1760–1767. https://doi.org/10.1056/NEJMoa0901053

Fisher CE, Hohl TM, Fan W et al (2017) Validation of single nucleotide polymorphisms in invasive aspergillosis following hematopoietic cell transplantation. Blood 129:2693–2701. https://doi.org/10.1182/blood-2016-10-743294

Fraser IP, Takahashi K, Koziel H et al (2000) *Pneumocystis carinii* enhances soluble mannose receptor production by macrophages. Microbes Infect 2:1305–1310. https://doi.org/10.1016/S1286-4579(00)01283-1

Furmonaviciene R, Ghaemmaghami AM, Boyd SE et al (2007) The protease allergen Derp1 cleaves cell surface DC-SIGN and DC-SIGNR: experimental analysis of in silico substrate identification and implications in allergic responses. Clin Exp Allergy 37:231–242. https://doi.org/10.1111/j.1365-2222.2007.02651.x

Galès A, Conduché A, Bernad J et al (2010) PPARγ controls Dectin-1 expression required for host antifungal defense against *Candida albicans*. PLoS Pathog 6. https://doi.org/10.1371/journal.ppat.1000714
Gamaletsou MN, Kontoyiannis DP, Sipsas NV et al (2012) *Candida* osteomyelitis: analysis of 207 pediatric and adult cases (1970–2011). Clin Infect Dis 55:1338–1351. https://doi.org/10.1093/cid/cis660
Garcia-Vallejo JJ, van Kooyk Y (2013) The physiological role of DC-SIGN: a tale of mice and men. Trends Immunol 34:482–486. https://doi.org/10.1016/j.it.2013.03.001
Gazi U, Rosas M, Singh S et al (2011) Fungal recognition enhances mannose receptor shedding through Dectin-1 engagement. J Biol Chem 286:7822–7829. https://doi.org/10.1074/jbc.M110.185025
Geijtenbeek TBH, Gringhuis SI (2009) Signalling through C-type lectin receptors: shaping immune responses. Nat Rev Immunol 9:465–479. https://doi.org/10.1038/nri2569
Gessner MA, Werner JL, Lilly LM et al (2012) Dectin-1-dependent interleukin-22 contributes to early innate lung defense against *Aspergillus fumigatus*. Infect Immun 80:410–417. https://doi.org/10.1128/IAI.05939-11
Giles SS, Dagenais TRT, Botts MR et al (2009) Elucidating the pathogenesis of spores from the human fungal pathogen *Cryptococcus neoformans*. Infect Immun 77:3491–3500. https://doi.org/10.1128/IAI.00334-09
Goodridge HS, Wolf AJ, Underhill DM (2009) β-glucan recognition by the innate immune system. Immunol Rev 230:38–50
Gour N, Lajoie S, Smole U et al (2018) Dysregulated invertebrate tropomyosin–Dectin-1 interaction confers susceptibility to allergic diseases. Sci Immunol. https://doi.org/10.1126/sciimmunol.aam9841
Gow NAR, Van De Veerdonk FL, Brown AJP, Netea MG (2012) *Candida albicans* morphogenesis and host defence: discriminating invasion from colonization. Nat Rev Microbiol 10:112–122. https://doi.org/10.1038/nrmicro2711
Griffiths JS, Thompson A, Stott M et al (2018) Differential susceptibility of Dectin-1 isoforms to functional inactivation by neutrophil and fungal proteases. FASEB J 32:3385–3397. https://doi.org/10.1096/fj.201701145R
Gringhuis SI, den Dunnen J, Litjens M et al (2007) C-type lectin DC-SIGN modulates toll-like receptor signaling via Raf-1 kinase-dependent acetylation of transcription factor NF-κB. Immunity 26:605–616. https://doi.org/10.1016/j.immuni.2007.03.012
Gringhuis SI, den Dunnen J, Litjens M et al (2009) Dectin-1 directs T helper cell differentiation by controlling noncanonical NF-κB activation through Raf-1 and Syk. Nat Immunol 10:203–213. https://doi.org/10.1038/ni.1692
Gugnani HC (2003) Ecology and taxonomy of pathogenic aspergilli. Front Biosci 8:1002. https://doi.org/10.2741/1002
Hagen F, Khayhan K, Theelen B et al (2015) Recognition of seven species in the *Cryptococcus gattii/Cryptococcus neoformans* species complex. Fungal Genet Biol 78:16–48. https://doi.org/10.1016/j.fgb.2015.02.009
Harris N, Super M, Rits M et al (1992) Characterization of the murine macrophage mannose receptor: demonstration that the downregulation of receptor expression mediated by interferon-gamma occurs at the level of transcription. Blood 80:2363–2373
Hattori T, Konno S, Hizawa N et al (2009) Genetic variants in the mannose receptor gene (*MRC1*) are associated with asthma in two independent populations. Immunogenetics. https://doi.org/10.1007/s00251-009-0403-x
Havlickova B, Czaika VA, Fredrich M (2008) Epidemiological trends in skin mycosis worldwide. Mycosis 51:2–15. https://doi.org/10.1111/j.1439-0507.2008.01606
Head MG, Fitchett JR, Cooke MK et al (2013) UK investments in global infectious disease research 1997–2010: a case study. Lancet Infect Dis 13:55–64. https://doi.org/10.1016/S1473-3099(12)70261-X

Head MG, Fitchett JR, Atun R, May RC (2014) Systematic analysis of funding awarded for mycology research to institutions in the UK, 1997–2010. BMJ Open 4. https://doi.org/10.1136/bmjopen-2013-004129

Hefter M, Lother J, Weiß E et al (2017) Human primary myeloid dendritic cells interact with the opportunistic fungal pathogen *Aspergillus fumigatus* via the C-type lectin receptor Dectin-1. Med Mycol 55:573–578. https://doi.org/10.1093/mmy/myw105

Heilmann CJ, Sorgo AG, Siliakus AR et al (2011) Hyphal induction in the human fungal pathogen *Candida albicans* reveals a characteristic wall protein profile. Microbiology 157:2297–2307. https://doi.org/10.1099/mic.0.049395-0

Heinsbroek SEM, Taylor PR, Martinez FO et al (2008) Stage-specific sampling by pattern recognition receptors during *Candida albicans* phagocytosis. PLoS Pathog 4. https://doi.org/10.1371/journal.ppat.1000218

Heinsbroek SE, Taylor PR, Rosas M et al (2006) Expression of functionally different Dectin-1 isoforms by murine macrophages. J Immunol 176:5513–5518. https://doi.org/10.4049/jimmunol.176.9.5513

Hernanz-Falcón P, Joffre O, Williams DL, Reis e Sousa C (2009) Internalization of Dectin-1 terminates induction of inflammatory responses. Eur J Immunol 39:507–513. https://doi.org/10.1002/eji.200838687

Huang HR, Li F, Han H et al (2018) Dectin-3 recognizes glucuronoxylomannan of *Cryptococcus neoformans* serotype AD and *Cryptococcus gattii* serotype B to initiate host defense against cryptococcosis. Front Immunol 9. https://doi.org/10.3389/fimmu.2018.01781

Idnurm A, Bahn YS, Nielsen K et al (2005) Deciphering the model pathogenic fungus *Cryptococcus neoformans*. Nat Rev Microbiol 3:753–764. https://doi.org/10.1038/nrmicro1245

Ifrim DC, Bain JM, Reid DM et al (2014) Role of Dectin-2 for host defense against systemic infection with *Candida glabrata*. Infect Immun 82:1064–1073. https://doi.org/10.1128/IAI.01189-13

Iliev ID, Leonardi I (2017) Fungal dysbiosis: immunity and interactions at mucosal barriers. Nat Rev Immunol 17:635–646. https://doi.org/10.1038/nri.2017.55

Iliev ID, Funari VA, Taylor KD et al (2012) Interactions between commensal fungi and the C-type lectin receptor Dectin-1 influence colitis. Science 336:1314–1317. https://doi.org/10.1126/science.1221789

Janeway CA (1992) The immune system evolved to discriminate infectious nonself from noninfectious self. Immunol Today 13:11–16. https://doi.org/10.1016/0167-5699(92)90198-G

Kashem SW, Igyártó BZ, Gerami-Nejad M et al (2015) *Candida albicans* morphology and dendritic cell subsets determine T helper cell differentiation. Immunity 42:356–366. https://doi.org/10.1016/j.immuni.2015.01.008

Kerrigan AM, Brown GD (2009) C-type lectins and phagocytosis. Immunobiology 214:562–575. https://doi.org/10.1016/j.imbio.2008.11.003

Kerrigan AM, Brown GD (2011) Syk-coupled C-type lectins in immunity. Trends Immunol 32:151–156. https://doi.org/10.1016/j.it.2011.01.002

Koppel EA, van Gisbergen KPJM, Geijtenbeek TBH, van Kooyk Y (2005) Distinct functions of DC-SIGN and its homologues L-SIGN (DC-SIGNR) and mSIGNR1 in pathogen recognition and immune regulation. Cell Microbiol 7:157–165

Kottom TJ, Hebrink DM, Jenson PE et al (2017) The interaction of *Pneumocystis* with the C-type lectin receptor mincle exerts a significant role in host defense against infection. J Immunol. https://doi.org/10.4049/jimmunol.1600744

Kottom TJ, Hebrink DM, Jenson PE et al (2018) Dectin-2 is a C-type lectin receptor that recognizes *Pneumocystis* and participates in innate immune responses. Am J Respir Cell Mol Biol 58:232–240. https://doi.org/10.1165/rcmb.2016-0186OC

Kousha M, Tadi R, Soubani AO (2011) Pulmonary aspergillosis: a clinical review. Eur Respir Rev 20:156–174

Koziel H, Eichbaum Q, Kruskal BA et al (1998) Reduced binding and phagocytosis of *Pneumocystis carinii* by alveolar macrophages from persons infected with HIV-1 correlates with mannose receptor downregulation. J Clin Invest 102:1332–1344. https://doi.org/10.1172/JCI560

Krutzik SR, Tan B, Li H et al (2005) TLR activation triggers the rapid differentiation of monocytes into macrophages and dendritic cells. Nat Med 11:653–660. https://doi.org/10.1038/nm1246

Lasbury ME, Lin P, Tschang D et al (2004) Effect of bronchoalveolar lavage fluid from *Pneumocystis carinii*-infected hosts on phagocytic activity of alveolar macrophages. Infect Immun 72:2140–2147. https://doi.org/10.1128/IAI.72.4.2140-2147.2004

Leal SM, Cowden S, Hsia YC et al (2010) Distinct roles for Dectin-1 and TLR4 in the pathogenesis of *Aspergillus fumigatus* keratitis. PLoS Pathog 6:1–16. https://doi.org/10.1371/journal.ppat.1000976

Lee SJ, Zheng NY, Clavijo M, Nussenzweig MC (2003) Normal host defense during systemic candidiasis in mannose receptor-deficient mice. Infect Immun 71:437–445. https://doi.org/10.1128/IAI.71.1.437-445.2003

Li X, Fleis RI, Shubitowski DM et al (2007) Fine mapping of murine asthma quantitative trait loci and analyses of *Ptgs1* and *Mrc1* as positional candidate genes. DNA Seq 18:190–195. https://doi.org/10.1080/10425170701207166

Li LY, Zhang HR, Jiang ZL et al (2018) Overexpression of dendritic cell-specific intercellular adhesion molecule-3-grabbing nonintegrin in dendritic cells protecting against Aspergillosis. Chin Med J (Engl) 131:2575–2582. https://doi.org/10.4103/0366-6999.244103

Lilly LM, Gessner MA, Dunaway CW et al (2012) The β-glucan receptor Dectin-1 promotes lung immunopathology during fungal allergy via IL-22. J Immunol 189:3653–3660. https://doi.org/10.4049/jimmunol.1201797

Lim J, Coates CJ, Seoane PI et al (2018) Characterizing the mechanisms of nonopsonic uptake of cryptococci by macrophages. J Immunol. https://doi.org/10.4049/jimmunol.1700790

Lima-Junior DS, Mineo TWP, Calich VLG, Zamboni DS (2017) Dectin-1 activation during *Leishmania amazonensis* phagocytosis prompts Syk-dependent reactive oxygen species production to trigger inflammasome assembly and restriction of parasite replication. J Immunol 199:2055–2068. https://doi.org/10.4049/jimmunol.1700258

Lobato-Pascual A, Saether PC, Fossum S et al (2013) Mincle, the receptor for mycobacterial cord factor, forms a functional receptor complex with MCL and FcεRI-γ. Eur J Immunol 43:3167–3174. https://doi.org/10.1002/eji.201343752

Loures FV, Röhm M, Lee CK et al (2015) Recognition of *Aspergillus fumigatus* hyphae by human plasmacytoid dendritic cells is mediated by Dectin-2 and results in formation of extracellular traps. PLoS Pathog 11:1–23. https://doi.org/10.1371/journal.ppat.1004643

Lowman DW, Greene RR, Bearden DW et al (2014) Novel structural features in *Candida albicans* hyphal glucan provide a basis for differential innate immune recognition of hyphae versus yeast. J Biol Chem 289:3432–3443. https://doi.org/10.1074/jbc.M113.529131

Ma L, Chen Z, Huang DW et al (2016) Genome analysis of three *Pneumocystis* species reveals adaptation mechanisms to life exclusively in mammalian hosts. Nat Commun 7. https://doi.org/10.1038/ncomms10740

Maldonado S, Dai J, Singh S et al (2015) Human pDCs express the C-type lectin receptor Dectin-1 and uptake and kill *Aspergillus fumigatus* spores *in vitro* (MPF4P.734). J Immunol 194

Marakalala MJ, Vautier S, Potrykus J et al (2013) Differential adaptation of *Candida albicans in vivo* modulates immune recognition by Dectin-1. PLoS Pathog 9:1–12. https://doi.org/10.1371/journal.ppat.1003315

Martinez-Pomares L (2012) The mannose receptor. J Leukoc Biol 92:1177–1186. https://doi.org/10.1189/jlb.0512231

Maruyama K, Takayama Y, Kondo T et al (2017) Nociceptors boost the resolution of fungal osteoinflammation via the TRP channel-CGRP-Jdp2 Axis. Cell Rep 19:2730–2742. https://doi.org/10.1016/j.celrep.2017.06.002

McGreal EP, Rosas M, Brown GD et al (2006) The carbohydrate-recognition domain of Dectin-2 is a C-type lectin with specificity for high mannose. Glycobiology 16:422–430. https://doi.org/10.1093/glycob/cwj077

Medrano FJ, Montes-Cano M, Conde M et al (2005) *Pneumocystis jirovecii* in general population. Emerg Infect Dis 11:245–250. https://doi.org/10.3201/eid1102.040487

Mezger M, Kneitz S, Wozniok I et al (2008) Proinflammatory response of immature human dendritic cells is mediated by Dectin-1 after exposure to *Aspergillus fumigatus* germ tubes. J Infect Dis 197:924–931. https://doi.org/10.1086/528694

Mora C, Tittensor DP, Adl S, et al (2011) How many species are there on earth and in the ocean? PLoS Biol 9:. https://doi.org/10.1371/journal.pbio.1001127

Mukaremera L, Lee KK, Wagener J et al (2018) Titan cell production in *Cryptococcus neoformans* reshapes the cell wall and capsule composition during infection. Cell Surf 1:15–24. https://doi.org/10.1016/j.tcsw.2017.12.001

Nakamura K, Kinjo T, Saijo S et al (2007) Dectin-1 is not required for the host defense to *Cryptococcus neoformans*. Microbiol Immunol 51:1115–1119. https://doi.org/10.1111/j.1348-0421.2007.tb04007.x

Nakamura Y, Sato K, Yamamoto H et al (2015) Dectin-2 deficiency promotes Th2 response and mucin production in the lungs after pulmonary infection with *Cryptococcus neoformans*. Infect Immun 83:671–681. https://doi.org/10.1128/IAI.02835-14

Netea MG, Gow NAR, Munro CA et al (2006) Immune sensing of *Candida albicans* requires cooperative recognition of mannans and glucans by lectin and toll-like receptors. J Clin Invest. https://doi.org/10.1172/JCI27114

Netea MG, Gow NAR, Joosten LAB et al (2010) Variable recognition of *Candida albicans* strains by TLR4 and lectin recognition receptors. Med Mycol 48:897–903. https://doi.org/10.3109/13693781003621575

Nollstadt KH, Powles MA, Fujioka H et al (1994) Use of β-1,3-glucan-specific antibody to study the cyst wall of *Pneumocystis carinii* and effects of pneumocandin B0 analog L-733,560. Antimicrob Agents Chemother 38:2258–2265. https://doi.org/10.1128/AAC.38.10.2258

Norimoto A, Hirose K, Iwata A et al (2014) Dectin-2 promotes house dust mite-induced T helper type 2 and type 17 cell differentiation and allergic airway inflammation in mice. Am J Respir Cell Mol Biol 51:201–209. https://doi.org/10.1165/rcmb.2013-0522OC

O'Riordan DM, Standing JE, Limper AH (1995) *Pneumocystis carinii* glycoprotein A binds macrophage mannose receptors. Infect Immun 63:779–784

Park CG, Takahara K, Umemoto E et al (2001) Five mouse homologues of the human dendritic cell C-type lectin, DC-SIGN. Int Immunol. https://doi.org/10.1093/intimm/13.10.1283

Parsons MW, Li L, Wallace AM et al (2014) Dectin-2 regulates the effector phase of house dust mite-elicited pulmonary inflammation independently from its role in sensitization. J Immunol 192:1361–1371. https://doi.org/10.4049/jimmunol.1301809

Patin EC, Orr SJ, Schaible UE (2017) Macrophage inducible C-type lectin as a multifunctional player in immunity. Front Immunol 8:1–10. https://doi.org/10.3389/fimmu.2017.00861

Perfect JR (2005) *Cryptococcus neoformans*: a sugar-coated killer with designer genes. FEMS Immunol Med Microbiol 45:395–404. https://doi.org/10.1016/j.femsim.2005.06.005

Perlroth J, Choi B, Spellberg B (2007) Nosocomial fungal infections: epidemiology, diagnosis, and treatment. Med Mycol 45:321–346. https://doi.org/10.1080/13693780701218689

Phair J, Muñoz A, Detels R et al (1990) The risk of *Pneumocystis carinii* pneumonia among men infected with human immunodeficiency virus type 1. Multicenter AIDS Cohort Study Group. N Engl J Med. https://doi.org/10.1056/nejm199001183220304

Pinke KH, de Lima HG, Cunha FQ, Lara VS (2016) Mast cells phagocyte *Candida albicans* and produce nitric oxide by mechanisms involving TLR2 and Dectin-1. Immunobiology 221:220–227. https://doi.org/10.1016/j.imbio.2015.09.004

Plantinga TS, van der Velden WJFM, Ferwerda B et al (2009) Early stop polymorphism in human Dectin-1 is associated with increased *Candida* colonization in hematopoietic stem cell transplant recipients. Clin Infect Dis 49:724–732. https://doi.org/10.1086/604714

Plantinga M, Guilliams M, Vanheerswynghels M et al (2013) Conventional and monocyte-derived CD11b + dendritic cells initiate and maintain T Helper 2 cell-mediated immunity to house dust mite allergen. Immunity 38:322–335. https://doi.org/10.1016/j.immuni.2012.10.016

Porcaro I, Vidal M, Jouvert S et al (2003) Mannose receptor contribution to *Candida albicans* phagocytosis by murine E-clone J774 macrophages. J Leukoc Biol 74:206–215. https://doi.org/10.1189/jlb.1202608

Powlesland AS, Ward EM, Sadhu SK et al (2006) Widely divergent biochemical properties of the complete set of mouse DC-SIGN-related proteins. J Biol Chem 281:20440–20449. https://doi.org/10.1074/jbc.M601925200

Rajaram MVS, Arnett E, Azad AK et al (2017) *M. tuberculosis*-initiated human mannose receptor signaling regulates macrophage recognition and vesicle trafficking by FcRγ-chain, Grb2, and SHP-1. Cell Rep. https://doi.org/10.1016/j.celrep.2017.09.034

Ritter M, Gross O, Kays S et al (2010) *Schistosoma mansoni* triggers Dectin-2, which activates the Nlrp3 inflammasome and alters adaptive immune responses. Proc Natl Acad Sci 107:20459–20464. https://doi.org/10.1073/pnas.1010337107

Rizzetto L, Weil T, Cavalieri D (2015) Systems level dissection of *Candida* recognition by dectins: a matter of fungal morphology and site of infection. Pathogens 4:639–661. https://doi.org/10.3390/pathogens4030639

Romani L (2011) Immunity to fungal infections. Nat Rev Immunol 11:275–288. https://doi.org/10.1038/nri2939

Royer PJ, Emara M, Yang C et al (2010) The mannose receptor mediates the uptake of diverse native allergens by dendritic cells and determines allergen-induced T cell polarization through modulation of IDO activity. J Immunol 185:1522–1531. https://doi.org/10.4049/jimmunol.1000774

Saijo S, Fujikado N, Furuta T et al (2007) Dectin-1 is required for host defense against *Pneumocystis carinii* but not against *Candida albicans*. Nat Immunol 8:39–46. https://doi.org/10.1038/ni1425

Saijo S, Ikeda S, Yamabe K et al (2010) Dectin-2 recognition of alpha-mannans and induction of Th17 cell differentiation is essential for host defense against *Candida albicans*. Immunity 32:681–691. https://doi.org/10.1016/j.immuni.2010.05.001

Sainz J, Lupiáñez CB, Segura-Catena J et al (2012) Dectin-1 and DC-SIGN polymorphisms associated with invasive pulmonary aspergillosis infection. PLoS ONE 7:1–10. https://doi.org/10.1371/journal.pone.0032273

Salazar F, Sewell HF, Shakib F, Ghaemmaghami AM (2013) The role of lectins in allergic sensitization and allergic disease. J Allergy Clin Immunol 132:27–36. https://doi.org/10.1016/j.jaci.2013.02.001

Sato K, Yang XL, Yudate T et al (2006) Dectin-2 is a pattern recognition receptor for fungi that couples with the Fc receptor γ chain to induce innate immune responses. J Biol Chem 281:38854–38866. https://doi.org/10.1074/jbc.M606542200

Sattler S, Reiche D, Sturtzel C et al (2012) The human C-Type lectin-like receptor CLEC-1 is upregulated by TGF-β and primarily localized in the endoplasmic membrane compartment. Scand J Immunol 75:282–292. https://doi.org/10.1111/j.1365-3083.2011.02665.x

Saville SP, Lazzell AL, Monteagudo C, Lopez-Ribot JL (2003) Engineered control of cell morphology *in vivo* reveals distinct roles for yeast and filamentous forms of *Candida albicans* during infection. Eukaryot Cell 2:1053–1060. https://doi.org/10.1128/EC.2.5.1053-1060.2003

Schlamm HT, Tarallo M, Baddley JW et al (2013) Aspergillosis in intensive care unit (ICU) patients: epidemiology and economic outcomes. BMC Infect Dis 13. https://doi.org/10.1186/1471-2334-13-29

Schop J (2007) Protective immunity against *Cryptococcus neoformans* infection. McGill J Med 10:35–43

Serrano-Gomez D, Dominguez-Soto A, Ancochea J et al (2004) Dendritic cell-specific intercellular adhesion molecule 3-grabbing nonintegrin mediates binding and internalization of *Aspergillus fumigatus* conidia by dendritic cells and macrophages. J Immunol 173:5635–5643. https://doi.org/10.4049/jimmunol.173.9.5635

Serrano-Gómez D, Leal JA, Corbí AL (2005) DC-SIGN mediates the binding of *Aspergillus fumigatus* and keratinophylic fungi by human dendritic cells. Immunobiology 210:175–183. https://doi.org/10.1016/j.imbio.2005.05.011

Shiokawa M, Yamasaki S, Saijo S (2017) C-type lectin receptors in anti-fungal immunity. Curr Opin Microbiol 40:123–130. https://doi.org/10.1016/j.mib.2017.11.004

Skalski JH, Kottom TJ, Limper AH (2015) Pathobiology of *Pneumocystis pneumonia*: life cycle, cell wall and cell signal transduction. FEMS Yeast Res 15:1–12. https://doi.org/10.1093/femsyr/fov046

Sobanov Y, Bernreiter A, Derdak S et al (2001) A novel cluster of lectin-like receptor genes expressed in monocytic, dendritic and endothelial cells maps close to the NK receptor genes in the human NK gene complex. Eur J Immunol 31:3493–3503. https://doi.org/10.1002/1521-4141(200112)31:12%3c3493:AID-IMMU3493%3e3.0.CO;2-9

Soilleux EJ, Barten R, Trowsdale J (2000) DC-SIGN; a related gene, DC-SIGNR and CD23 form a cluster on 19p13. J Immunol 165:2937–2942. https://doi.org/10.4049/jimmunol.165.6.2937

Stappers MHT, Clark AE, Aimanianda V et al (2018) Recognition of DHN-melanin by a C-type lectin receptor is required for immunity to *Aspergillus*. Nature 555:382–386. https://doi.org/10.1038/nature25974

Steele C, Marrero L, Swain S et al (2003) Alveolar macrophage-mediated killing of *Pneumocystis carinii* f. sp. muris involves molecular recognition by the Dectin-1 beta-glucan receptor. J Exp Med 198:1677–1688. https://doi.org/10.1084/jem.20030932

Stehle SE, Rogers RA, Harmsen AG, Ezekowitz RA (2000) A soluble mannose receptor immunoadhesin enhances phagocytosis of *Pneumocystis carinii* by human polymorphonuclear leukocytes *in vitro*. Scand J Immunol 52:131–137. https://doi.org/10.1046/j.1365-3083.2000.00755.x

Strasser D, Neumann K, Bergmann H et al (2012) Syk kinase-coupled C-type lectin receptors engage protein kinase C-δ to elicit Card9 adaptor-mediated innate immunity. Immunity 36:32–42. https://doi.org/10.1016/j.immuni.2011.11.015

Sun H, Xu XY, Shao HT et al (2013) Dectin-2 is predominately macrophage restricted and exhibits conspicuous expression during *Aspergillus fumigatus* invasion in human lung. Cell Immunol 284:60–67. https://doi.org/10.1016/j.cellimm.2013.06.013

Swain SD, Lee SJ, Nussenzweig MC, Harmsen AG (2003) Absence of the macrophage mannose receptor in mice does not increase susceptibility to *Pneumocystis carinii* infection *in vivo*. Infect Immun 71:6213–6221. https://doi.org/10.1128/IAI.71.11.6213-6221.2003

Swidergall M, Filler SG (2017) Oropharyngeal candidiasis: fungal invasion and epithelial cell responses. PLoS Pathog 13:1–7. https://doi.org/10.1371/journal.ppat.1006056

Syme RM, Spurrell JCL, Amankwah EK et al (2002) Primary dendritic cells phagocytose *Cryptococcus neoformans* via mannose receptors and Fcγ receptor II for presentation to T lymphocytes. Infect Immun 70:5972–5981. https://doi.org/10.1128/IAI.70.11.5972

Taylor PR, Tsoni SV, Willment JA et al (2007) Dectin-1 is required for β-glucan recognition and control of fungal infection. Nat Immunol 8:31–38. https://doi.org/10.1038/ni1408

Taylor PR, Roy S, Leal SM et al (2014) Activation of neutrophils by autocrine IL-17A-IL-17RC interactions during fungal infection is regulated by IL-6, IL-23, RORγt and Dectin-2. Nat Immunol 15:143–151. https://doi.org/10.1038/ni.2797

Thiagarajan PS, Yakubenko VP, Elsori DH et al (2013) Vimentin is an endogenous ligand for the pattern recognition receptor Dectin-1. Cardiovasc Res 99:494–504. https://doi.org/10.1093/cvr/cvt117

Thomas CF, Limper AH (2004) *Pneumocystis* pneumonia. N Engl J Med 107:830–842. https://doi.org/10.1016/S0929-6646(08)60199-0

van de Veerdonk FL, Marijnissen RJ, Kullberg BJ et al (2009) The macrophage mannose receptor induces IL-17 in response to *Candida albicans*. Cell Host Microbe 5:329–340. https://doi.org/10.1016/j.chom.2009.02.006

Van De Veerdonk FL, Gresnigt MS, Romani L et al (2017) *Aspergillus fumigatus* morphology and dynamic host interactions. Nat Rev Microbiol 15:661–674. https://doi.org/10.1038/nrmicro.2017.90

Vautier S, Drummond RA, Redelinghuys P et al (2012) Dectin-1 is not required for controlling *Candida albicans* colonization of the gastrointestinal tract. Infect Immun 80:4216–4222. https://doi.org/10.1128/IAI.00559-12

Wächtler B, Wilson D, Haedicke K et al (2011) From attachment to damage: defined genes of *Candida albicans* mediate adhesion, invasion and damage during interaction with oral epithelial cells. PLoS ONE 6. https://doi.org/10.1371/journal.pone.0017046

Wagener M, Hoving JC, Ndlovu H, Marakalala MJ (2018) Dectin-1-Syk-CARD9 signaling pathway in TB immunity. Front Immunol 9:1–7. https://doi.org/10.3389/fimmu.2018.00225

Walsh NM, Wuthrich M, Wang H et al (2017) Characterization of C-type lectins reveals an unexpectedly limited interaction between *Cryptococcus neoformans* spores and Dectin-1. PLoS ONE 12:1–21. https://doi.org/10.1371/journal.pone.0173866

Wang Q, Zhao G, Lin J et al (2016) Role of the mannose receptor during *Aspergillus fumigatus* infection and interaction with Dectin-1 in corneal epithelial cells. Cornea. https://doi.org/10.1097/ICO.0000000000000710

Wang H, Lee TJ, Fites SJ et al (2017) Ligation of Dectin-2 with a novel microbial ligand promotes adjuvant activity for vaccination. PLoS Pathog 13:1–20. https://doi.org/10.1371/journal.ppat.1006568

Weis WI, Taylor ME, Drickamer K (1998) The C-type lectin superfamily in the immune system. Immunol Rev 163:19–34. https://doi.org/10.1111/j.1600-065X.1998.tb01185.x

Werner JL, Metz AE, Horn D et al (2009) Requisite role for the dectin-1 β-glucan receptor in pulmonary defense against *Aspergillus fumigatus*. J Immunol 182:4938–4946. https://doi.org/10.4049/jimmunol.0804250

Williams SJ (2017) Sensing lipids with mincle: structure and function. Front Immunol 8. https://doi.org/10.3389/fimmu.2017.01662

Xu Q, Zhao G, Lin J et al (2015) Role of dectin-1 in the innate immune response of rat corneal epithelial cells to *Aspergillus fumigatus* cornea and external eye diseases. BMC Ophthalmol 15:1–7. https://doi.org/10.1186/s12886-015-0112-1

Yang AM, Inamine T, Hochrath K et al (2017) Intestinal fungi contribute to development of alcoholic liver disease. J Clin Invest 127:2829–2841. https://doi.org/10.1172/JCI90562

Yuan K, Zhao G, Che C et al (2017) Dectin-1 is essential for IL-1β production through JNK activation and apoptosis in *Aspergillus fumigatus* keratitis. Int Immunopharmacol 52:168–175. https://doi.org/10.1016/j.intimp.2017.09.008

Zaragoza O, Rodrigues ML, De Jesus M et al (2010) The capsule of the fungal pathogen *Cryptococcus neoformans*. Adv Appl Microbiol 2164:1–64. https://doi.org/10.1016/S0065-2164(09)01204-0

Zelensky AN, Gready JE (2005) The C-type lectin-like domain superfamily. FEBS J 272:6179–6217

Zhu LL, Zhao XQ, Jiang C et al (2013) C-type lectin receptors Dectin-3 and Dectin-2 form a heterodimeric pattern-recognition receptor for host defense against fungal infection. Immunity 39:324–334. https://doi.org/10.1016/j.immuni.2013.05.017

Chapter 2
Immune Recognition of Pathogen-Derived Glycolipids Through Mincle

Yasunobu Miyake and Sho Yamasaki

Abstract Mincle (macrophage inducible C-type lectin, Clec4e, Clecsf9) was originally identified as a member of the C-type lectin receptor family in 1999. Then, the function of Mincle to control antifungal immunity by binding to *Candida albicans* was reported in 2008. Around the same time, it was reported that Mincle recognized damaged cells and induced sterile inflammation by coupling with the ITAM-adaptor molecule FcRγ. In the following year, a breakthrough discovery reported that Mincle was an essential receptor for mycobacterial cord factor (trehalose-6,6′-dimycolate, TDM). Mincle gained increasing attention immediately after this critical finding. Although our understanding of the recognition of *Mycobacteria* has been advanced significantly, it was also revealed that Mincle interacts with pathogens other than *Mycobacteria*. In addition, endogenous ligands of Mincle were identified recently. Therefore, Mincle is now considered a danger receptor both for self and non-self ligands, so-called damage-associated molecular patterns (DAMPs) and pathogen-associated molecular patterns (PAMPs). This chapter will give an overview of the accumulated knowledge of the multi-task danger receptor Mincle from its discovery to the latest findings.

Keywords C-type lectin receptor · Danger signal · Glycolipids

Y. Miyake (✉)
Division of Molecular and Cellular Immunoscience, Department of Biomolecular Sciences, Faculty of Medicine, Saga University, Saga, Japan
e-mail: ymiyake@cc.saga-u.ac.jp

S. Yamasaki
Department of Molecular Immunology, Research Institute for Microbial Diseases, Osaka University, Suita 565-0871, Japan
e-mail: yamasaki@biken.osaka-u.ac.jp

Laboratory of Molecular Immunology, Immunology Frontier Research Center, Osaka University, Osaka, Japan

S.-L. Hsieh (ed.), *Lectin in Host Defense Against Microbial Infections*, Advances in Experimental Medicine and Biology 1204,
https://doi.org/10.1007/978-981-15-1580-4_2

2.1 Introduction

Mincle was originally identified as a transcriptional target of NF-IL6 (C/EBPβ) in murine peritoneal macrophages following stimulation with lipopolysaccharide plus interferon (IFN)-γ (Matsumoto et al. 1999). The Mincle gene was mapped to murine chromosome 6F2 and human chromosome 12p13 in the antigen presenting cell lectin-like receptor gene complex (APLEC) locus, which consists of evolutionary conserved lectin receptors including macrophage C-type lectin (MCL, clec4d), dendritic cell-associated C-type lectin-2 (Dectin-2, clec6a) and dendritic cell immunoreceptor (DCIR, clec4a) (Flornes et al. 2004). The 5′-flanking region of Mincle contains a number of potential binding sites for various transcription factors including NF-IL6 and NF-κB. Consistently, the mRNA expression of Mincle is robustly upregulated after exposure to various stimuli such as inflammatory cytokines and Toll-like receptor (TLR) ligands despite its negligible expression under steady-state conditions. Mincle ligands also upregulate Mincle expression through an NF-IL/NF-κB dependent auto-amplification loop (Zhang et al. 2014; Schoenen et al. 2014). In contrast, IL-4 downregulates Mincle expression in a STAT6-dependent manner in mouse and human antigen presenting cells (Hupfer et al. 2016). Mincle is a single-pass type II transmembrane protein with a typical Ca^{2+}-dependent carbohydrate-recognition domain (CRD) (Weis et al. 1998) in its extracellular region and thus belongs to the C-type lectin receptor (CLR) family. Mincle is mainly expressed on myeloid cell lineages including macrophages (Yamasaki et al. 2008), monocytes (Vijayan et al. 2012), neutrophils (Schoenen et al. 2010) and dendritic cells (Miyake et al. 2013). It was also reported that nonmyeloid cells such as B cells (Kawata et al. 2012) and microglia cells (Xie et al. 2017) express Mincle.

2.2 Downstream Signals of Mincle

2.2.1 Inflammatory Signals

Although no typical signaling motif has been found in the cytoplasmic region of Mincle, it has potential serine-threonine phosphorylation sites. Deletion of this cytoplasmic portion has no impact on the signaling capacity of Mincle, suggesting that Mincle by itself has no potential for signal transduction. Instead, Mincle is constitutively associated with a signal adaptor molecule, Fc receptor γ chain (FcRγ), which possesses an immunoreceptor tyrosine-based activation motif (ITAM) in its cytoplasmic region (Yamasaki et al. 2008). A positively charged arginine residue in the transmembrane region of Mincle mediates the interaction with FcRγ. This interaction is specific because Mincle does not bind to other ITAM-adaptor molecules such as DAP12 and CD3ζ. Upon ligand binding to Mincle, Src-family kinases phosphorylate the tyrosine residues of the ITAM sequence of FcRγ, followed by the recruitment of spleen tyrosine kinase (Syk) and activation of a downstream signaling cascade including PLCγ, PKCδ, CARD9/Bcl-10/MALT1 and IKKs (Fig. 2.1). Finally, the transcription factor NF-κB is activated and proinflammatory cytokines and chemokines such as TNF-α, MIP-2 (CXCL2), KC (CXCL1) and IL-6 are produced. This is a major signaling pathway of Mincle leading to both the activation of host immunity against infection and the induction of sterile inflammation during endogenous danger such as tissue damage.

2.2.2 Anti-inflammatory Signals

In addition to its proinflammatory role, Mincle also transduces negative signals in several settings (Fig. 2.1). The engagement of Mincle by TDM or *M. bovis* BCG promotes IL-10 production in bone marrow-derived macrophages (BMDMs), which downregulates IL-12p40 secretion (Patin et al. 2016). The siRNA mediated knockdown of Mincle in human macrophages downregulated IL-10 production during *Helicobacter pylori* infection in vitro (Devi et al. 2015). In contrast, it was reported that TDM has negligible potency to produce IL-10 and that the lipoarabinomannan (LAM)/Dectin-2 axis has a crucial role in IL-10 production in bone marrow-derived dendritic cells (BMDCs) following mycobacterial stimulation (Yonekawa et al. 2014). Mincle inhibits IL-1β production through the iNOS-dependent nitrosylation of the NLRP3 inflammasome, and thus contributes to the resolution of TDM-induced granuloma formation (Lee WB. et al. Mincle-mediated translational regulation is required for strong nitric oxide production and inflammation resolution. Nat Commun 2016). TDM delays phagosome fusion with lysosomes and thus promotes the survival of mycobacteria inside host macrophages (Indrigo et al. 2003). TDM interferes with FcγR-mediated phagosome maturation through SHP-1 (Patin 2017). TLR4 signaling is suppressed by Mincle through two distinct mechanisms (Greco et al. 2016). Mincle negatively regulates expression of the TLR co-receptor CD14 in a SOCS1-dependent manner. Additionally, Mincle is responsible for the A20/ABIN3-dependent degradation of TRAF6 and Mal that are downstream signaling molecules of TLR4. Mincle inhibits the apoptosis of neutrophils and macrophages by suppressing the expression of apoptosis-related molecules such as Fas, Fas ligand and caspase-3 in *Aspergillus fumigatus*-induced keratitis (Lin et al. 2017). In addition, the antibody neutralization of Mincle exacerbated fungal keratitis induced by *A. fumigatus* (Yu et al. 2018).

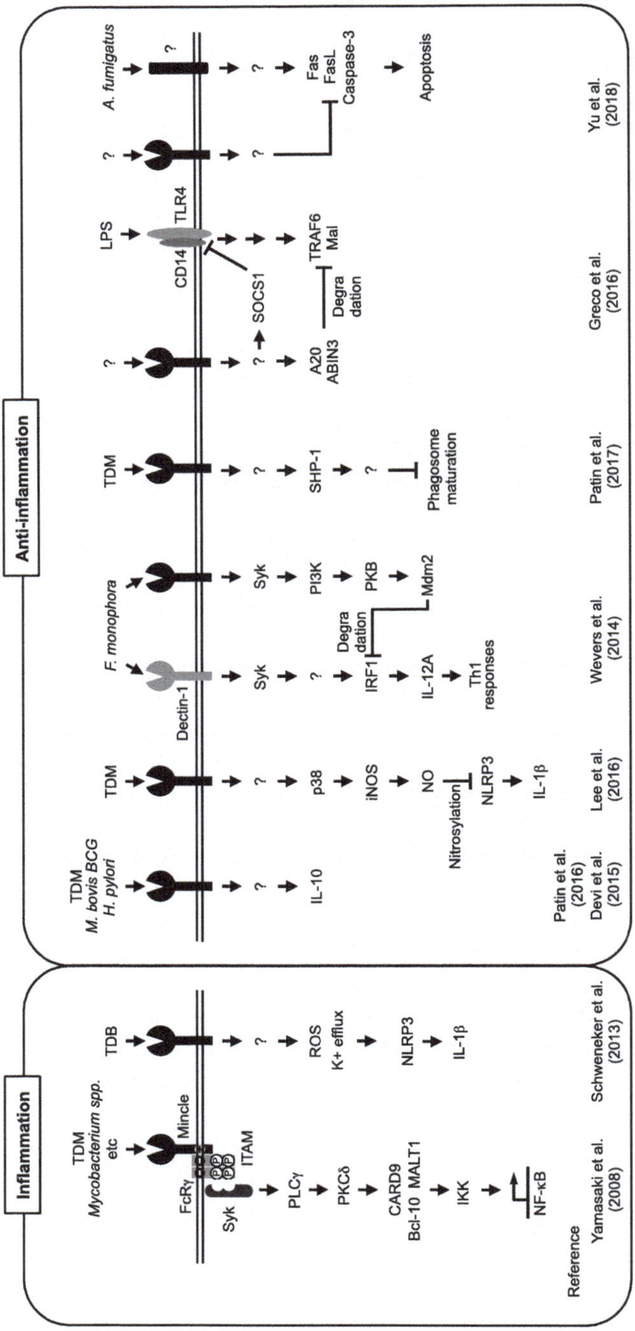

Fig. 2.1 Downstream signals of Mincle

2.2.3 *Inhibition of Mincle Signaling*

Integrin CD11b is a negative regulator of TDM-induced Mincle signaling (Zhang et al. 2018). Activated CD11b recruits Lyn, SIRPα and SHP1, which dephosphorylate Syk to inhibit Mincle-mediated inflammation. SAP130 is released from hypoxia-treated neurons to activate microglial Mincle, which induces inflammation. Interestingly, human serum albumin binds to and masks Mincle, which attenuates microglial inflammatory responses (Xie et al. 2017). Some plant flavonoids suppress Mincle signaling. For example, *Glycyrrhiza uralensis* is a herbal medicine with multiple biological effects including anti-allergic, antiangiogenesis, and antitumor growth activities. Isoliquiritigenin derived from *G. uralensis* inhibited the Mincle-stimulated expression of fibrosis-related genes in a stromal vascular fraction isolated from obese adipose tissue and macrophages (Watanabe et al. 2016). *Sophora flavescens* is a traditional Chinese medicine that exhibits bactericidal activity against *Mycobacterium tuberculosis* (MTB). Furthermore, alkaloid-free flavonoid extracts from *S. flavescens* suppressed proinflammatory mediators released from mouse lung alveolar macrophages following stimulation with TDM (Liu et al. 2018). The oral administration of these flavonoids suppresses inflammation and granuloma formation in the lung following TDM administration in vivo.

2.3 Mincle Senses PAMPs

Mincle recognizes a wide variety of PAMPs and DAMPs and mediates physiological and pathological processes in our body (Table 2.1).

2.3.1 Mycobacteria

Tuberculosis, once termed consumption, is a fatal infectious disease caused by MTB. Mincle recognizes *Mycobacteria* including MTB, *M. bovis* BCG and *M. smegmatis* (Ishikawa et al. 2009). The mycobacterial cell wall contains various unique components that are recognized by host immune receptors such as TLRs, CLRs and scavenger receptors (Stamm et al. 2015). One well-studied component is TDM, which is abundant on the cell surface of mycobacteria. TDM is a trehalose ester with two mycolic acids, a unique very long chain fatty acid of mycobacteria, which possesses strong immunostimulatory and granuloma-inducing capacities (Ryll et al. 2001). Mincle is a dominant receptor for TDM because TDM-induced inflammatory responses including the robust production of proinflammatory mediators and lung granuloma formation disappear in Mincle-deficient mice (Ishikawa et al. 2009; Schoenen et al. 2010). However, Mincle has a dispensable role in the uptake of TDM or TDM-coated particles by macrophages (Kodar et al. 2015). Furthermore, Mincle

Table 2.1 Mincle ligands

Origin			Ligand	Function	References
Non-self	Bacteria	*Mycobacterium* spp.	TDM (trehalose dimycolate)	Inflammation	Ishikawa et al. (2009)
			TMM (trehalose monomycolate)	Inflammation	Ishikawa et al. (2009) and Stocker et al. (2014)
			GroMM (glycerol monomycolate)	Inflammation	Hattori et al. (2011)
			GMM (glucose monomycolate)	Inflammation	Matsunaga et al. (2008)
			β-gentiobiosyl diglyceride	Inflammation	Richardson et al. (2015)
		Corynebacterium spp.	Corynomycolic acid-containing glycolipids	Inflammation	Van der Peet et al. (2015)
		Streptococcus pneumoniae	α-Glucosyl-diacylglycerol	Inflammation	Behler-Janbeck et al. (2016)
		Lactobacillus plantarum		Inflammation	Schick et al. (2017)
		Nonomuraea spp.	Brartemicin	Inflammation	Jacobsen et al. (2015)
		Tannerella forsythia	S-layer	Inflammation	Chinthamani et al. (2017)
	Fungi	*Candida albicans*	?	Inflammation	Bugarcic et al. (2008), Wells et al. (2008)
		Malassezia spp.	Gentiobiosyl-glycerol Mannosyl fatty acids containing mannitol	Inflammation	Yamasaki et al. (2009) and Ishikawa et al. (2013)

(continued)

is unlikely to act as an endocytic receptor because no internalization signals have been identified to date.

Mincle also recognizes TDM-related glycolipids. A series of synthetic TDM analogues with symmetrically shortened acyl chains also bind to Mincle to induce G-CSF and NO production in macrophages (Huber et al. 2016). Therefore, Mincle might also function as an adjuvant (Kallerup et al. 2015). Trehalose monomycolate (TMM), a precursor of TDM biosynthesis, is weakly associated with Mincle (Ishikawa et al. 2009; Stocker et al. 2014). Trehalose monoesters and diesters with a blanched acyl chain show altered ligand activity on human and murine Mincle (Bird et al. 2018). Mycobacteria converts the biosynthetic pathway from TDM into glucose monomycolate (GMM) in the host cell environment (Matsunaga et al. 2008) and Mincle also recognizes GMM (Decout et al. 2017). Arabinose monomycolate (AraMM), a minor component of the mycobacterial cell wall, is recognized by Mincle, but to a lesser degree compared with TMM and GMM (Tima et al. 2017). Glycerol monomycolate (GroMM) is another mycolate-containing glycolipid derived from mycobacteria.

Table 2.1 (continued)

	Origin		Ligand	Function	References
		Fonsecaea spp.	?	Anti-inflammation	Wevers et al. (2014)
				Inflammation	Sousa et al. (2011)
		Pneumocystis carinii	Major surface glycoprotein	Inflammation	Kottom et al. (2017)
	Parasite	*Leishmania major*	Unidentified soluble factor	Anti-inflammation	Iborra et al. (2016)
	Synthesis		TDB (trehalose dibehenate)	Inflammation	Ishikawa et al. (2009)
			GroMB (glycerol monobehenate)	Inflammation	Hattori et al. (2014)
			Trehalose mono/diesters	Inflammation	Bird et al. (2018)
			AraMM (arabinose monomycolates)	Inflammation	Tima et al. (2017)
Self	Damaged cells		SAP130	Inflammation	Yamasaki et al. (2008)
			β-glucosylceramide	Inflammation	Nagata et al. (2017)
	Liver		Cholesterol crystal	Inflammation	Kiyotake et al. (2015)
	Skin		Cholesterol sulfate	Inflammation	Kostarnoy et al. (2017)
	Serum		Human serum albumin	Anti-inflammation	Kawata et al. (2012)

Given that glycerol is abundant in host cells, mycobacteria utilize host glycerol to produce GroMM during the course of infections (Hattori et al. 2011). Interestingly, human but not murine Mincle recognizes GroMM and the synthetic analogue glycerol monobehenate (GroMB) (Hattori et al. 2014). Consistently, GroMM liposome treatment has no impact on WT mice, although it induced skin inflammation in human Mincle Tg mice. Mycobacteria produce a wide range of immunogenic β-gentiobiosyl diglycerides that are weakly recognized by murine Mincle but not human Mincle (Richardson et al. 2015). Glycan array analysis of major carbohydrate structures on the cell wall of mycobacteria confirmed that Mincle has a unique binding specificity that is exclusive to trehalose-containing structures such as TDM (Zheng et al. 2017).

Trehalose dibehenate (TDB) is a synthetic analogue of TDM with shorter acyl chains designed as a vaccine adjuvant (Pimm et al. 1979). Stimulation of BMDCs with TDB induced Nlrp3 inflammasome-dependent IL-1β secretion (Schweneker et al. 2013). Although CARD9 is required for NF-κB activation by TDB, it is dispensable for the TDB-induced activation of the Nlrp3 inflammasome. Additionally, the efflux of intracellular potassium, lysosomal rupture, and oxygen radical (ROS) production are crucial for caspase-1 processing and IL-1β secretion by TDB. In addition, TDB is a strong activator of granulocyte macrophage colony-stimulating factor (GM-CSF) derived BMDMs but is less effective for macrophage colony-stimulating factor (M-CSF)/IL-4 derived BMDMs, suggesting that TDB differentially regulates M1- and M2-like macrophages (Kodar et al. 2017). Mincle is induced specifically on M1 macrophages, where Mincle-Syk signaling promotes and maintains the inflammatory phenotype of M1 macrophages. The adaptive transfer of Mincle-expressing M1 macrophages promoted cisplatin-induced renal inflammation, which was prevented by the knockdown of Mincle (Lv et al. 2017). TDB enhances Th2 cytokine production in RBL-2H3 cells suggesting the Mincle/TDB axis is involved in mast cell activation (Honjoh et al. 2017). Indeed, Mincle is expressed on primary isolated human mast cells (Ma et al. 2011).

Several groups reported the contribution of Mincle in host defense against mycobacterial infection in vivo; however these results are complicated and are not always consistent between studies (Lee et al. 2012; Behler et al. 2012; Heitmann et al. 2013; Behler et al. 2015) (Table 2.2). Different experimental conditions, including mycobacterial strain, infection route/dose and analysis tissue, may affect the results of these previous studies. Bacterial clearance and inflammatory responses are compromised to varying degrees in Mincle-deficient mice suggesting the importance of Mincle in host defense against mycobacterial infection. Nevertheless, these defects are not so severe in all cases. Other receptors such as TLRs, scavenger receptors and other CLRs also contribute to host defense against mycobacteria (Stamm et al. 2015).

Multiple C-type lectin receptors, which share high homology with Mincle, are clustered in the vicinity of the Mincle gene locus and are part of the Dectin-2 family (Fig. 2.2) (Kerscher et al. 2013). Some of these receptors are involved in host defense against mycobacteria. Mincle and MCL cooperatively recognize TDM and induce host immune responses and granuloma formation (see Sect. 2.6) (Miyake et al. 2013). Dectin-2 binds to mannose-capped LAM and controls antimycobacterial host immune responses through the production of both pro- and anti-inflammatory cytokines (Yonekawa et al. 2014). Phosphatidyl-inositol mannosides (PIMs) recognized by dendritic cell immunoactivating receptors (DCAR, clec4b1) expressed on monocyte-derived inflammatory cells induce chemokine production to promote Th1 responses (Toyonaga et al. 2016), although a human homologue of DCAR has not been identified to date. DCIR deficiency impairs STAT1-mediated type I IFN signaling in dendritic cells, leading to enhanced host immune responses against mycobacteria (Troegeler et al. 2017). Furthermore, DCIR may suppress excessive inflammation and prevent tissue damage during mycobacterial infection. Although direct interactions between DCIR and mycobacteria have not been demonstrated, it was reported that DCIR weakly interacted with PIMs (Toyonaga et al. 2016). Taken together, Mincle, MCL Dectin-2 and DCAR form a "mycobacterial receptor cluster", which sense distinct ligands on the mycobacterial cell wall both independently and cooperatively to trigger various types of host immune responses. Our immune system has acquired these mycobacterial receptors by gene duplication through long-standing struggles against mycobacteria. Other Dectin-2 family receptors may also be involved in the recognition of mycobacteria.

Table 2.2 Phenotype of Mincle-deficient mice against mycobacterial infection

Strain	Infection dose(CFU)	Infection route	Infection period (days)	Analysis tissue	Analysis factor	Reference
M. tuberculosis Erdman	100	Aerosol	0–60	Lung	TNFα, IL-6, IFNγ, IL-1β (protein) ↑	Lee et al. (2012)
					Granuloma →	
					CFU ↑	
M. bovis BCG	2–300,000	Intratracheally	0–120	BALF	TNFα, CCL2, KC, IFNγ, Rantes (protein) ↓	Behler et al. (2012)
					Infiltration of macrophages, neutrophils and lymphocytes ↓	
					CFU →	
				Lung	CFU ↑	
				Lung dLN	CFU ↑	
				Spleen	CFU ↑	
	800,000	Intravenously	0–56	BALF	CFU ↑	
				Lung	CFU ↑	
				Lung dLN	CFU ↑	
				Spleen	CFU ↑	

(continued)

Table 2.2 (continued)

Strain	Infection dose(CFU)	Infection route	Infection period (days)	Analysis tissue	Analysis factor	Reference
M. tuberculosis H37Rv	60–150	Aerosol	0–105	Lung	CCL3, CCL4, G-CSF, TNF, IL-1β, IL-6, IFNγ, IL12β, IL17a, Nos2, Irg47 (mRNA) →	Heitmann et al. (2013)
					CCL3, G-CSF, IL-1β, IL-6 (protein) →	
					Granuloma →	
					CFU →	
					CCL4, TNF (protein) ↑	
	2000		0–145	Lung	Granuloma →	
					CFU →	
				Spleen	CFU →	
				Liver	CFU →	

(continued)

Table 2.2 (continued)

Strain	Infection dose(CFU)	Infection route	Infection period (days)	Analysis tissue	Analysis factor	Reference
M. bovis BCG	800,000	Intravenously	0–28	Spleen	CFU ↑	Behler et al. (2015)
					Granuloma ↓	
					Accumulation of macrophages ↓	
					TNFα, IFNg, IL-1β, CCL5 (protein) ↓	
					NO ↓	
				Splenic macrophage	TNF, IFNγ (mRNA) ↓	
					IL-10 Nos2 (mRNA) →	
				Splenic neutrophil	IFNγ, IL-1, Nos2 (mRNA) ↓	
					TNF (mRNA) →	
				Splenic dendritic cell	TNF, IFNγ, IL-1 (mRNA) →	
				Liver	CFU ↑	
					Granuloma →	

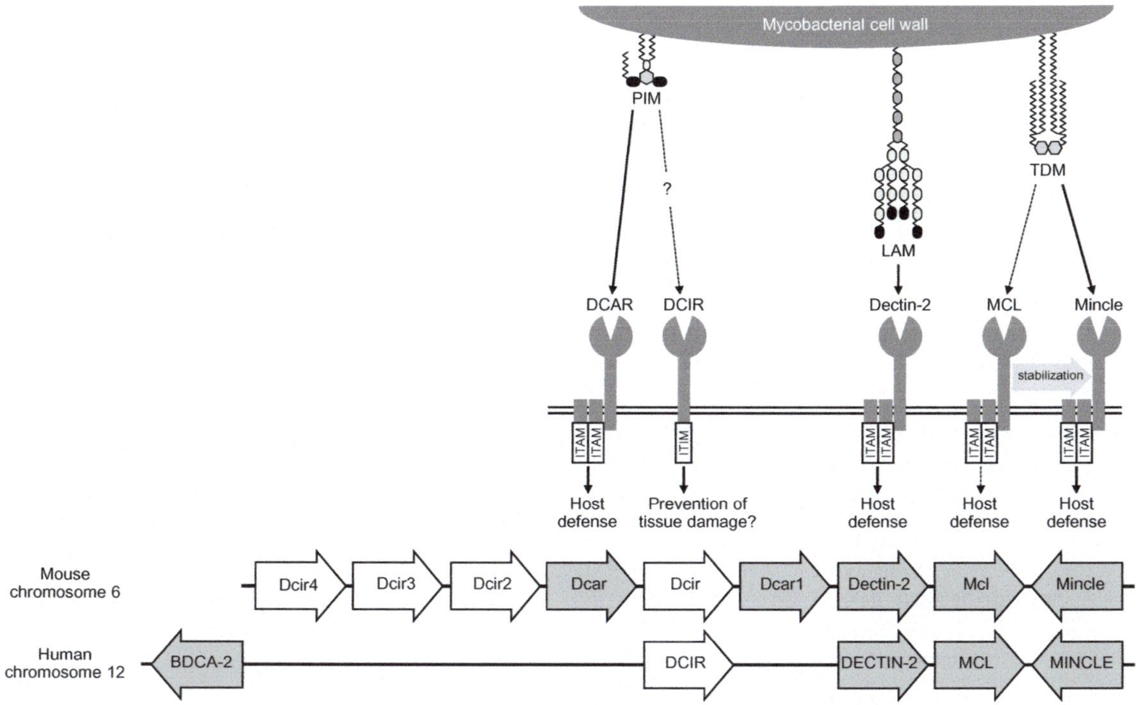

Fig. 2.2 Dectin-2 family receptors for recognition of mycobacterial components. Gray and white arrows indicate activating and inhibitory receptor, respectively

2.3.2 Other Bacteria

Mincle also recognizes pathogenic bacteria other than mycobacteria and is involved in contributing to host defense against these infections.

Streptococcus pneumoniae is a gram-positive bacterium that causes pneumonia, sepsis and meningitis. Binding analysis using a library of recombinantly expressed CLR-Fc fusion proteins showed that Mincle specifically binds to *S. pneumonia* in a serotype dependent manner (Rabes et al. 2015). α-glucosyl-diacylglycerol (αGlc-DAG) is a ligand of Mincle in *S. pneumoniae* (Behler-Janbeck et al. 2016). Purified αGlc-DAG induced inflammatory cytokine release by alveolar macrophages from WT but not Mincle KO mice. Mincle deficiency in mice challenged with serotype 19F *S. pneumoniae* led to increased bacterial loads and decreased survival together with strongly dysregulated cytokine responses (Behler-Janbeck et al. 2016). However, Mincle deficiency does not influence the inflammatory responses of alveolar macrophages and bacterial clearance during infection with serotype 2 and 3 *S. pneumoniae*, even though Mincle binds to both serotypes (Rabes et al. 2015; Behler-Janbeck et al. 2016). Taken together, Mincle contributes to host defense against *S. pneumoniae* in a serotype dependent manner. The gram-positive nonpathogenic bacterium *Lactobacillus plantarum*, commonly found in fermented food products, also produces α-glucosyl diglyceride, termed GL-1 and Mincle recognizes synthetic GL-1 and its derivatives (Schick et al. 2017). Direct interactions between Mincle and *Lactobacillus plantarum* have not been reported to date.

Klebsiella pneumoniae is an anaerobic, gram-negative bacterium indigenous to the oral cavity and intestinal tract; however, it often causes severe respiratory and urinary tract infections. Mincle-deficient mice showed delayed bacterial clearance and compromised host defense including inflammatory cytokine production, phagocytosis of bacteria and the formation of neutrophil extracellular traps (NETs) during infection by *K. pneumoniae* (Sharma et al. 2014). Mincle-mediated NETs formation by regulating autophagy but not ROS generation (Sharma et al. 2017). The direct ligand of Mincle in *K. pneumoniae* is still unknown.

Tannerella forsythia is an anaerobic, gram-negative bacterium in the oral cavity that is associated with the development of periodontal disease. Mincle binds to the S-layer of *T. forsythia* (Chinthamani et al. 2017), which is a monolayer structure of the cell envelope commonly found in bacteria. It is composed of identical glycoproteins; however, S-layer proteins are not conserved and vary between species. Thus, it is unclear whether Mincle recognizes the S-layer of other species. *T. forsythia* is the only gram-negative bacterium to possess an O-glycosylated S-layer. Mincle/S-layer interactions are required to induce both pro- and anti-inflammatory cytokine secretion in differentiated THP-1 cells stimulated with *T. forsythia* as well as its S-layer.

Corynebacterium spp. belong to the order *Corynebacterineae* as well as *Mycobacterium* spp. and contain corynomycolic acid, a shorter form of mycolic acid. Mincle recognizes corynomycolic acid-containing glycolipids (Van der Peet et al. 2015). G-CSF production induced by *Corynebacterium* spp. was compromised in Mincle-deficient macrophages (Shah et al. 2016).

Brartemicin is a natural product isolated from actinomycete of the genus *Nonomuraea* that was reported to inhibit cancer cell invasion (Igarashi et al. 2009). Structurally, brartemicin has a doubly esterified α,α-trehalose core similar to TDM. Brartemicin and its related-analogues are high affinity ligands of Mincle (Jacobsen et al. 2015) and lipidated bratemicin functions as an adjuvant by activating antigen presenting cells (Foster et al. 2018).

2.3.3 Fungi

Candida albicans is a commensal fungus present on the skin and in mucosal tissues that causes candidiasis during immunosuppressive conditions. Mincle binds to *C. albicans* via an unknown ligand (Bugarcic et al. 2008) and promotes TNF production and bacterial clearance after *C. albicans* challenge (Wells et al. 2008). Mincle expression on human peripheral blood neutrophils was associated with phagocytosis and yeast killing activity (Vijayan et al. 2012). In contrast, Mincle expression on peripheral blood monocytes was inversely correlated with candidacidal activity but was necessary for the induction of inflammatory cytokines upon *Candida* challenge.

Malassezia spp. causes skin infections in humans and dogs and is associated with the disease incidence and severity of atopic dermatitis and acne. Mincle broadly recognizes *Malassezia* spp. such as *M. pachydermatis*, *M. japonica*, *M. slooffiae*, and *M. furfur* (Yamasaki et al. 2009). Two unique glycolipids were identified as Mincle ligands, gentiobiosyl-glycerol and mannosyl fatty acids containing mannitol (Ishikawa et al. 2013). These glycolipids induce Mincle-dependent TNF-α production in macrophages and have adjuvant activity that induces Th1/Th17 responses. Mincle-deficient mice showed decreased proinflammatory responses against *M. pachydermatis* (Yamasaki et al. 2009).

Pneumocystis carinii causes pneumocystis pneumonia in rodents. Mincle binds to the major surface glycoprotein (MSG) of *P. carinii* through its CRD region (Kottom et al. 2017). Mincle-deficient mice exhibited increased micro-organism burden and inflammation in the lung during *P. carinii* infection under CD4 lymphocyte-depleted conditions. Because MSG is an abundant protein of *Pneumocystis* spp., Mincle may also recognize the human pathogen *P. jirovecii*, which is the most common opportunistic pathogen of HIV patients.

Fonsecaea spp. are the major causative agents of chromoblastomycosis. Mincle binds to *F. monophora* via an unidentified ligand to suppress antifungal immunity induced by Dectin-1 (Wevers et al. 2014), which recognizes *F. monophora* and activates the transcription factor IRF1 that controls the Th1-promoting cytokine IL-12A. However, *F. monophora* binding to Mincle induces the E3 ubiquitin ligase Mdm2-dependent degradation of IRF1, thereby blocking IL-12A transcription. The absence of IL-12 leads to impaired Th1 responses and promotes Th2 polarization. Mincle also binds to other pathogenic chromoblastomycosis-related fungi such as *F. pedrosoi*, *F. compact* and *Cladophialophora carrionii*. Furthermore, Mincle and TLR ligands coordinately activate inflammatory responses against *F. pedrosoi* (Sousa et al. 2011).

2.3.4 *Parasites*

Leishmania major is a parasitic eukaryote that causes leishmaniasis. *L. major* is a poor inducer of dendritic cell activation and thus escapes Th1-mediated host defense (Ribeiro-Gomes et al. 2012). Mincle-deficient mice showed enhanced parasite clearance and reduced pathology during *L. major* skin infection (Iborra et al. 2016). *L. major* releases soluble Mincle ligand that dampens the activation of Mincle-expressing dendritic cells by recruiting SHP1 to the ITAM-adaptor molecule FcRγ. Thus, numbers of $CD4^+$ $IFN\gamma^+$ cells increased in the infected skin lesions of Mincle-deficient mice. *L. major* escapes host immune responses by shifting Mincle to an inhibitory ITAM configuration (Blank et al. 2009).

2.4 Mincle Senses DAMPs

2.4.1 *Damaged Self*

Mincle recognizes damaged cells and initiates sterile inflammation. Whole-body irradiation or dexamethasone administration in mice causes massive cell death and sterile inflammation in the thymus. Mincle blocking antibody reduces MIP-2 production and infiltration of neutrophils (Yamasaki et al. 2008). SAP130 (Sf3b3), a component of U2 snRNP, was selectively bound to Mincle-Ig fusion protein (Yamasaki et al. 2008). Although SAP130 is localized to the nucleus in healthy living cells, it leaks into the extracellular space during the course of necrotic cell death. The damaged cells/Mincle axis is involved in several pathological settings. SAP130 released from ethanol-treated hepatocytes (Zhou et al. 2016) induces IL-1β production in Kupffer cells in a Mincle-dependent manner (Kim et al. 2018). Mincle KO mice had significantly reduced hepatic steatosis, inflammation with neutrophil infiltration, and fibrosis compared with WT mice in an alcohol-induced liver injury model. The expression of Mincle and SAP130 was elevated in brain injury models including traumatic brain injury (de Rivero Vaccari et al. 2015), ischemic stroke (Suzuki et al. 2013) and subarachnoid hemorrhage (He et al. 2015). Furthermore, the knock-down of Mincle reduced IL-1β, the infiltration of myeloperoxidase positive cells and brain water content, and improved neurological functions in a rat subarachnoid hemorrhage model. The administration of recombinant SAP130 exacerbated inflammation and brain damage (He et al. 2015). Taken together, Mincle and/or SAP130 are at least partly involved in sterile inflammation caused by tissue damages.

β-glucosylceramide (GlcCer) is another endogenous ligand released from dead cells (Nagata et al. 2017). Although β-GlcCer is a ubiquitous intracellular metabolite, the accumulation of β-GlcCer leads to Gaucher disease, a disorder characterized mainly by systemic inflammation. β-GlcCer is detected in the supernatants of dead cells and induces proinflammatory cytokine production in BMDMs in a Mincle-dependent manner. GBA1 deficient mice, which are a model animal for Gaucher disease, show exacerbated inflammation in the thymus upon whole-body irradiation. Mincle-deficiency tends to decrease this hyper-inflammation, suggesting that Mincle may be involved in the disease progression of Gaucher disease.

2.4.2 Cholesterol

Human, but not murine, Mincle functions as a direct receptor for cholesterol crystals and induces inflammation (Kiyotake et al. 2015). This species-specific interaction can be explained by the existence of a cholesterol recognition/interaction amino acid consensus (CRAC) motif (Li et al. 1998) that is only present in human Mincle. Human Mincle also recognizes sitosterol and desmosterol but not ergosterol, cholesterol esters or steroid hormones. Cholesterol crystals are not recognized by Mincle-related CLRs including MCL, Dectin-2 and Dectin-1. However, murine Mincle recognizes cholesterol sulfate, present at relatively high concentrations in the epithelial layer of skin, and which induces the secretion of proinflammatory mediators to promote skin allergy in a murine model of allergic contact dermatitis (Kostarnoy et al. 2017).

2.5 Ligand Recognition Mode of Mincle

Crystal structures of CRD have been reported for human and bovine Mincle (Furukawa et al. 2013; Feinberg et al. 2013). Mincle CRD contains a conserved EPN motif, which is a potential sugar-binding site for mannose and fucose that is essential for the recognition of glycolipid ligands such as TDM (Ishikawa et al. 2009). The overall structure around this sugar-binding site is similar to other CLRs such as DC-SIGNR. One glucose residue of trehalose is closely associated with this primary sugar-binding site. The second glucose residue of trehalose is accommodated in a unique secondary sugar-binding site near the primary sugar-binding site, which further stabilizes the interaction (Feinberg et al. 2013). Long acyl chains of TDM facilitate efficient immune responses in macrophages (Khan et al. 2011). Unique hydrophobic stretches are found in Mincle CRD (Furukawa et al. 2013; Feinberg et al. 2013) and these form shallow grooves adjacent to the sugar-binding site and associate with the acyl chain of TDM. The flexibility of long acyl chains of TDM allows multiple interactions with the hydrophobic grooves of Mincle CRD (Söldner et al. 2018). These interactions further stabilize the association of Mincle and TDM. A number of TDM-related molecules have been chemically synthesized and their ligand activities were extensively analyzed. More detailed information on the structure-activity relationship of glycolipid ligands has been summarized in excellent reviews elsewhere (Williams 2017; Braganza et al. 1940).

2.6 MCL, a Key Partner of Mincle

MCL is expressed on myeloid cells and shows high homology to Mincle (Balch et al. 1998; Arce et al. 2004). The genomic annotation of the Mincle and MCL genes of various mammalian species reveals that monotremes and marsupialia possess Mincle but not MCL. However, placental mammals possess both Mincle and MCL, which are located next to each other in the genome (Flornes et al. 2004). This suggests that MCL may arise from the gene duplication of Mincle (Miyake et al. 2013). Consistently, MCL recognizes TDM; however, its binding affinity is much weaker compared with Mincle. In contrast to Mincle, MCL is constitutively expressed in the absence of stimulation but is retained intracellularly (Kerscher et al. 2016). Immediately after Mincle induction, MCL forms a functional complex (Yamasaki 2013) that is stabilized on the cell surface (Miyake et al. 2015). MCL is indispensable for efficient Mincle expression and signaling in both mice and humans (Miyake et al. 2013; Kerscher et al. 2016; Ostrop et al. 2015). Notably, MCL-deficient mice show significant defects in host defense against infection by *M. bovis* BCG or *M. tuberculosis* H37Rv (Wilson et al. 2015). Furthermore, single nucleotide polymorphisms (SNPs) of human MCL (rs4304840) were associated with increased susceptibility to pulmonary tuberculosis. This polymorphism causes an amino acid change in the transmembrane region of MCL (Ser32Gly), which compromises the interaction of MCL with FcRγ and affects the ability of MCL to be transported to the cell surface. Thus, the rs4304840 polymorphism might reduce the surface expression of MCL and its partner, Mincle.

2.7 Pathological Relevance

Regarding the susceptibility of tuberculosis, four tagging SNPs (rs10841845, rs10841847, rs10841856 and rs4620776) were genotyped in 416 tuberculosis cases and 405 healthy controls; however, no significant association was detected with any of the SNPs analyzed (Bowker et al. 2016). A gender specific association between Mincle SNPs and anti-cyclic citrullinated peptide (CCP)-positive rheumatoid arthritis was reported (Wu et al. 2012). The SNP rs10841845 G allele was reported to have a protective effect against anti-CCP-positive rheumatoid arthritis and confer reduced rheumatoid arthritis activity in men. Enhanced or dysregulated Mincle expression was reported in rheumatoid arthritis (Nakamura et al. 2006), obesity (Ichioka et al. 2011) and recurrent cholesteatoma (Kim et al. 2017). Importantly, it remains unclear whether such upregulation of Mincle is the cause or effect, because Mincle expression is highly susceptible to environmental stress.

More direct evidence has indicated the involvement of Mincle in pathological progression. Mincle-deficient mice are resistant to experimental autoimmune uveitis (EAU) induced by complete Freund's adjuvant (CFA), which contains Mincle ligand mycobacteria (Lee et al. 2016). In addition, Mincle activation by TDB was sufficient to generate the EAU phenotype. However, although Mincle is not involved in fungal-derived EAU (Brown et al. 2017), it is involved in necrosome-promoted pancreatic oncogenesis (Seifert et al. 2016). Mincle was expressed on infiltrating monocytes in human pancreatic ductal adenocarcinoma (PDA), and was associated with SAP130. Ligation of Mincle by SAP130 promotes oncogenesis, whereas the deletion of Mincle protects against oncogenesis through the regulation of myeloid cell-induced adaptive immune suppression. Mincle-deficient mice exhibited lower survival rates in experimental sepsis induced by cecal ligation and puncture and *E. coli*–induced peritonitis (Lee et al. 2017). Mincle enhances neutrophil migration from the blood stream into inflammatory sites. In addition, Mincle deletion or blockade protected against concanavalin A-induced hepatitis, whereas Mincle ligation exacerbated disease (Greco et al. 2016). Mincle exacerbated neuronal loss following ischemic but not traumatic spinal cord injury (Arumugam et al. 2017). The loss of Mincle was beneficial in a model of transient middle cerebral artery occlusion but did not alter outcomes following heart or gut ischemia. Of note, there is no evidence for a direct role of Mincle in microglia or neural activation, although Mincle is expressed in a subset of macrophages resident in the perivascular niche. Furthermore, Mincle underlies obesity-induced adipose tissue inflammation and fibrosis (Tanaka et al. 2014).

The APLEC locus, which contains the Mincle gene, was associated with sensitivity of sepsis induced by *S. aureus* infection, herpes simplex virus-induced encephalitis (Guo et al. 2009) and experimental autoimmune encephalomyelitis (Flytzani et al. 2013) rat models. A correlation of the APLEC locus and rheumatoid arthritis was reported for humans and rats (Lorentzen et al. 2007; Guo et al. 2008).

2.8 Mincle as a Target for Adjuvant

TDM is a prominent component of CFA and Ribi adjuvant. TDM alone shows strong adjuvant activity; therefore, Mincle might be a promising target for adjuvant in vaccine development.

CAF01 is a TDB-containing liposomal adjuvant consistent with a cationic surfactant, dimethyl dioctadecyl ammonium bromide (Christensen et al. 2009). Strong Th1-type cellular immunity is induced by vaccination with mycobacterial antigens such as ESAT-6 and Ag85B in combination with CAF01 adjuvant (Holten-Andersen et al. 2004; Davidsen et al. 2005; Jensen et al. 2017). CAF01 provides long-lasting CD4 memory responses (Lindenstrøm et al. 2009) and elicits Th17 responses (Kamath et al. 2012; Lindenstrøm et al. 2012). Therefore, it markedly potentiates vaccine effects. IL-1 but not IL-18 and IL-33 was required for TDB-induced Th1/17 responses in vivo (Desel et al. 2013). Although neonates and infants showed a reduced efficacy of vaccination (Thakur et al. 2013), CAF01 elicited significantly higher and sustained immune responses compared with other TLR agonists in neonates (Vono et al. 2018). The degree of synergy between Mincle and TLRs varies dramatically with age; it is the greatest in newborns and infants and has less synergy in adults, indicating that a combination adjuvant system might enhance the efficacy of early life vaccines (Van Haren et al. 2016). The efficient adjuvant potential of CAF01 in animal models has been shown by the vaccination of a wide variety of infectious pathogens including HIV (Gram et al. 2009; Fomsgaard et al. 2011), influenza virus (Martel CJ. et al. CAF01 potentiates immune responses and efficacy of an inactivated influenza vaccine in ferrets. PLoS One 2011; Rosenkrands et al. 2011; Christensen et al. 1928), polio virus (Dietrich et al. 2014), group A streptococcus (Mortensen et al. 2017), chlamydia (Yu et al. 2010; Olsen et al. 2010, 2017), *Helicobacter pylori* (Hitzler et al. 2011), leishmania (Leal et al. 2015), and malaria (Pinto et al. 2012; Agger et al. 2008). The clinical safety and efficacy of CAF01 was demonstrated by the vaccination for MTB (van Dissel et al. 2014) and HIV (Karlsson et al. 2013; Román et al. 2013) in humans.

Modification of CAF01 alters the potency of its immune-activating properties. CAF04, which replaces TDB with monomycolyl glycerol, had greater efficient adjuvant activity than CAF01 (Andersen et al. 2009; Billeskov et al. 2016). Whereas CAF01 potently adjuvants helper T cell responses, the stimulation of cytotoxic T cell responses is weak or limited (Agger et al. 2008). The combination of CAF01 plus TLR4-ligand monophosphoryl lipid A improved $CD8^+$ T cell inducing properties without affecting humoral immunity (Nordly et al. 2011). Likewise, TLR3-ligand polyinosinic-polycytidylic acid plus CAF01 or CAF04 (designated as CAF05 and CAF09, respectively) had superior potential to induce $CD8^+$ T cell responses (Nordly et al. 2011; Hansen et al. 2012; Korsholm et al. 2014). The pegylation of CAF01 enhanced IL-5 production and decreased IFN-γ production, thus altering Th1/Th2 immune responses (Kaur et al. 2012).

2.9 Mincle in Other Species

Mincle homologous genes in the genomic sequence database are widely found from fish to mammalian species. Alignment of the CRD region of Mincle across mammalian species demonstrates the strong conservation of key residues involved in its binding to trehalose and acyl chains of TDM (Rambaruth et al. 2015), suggesting Mincle might function as a TDM receptor across species. While human and murine Mincle has been extensively analyzed, several reports have demonstrated the function of Mincle in other species. The guinea pig is a widely used animal for the research and development of tuberculosis vaccines because its pathological disease process is similar to that in humans. Guinea pig Mincle recognizes TDM to the same degree as human and murine Mincle (Toyonaga et al. 2014). Furthermore, guinea pig Mincle is expressed on thioglycollate-elicited peritoneal macrophages and the TDM/Mincle axis is involved in host immune responses against mycobacterial challenge in guinea pigs. Notably, MCL lacks a C-terminus region and fails to bind to TDM in guinea pigs. Bovine Mincle binds to acylated derivatives of trehalose (Feinberg et al. 2016) and brartemicin (Jacobsen et al. 2015). Mincle expression was detected in papillomavirus-associated urothelial tumors in cattle (Roperto et al. 2015). Fish Mincle (barramundi) was reported to act as a partial replacement for

TLR4 in lipopolysaccharide-induced inflammation (Zoccola et al. 2017); however, its mycobacterial recognition activity is unknown.

2.10 Perspectives

The diverse functions of Mincle have become clear following extensive research over the last decade. Several subjects remain to be investigated. Because Mincle recognizes a wide variety of bacteria and fungi, there is a very great likelihood that Mincle may interact with enteric bacteria. There are several possibilities whereby Mincle might contribute to host defense against gut flora, the exacerbation of inflammatory bowel disease, and tissue repair/maintenance of the intestinal tract; however, the function of Mincle in the gut remains unknown. The function of Mincle in the central nervous system (CNS), a glycolipid-rich tissue, is also unknown. Although several reports have indicated involvement of the Mincle/SAP130 axis in brain injury, it is still unclear whether Mincle recognizes endogenous glycolipids in the CNS. It will be tempting to clarify the function of Mincle in the process of sensor nerve transmission or the development/differentiation of the CNS.

References

Agger EM et al (2008) Cationic liposomes formulated with synthetic mycobacterial cordfactor (CAF01): a versatile adjuvant for vaccines with different immunological requirements. PLoS ONE 3:e3116

Andersen CS et al (2009) A simple mycobacterial monomycolated glycerol lipid has potent immunostimulatory activity. J Immunol 182:424–432

Arce I et al (2004) The human C-type lectin CLECSF8 is a novel monocyte/macrophage endocytic receptor. Eur J Immunol 34:210–220

Arumugam TV et al (2017) An atypical role for the myeloid receptor Mincle in central nervous system injury. J Cereb Blood Flow Metab 37:2098–2111

Balch SG et al (1998) Cloning of a novel c-type lectin expressed by murine macrophages. J Biol Chem 273:18656–18664

Behler F et al (2012) Role of Mincle in alveolar macrophage-dependent innate immunity against mycobacterial infections in mice. J Immunol 189:3121–3129

Behler F et al (2015) Macrophage-inducible C-type lectin Mincle-expressing dendritic cells contribute to control of splenic *Mycobacterium bovis* BCG infection in mice. Infect Immun 83:184–196

Behler-Janbeck F et al (2016) C-type lectin Mincle recognizes glucosyl-diacylglycerol of *Streptococcus pneumoniae* and plays a protective role in pneumococcal pneumonia. PLoS Pathog 12:e1006038

Billeskov R et al (2016) Testing the H56 vaccine delivered in 4 different adjuvants as a BCG-booster in a non-human primate model of tuberculosis. PLoS ONE 11:e0161217

Bird JH et al (2018) Synthesis of branched trehalose glycolipids and their Mincle agonist activity. J Org Chem 83:7593–7605

Blank U et al (2009) Inhibitory ITAMs as novel regulators of immunity. Immunol Rev 232:59–71

Bowker N et al (2016) Polymorphisms in the pattern recognition receptor Mincle gene (CLEC4E) and association with tuberculosis. Lung 194:763–767

Braganza CD et al (1940) Identification and biological activity of synthetic macrophage inducible C-type lectin ligands. Front Immunol 2018:8

Brown BR et al (2017) Fungal-derived cues promote ocular autoimmunity through a Dectin-2/Card9-mediated mechanism. Clin Exp Immunol 190:293–303

Bugarcic A et al (2008) Human and mouse macrophage-inducible C-type lectin (Mincle) bind *Candida albicans*. Glycobiology 18:679–685

Chinthamani S et al (2017) Macrophage inducible C-type lectin (Mincle) recognizes glycosylated surface (S)-layer of the periodontal pathogen *Tannerella forsythia*. PLoS ONE 12:e0173394

Christensen D et al (1928) Seasonal influenza split vaccines confer partial cross-protection against heterologous influenza virus in ferrets when combined with the CAF01 adjuvant. Front Immunol 2018:8

Christensen D et al (2009) Liposome-based cationic adjuvant formulations (CAF): past, present, and future. J Liposome Res 19:2–11

Davidsen J et al (2005) Characterization of cationic liposomes based on dimethyldioctadecylammonium and synthetic cord factor from M. tuberculosis (trehalose 6,6'-dibehenate)-a novel adjuvant inducing both strong CMI and antibody responses. Biochim Biophys Acta 1718:22–31

de Rivero Vaccari JC et al (2015) Mincle signaling in the innate immune response after traumatic brain injury. J Neurotrauma 32:228–236

Decout A et al (2017) Rational design of adjuvants targeting the C-type lectin Mincle. Proc Natl Acad Sci U S A 114:2675–2680

Desel C et al (2013) The Mincle-activating adjuvant TDB induces MyD88-dependent Th1 and Th17 responses through IL-1R signaling. PLoS ONE 8:e53531

Devi S et al (2015) Induction of Mincle by *Helicobacter pylori* and consequent anti-inflammatory signaling denote a bacterial survival strategy. Sci Rep 5:15049

Dietrich J et al (2014) Inducing dose sparing with inactivated polio virus formulated in adjuvant CAF01. PLoS ONE 9:e100879

Feinberg H et al (2013) Mechanism for recognition of an unusual mycobacterial glycolipid by the macrophage receptor Mincle. J Biol Chem 288:28457–28465

Feinberg H et al (2016) Binding sites for acylated trehalose analogs of glycolipid ligands on an extended carbohydrate recognition domain of the macrophage receptor Mincle. J Biol Chem 291:21222–21233

Flornes LM et al (2004) Identification of lectin-like receptors expressed by antigen presenting cells and neutrophils and their mapping to a novel gene complex. Immunogenetics 56:506–517

Flytzani S et al (2013) Anti-MOG antibodies are under polygenic regulation with the most significant control coming from the C-type lectin-like gene locus. Genes Immun 14(7):409–419

Fomsgaard A et al (2011) Development and preclinical safety evaluation of a new therapeutic HIV-1 vaccine based on 18 T-cell minimal epitope peptides applying a novel cationic adjuvant CAF01. Vaccine 29:7067–7074

Foster AJ et al (2018) Lipidated brartemicin analogues are potent Th1-stimulating vaccine adjuvants. J Med Chem 61:1045–1060

Furukawa A et al (2013) Structural analysis for glycolipid recognition by the C-type lectins Mincle and MCL. Proc Natl Acad Sci U S A 110:17438–17443

Gram GJ et al (2009) A novel liposome-based adjuvant CAF01 for induction of CD8(+) cytotoxic T-lymphocytes (CTL) to HIV-1 minimal CTL peptides in HLA-A*0201 transgenic mice. PLoS ONE 4:e6950

Greco SH et al (2016a) Mincle suppresses Toll-like receptor 4 activation. J Leukoc Biol 100:185–194

Greco SH et al (2016b) Mincle signaling promotes con A Hepatitis. J Immunol 197:2816–2827

Guo JP et al (2008) Profound and paradoxical impact on arthritis and autoimmunity of the rat antigen-presenting lectin-like receptor complex. Arthritis Rheum 58(5):1343–1353

Guo JP et al (2009) The rat antigen-presenting lectin-like receptor complex influences innate immunity and development of infectious diseases. Genes Immun 10:227–236

Hansen J et al (2012) CAF05: cationic liposomes that incorporate synthetic cord factor and poly(I:C) induce CTL immunity and reduce tumor burden in mice. Cancer Immunol Immunother 61:893–903

Hattori Y et al (2011) Glycerol monomycolate, a latent tuberculosis-associated mycobacterial lipid, induces eosinophilic hypersensitivity responses in guinea pigs. Biochem Biophys Res Commun 409:304–307

Hattori Y et al (2014) Glycerol monomycolate is a novel ligand for the human, but not mouse macrophage inducible C-type lectin, Mincle. J Biol Chem 289:15405–15412

He Y et al (2015) Macrophage-inducible C-type lectin/spleen tyrosine kinase signaling pathway contributes to neuroinflammation after subarachnoid hemorrhage in rats. Stroke 46:2277–2286

Heitmann L et al (2013) Mincle is not essential for controlling *Mycobacterium tuberculosis* infection. Immunobiology 218:506–516

Hitzler I et al (2011) Dendritic cells prevent rather than promote immunity conferred by a helicobacter vaccine using a mycobacterial adjuvant. Gastroenterology 141:186–196

Holten-Andersen L et al (2004) Combination of the cationic surfactant dimethyl dioctadecyl ammonium bromide and synthetic mycobacterial cord factor as an efficient adjuvant for tuberculosis subunit vaccines. Infect Immun 72:1608–1617

Honjoh C et al (2017) Association of C-type lectin Mincle with FcεRIβγ subunits leads to functional activation of RBL-2H3 cells through Syk. Sci Rep 7:46064

Huber A et al (2016) Trehalose diester glycolipids are superior to the monoesters in binding to Mincle, activation of macrophages in vitro and adjuvant activity in vivo. Innate Immun 22:405–418

Hupfer T et al (2016) Stat6-dependent inhibition of Mincle expression in mouse and human antigen-presenting cells by the Th2 cytokine IL-4. Front Immunol 7:423

Iborra S et al (2016) Leishmania uses Mincle to target an inhibitory ITAM signaling pathway in dendritic cells that dampens adaptive immunity to infection. Immunity 45:788–801

Ichioka M et al (2011) Increased expression of macrophage-inducible C-type lectin in adipose tissue of obese mice and humans. Diabetes 60:819–826

Igarashi Y et al (2009) Brartemicin, an inhibitor of tumor cell invasion from the actinomycete Nonomuraea sp. J Nat Prod 72:980–982

Indrigo J et al (2003) Cord factor trehalose 6,6'-dimycolate (TDM) mediates trafficking events during mycobacterial infection of murine macrophages. Microbiology 149:2049–2059

Ishikawa E et al (2009) Direct recognition of the mycobacterial glycolipid, trehalose dimycolate, by C-type lectin Mincle. J Exp Med 206:2879–2888

Ishikawa T et al (2013) Identification of distinct ligands for the C-type lectin receptors Mincle and Dectin-2 in the pathogenic fungus Malassezia. Cell Host Microbe 13:477–488

Jacobsen KM et al (2015) The natural product brartemicin is a high affinity ligand for the carbohydrate-recognition domain of the macrophage receptor mincle. Medchemcomm 6:647–652

Jensen C et al (2017) Optimisation of a murine splenocyte mycobacterial growth inhibition assay using virulent *Mycobacterium tuberculosis*. Sci Rep 7:2830

Kallerup RS et al (2015) Influence of trehalose 6,6'-diester (TDX) chain length on the physicochemical and immunopotentiating properties of DDA/TDX liposomes. Eur J Pharm Biopharm 90:80–89

Kamath AT et al (2012) Synchronization of dendritic cell activation and antigen exposure is required for the induction of Th1/Th17 responses. J Immunol 188:4828–4837

Karlsson I et al (2013) Adjuvanted HLA-supertype restricted subdominant peptides induce new T-cell immunity during untreated HIV-1-infection. Clin Immunol 146:120–130

Kaur R et al (2012) Pegylation of DDA: TDB liposomal adjuvants reduces the vaccine depot effect and alters the Th1/Th2 immune responses. J Control Release 158:72–77

Kawata K et al (2012) Mincle and human B cell function. J Autoimmun 39:315–322

Kerscher B et al (2013) The Dectin-2 family of C-type lectin-like receptors: an update. Int Immunol 25:271–277

Kerscher B et al (2016a) Signalling through MyD88 drives surface expression of the mycobacterial receptors MCL (Clecsf8, Clec4d) and Mincle (Clec4e) following microbial stimulation. Microbes Infec 18:505–509

Kerscher B et al (2016b) Mycobacterial receptor, Clec4d (CLECSF8, MCL), is coregulated with Mincle and upregulated on mouse myeloid cells following microbial challenge. Eur J Immunol 46:381–389

Khan AA et al (2011) Long-chain lipids are required for the innate immune recognition of trehalose diesters by macrophages. ChemBioChem 12:2572–2576

Kim SH et al (2017) Expression of C-type lectin receptor mRNA in chronic otitis media with cholesteatoma. Acta Otolaryngol 137:581–587

Kim JW et al (2018) Spliceosome-associated protein 130 exacerbates alcohol-induced liver injury by inducing NLRP3 inflammasome-mediated IL-1β in mice. Am J Pathol 188:967–980

Kiyotake R et al (2015) Human Mincle binds to cholesterol crystals and triggers innate immune responses. J Biol Chem 290:25322–25332

Kodar K et al (2015) The uptake of trehalose glycolipids by macrophages is independent of Mincle. ChemBioChem 16:683–693

Kodar K et al (2017) The Mincle ligand trehalose dibehenate differentially modulates M1-like and M2-like macrophage phenotype and function via Syk signaling. Immun Inflamm Dis 5:503–514

Korsholm KS et al (2014) Induction of CD8+ T-cell responses against subunit antigens by the novel cationic liposomal CAF09 adjuvant. Vaccine 32:3927–3935

Kostarnoy AV et al (2017) Receptor Mincle promotes skin allergies and is capable of recognizing cholesterol sulfate. Proc Natl Acad Sci U S A 114:E2758–E2765

Kottom TJ et al (2017) The interaction of pneumocystis with the C-type lectin receptor Mincle exerts a significant role in host defense against infection. J Immunol 198:3515–3525

Leal JM et al (2015) Intranasal vaccination with killed *Leishmania amazonensis* promastigotes antigen (LaAg) associated with CAF01 adjuvant induces partial protection in BALB/c mice challenged with Leishmania (infantum) chagasi. Parasitology 142:1640–1646

Lee WB et al (2012) Neutrophils promote mycobacterial trehalose dimycolate-induced lung inflammation via the Mincle pathway. PLoS Pathog 8:e1002614

Lee WB et al (2016) Mincle-mediated translational regulation is required for strong nitric oxide production and inflammation resolution. Nat Commun 7:11322

Lee WB et al (2017) Mincle activation enhances neutrophil migration and resistance to polymicrobial septic peritonitis. Sci Rep 7:41106

Lee EJ et al (2016) Mincle activation and the Syk/Card9 signaling axis are central to the development of autoimmune disease of the eye. J Immunol 196:3148–3158

Li H et al (1998) Peripheral-type benzodiazepine receptor function in cholesterol transport. Identification of a putative cholesterol recognition/interaction amino acid sequence and consensus pattern. Endocrinology 139:4991–4997

Lin J et al (2017) Mincle inhibits neutrophils and macrophages apoptosis in *A. fumigatus* keratitis. Int Immunopharmacol 52:101–109

Lindenstrøm T et al (2009) Tuberculosis subunit vaccination provides long-term protective immunity characterized by multifunctional CD4 memory T cells. J Immunol 182:8047–8055

Lindenstrøm T et al (2012) Vaccine-induced th17 cells are maintained long-term postvaccination as a distinct and phenotypically stable memory subset. Infect Immun 80:3533–3544

Liu D et al (2018) *Sophora flavescens* protects against mycobacterial trehalose dimycolate-induced lung granuloma by inhibiting inflammation and infiltration of macrophages. Sci Rep 8:3903

Lorentzen JC et al (2007) Association of arthritis with a gene complex encoding C-type lectin-like receptors. Arthritis Rheum 56(8):2620–2632

Lv LL et al (2017) The pattern recognition receptor, Mincle, is essential for maintaining the M1 macrophage phenotype in acute renal inflammation. Kidney Int 91:587–602

Ma D et al (2011) Purification and characterization of two new allergens from the salivary glands of the horsefly, Tabanus yao. Allergy 66:101–109

Martel CJ et al (2011) CAF01 potentiates immune responses and efficacy of an inactivated influenza vaccine in ferrets. PLoS ONE 6:e22891

Matsumoto M et al (1999) A novel LPS-inducible C-type lectin is a transcriptional target of NF-IL6 in macrophages. J Immunol 163:5039–5048

Matsunaga I et al (2008) Mycolyltransferase-mediated glycolipid exchange in mycobacteria. J Biol Chem 283:28835–28841

Miyake Y et al (2013) C-type lectin MCL is an FcRγ-coupled receptor that mediates the adjuvanticity of mycobacterial cord factor. Immunity 1050–1062

Miyake Y et al (2015) C-type lectin receptor MCL facilitates Mincle expression and signaling through complex formation. J Immunol 194:5366–5374

Mortensen R et al (2017) Local Th17/IgA immunity correlate with protection against intranasal infection with *Streptococcus pyogenes*. PLoS ONE 12:e0175707

Nagata M et al (2017) Intracellular metabolite β-glucosylceramide is an endogenous Mincle ligand possessing immunostimulatory activity. Proc Natl Acad Sci U S A 114:E3285–E3294

Nakamura N et al (2006) Isolation and expression profiling of genes upregulated in bone marrow-derived mononuclear cells of rheumatoid arthritis patients. DNA Res 13(4):169–183

Nordly P et al (2011a) Incorporation of the TLR4 agonist monophosphoryl lipid A into the bilayer of DDA/TDB liposomes: physico-chemical characterization and induction of CD8+ T-cell responses in vivo. Pharm Res 28:553–562

Nordly P et al (2011b) Immunity by formulation design: induction of high CD8+ T-cell responses by poly(I:C) incorporated into the CAF01 adjuvant via a double emulsion method. J Control Release 150:307–317

Olsen AW et al (2010) Protection against Chlamydia promoted by a subunit vaccine (CTH1) compared with a primary intranasal infection in a mouse genital challenge model. PLoS ONE 5:e10768

Olsen AW et al (2017) Protective effect of vaccine promoted neutralizing antibodies against the intracellular pathogen *Chlamydia trachomatis*. Front Immunol 8:1652

Ostrop J et al (2015) Contribution of MINCLE-SYK signaling to activation of primary human APCs by mycobacterial cord factor and the novel adjuvant TDB. J Immunol 195:2417–2428

Patin EC et al (2017) Trehalose dimycolate interferes with FcγR-mediated phagosome maturation through Mincle, SHP-1 and FcγRIIB signalling. PLoS ONE 12:e0174973

Patin EC et al (2016) Mincle-mediated anti-inflammatory IL-10 response counter-regulates IL-12 in vitro. Innate Immun 22:181–185

Pimm MV et al (1979) Immunotherapy of an ascitic rat hepatoma with cord factor (trehalose-6, 6′-dimycolate) and synthetic analogues. Int J Cancer 24:780–785

Pinto VV et al (2012) The effect of adjuvants on the immune response induced by a DBL4ε-ID4 VAR2CSA based *Plasmodium falciparum* vaccine against placental malaria. Vaccine 30:572–579

Rabes A et al (2015) The C-type lectin receptor Mincle binds to *Streptococcus pneumoniae* but plays a limited role in the anti-pneumococcal innate immune response. PLoS ONE 10:e0117022

Rambaruth ND et al (2015) Mouse Mincle: characterization as a model for human Mincle and evolutionary implications. Molecules 20:6670–6682

Ribeiro-Gomes FL et al (2012) Efficient capture of infected neutrophils by dendritic cells in the skin inhibits the early anti-leishmania response. PLoS Pathog 8:e1002536

Richardson MB et al (2015) Mycobacterium tuberculosis β-gentiobiosyl diacylglycerides signal through the pattern recognition receptor Mincle: total synthesis and structure activity relationships. Chem Commun (Camb) 51:15027–15030

Román VR et al (2013) Therapeutic vaccination using cationic liposome-adjuvanted HIV type 1 peptides representing HLA-supertype-restricted subdominant T cell epitopes: safety, immunogenicity, and feasibility in Guinea-Bissau. AIDS Res Hum Retroviruses 29:1504–1512

Roperto S et al (2015) Mincle, an innate immune receptor, is expressed in urothelial cancer cells of papillomavirus-associated urothelial tumors of cattle. PLoS ONE 10:e0141624

Rosenkrands I et al (2011) Enhanced humoral and cell-mediated immune responses after immunization with trivalent influenza vaccine adjuvanted with cationic liposomes. Vaccine 29:6283–6291

Ryll R et al (2001) Immunological properties of trehalose dimycolate (cord factor) and other mycolic acid-containing glycolipids—a review. Microbiol Immunol 45:801–811

Schick J et al (2017) Toll-like receptor 2 and Mincle cooperatively sense corynebacterial cell wall glycolipids. Infect Immun 85:e00075–17

Schoenen H et al (2010) Cutting edge: Mincle is essential for recognition and adjuvanticity of the mycobacterial cord factor and its synthetic analog trehalose-dibehenate. J Immunol 184:2756–2760

Schoenen H et al (2014) Differential control of Mincle-dependent cord factor recognition and macrophage responses by the transcription factors C/EBPβ and HIF1α. J Immunol 193:3664–3675

Schweneker K et al (2013) The mycobacterial cord factor adjuvant analogue trehalose-6,6'-dibehenate (TDB) activates the Nlrp3 inflammasome. Immunobiology 218:664–673

Seifert L et al (2016) The necrosome promotes pancreatic oncogenesis via CXCL1 and MINCLE-induced immune suppression. Nature 532:245–249

Shah S et al (2016) Total synthesis of a cyclopropane-fatty acid α-glucosyl diglyceride from *Lactobacillus plantarum* and identification of its ability to signal through Mincle. Chem Commun (Camb) 52:10902–10905

Sharma A et al (2014) Protective role of Mincle in bacterial pneumonia by regulation of neutrophil mediated phagocytosis and extracellular trap formation. J Infect Dis 209:1837–1846

Sharma A et al (2017) Mincle-mediated neutrophil extracellular trap formation by regulation of autophagy. J Infect Dis 215:1040–1048

Söldner CA et al (2018) Interaction of glycolipids with the macrophage surface receptor Mincle—a systematic molecular dynamics study. Sci Rep 8:5374

Sousa Mda G et al (2011) Restoration of pattern recognition receptor costimulation to treat chromoblastomycosis, a chronic fungal infection of the skin. Cell Host Microbe 9:436–443

Stamm CE et al (2015) Sensing of *Mycobacterium tuberculosis* and consequences to both host and bacillus. Immunol Rev 264:204–219

Stocker BL et al (2014) On one leg: trehalose monoesters activate macrophages in a Mincle-dependant manner. ChemBioChem 15:382–388

Suzuki Y et al (2013) Involvement of Mincle and Syk in the changes to innate immunity after ischemic stroke. Sci Rep 3:3177

Tanaka M et al (2014) Macrophage-inducible C-type lectin underlies obesity-induced adipose tissue fibrosis. Nat Commun 5:4982

Thakur A et al (2013) Cell-mediated and humoral immune responses after immunization of calves with a recombinant multiantigenic *Mycobacterium avium* subsp. paratuberculosis subunit vaccine at different ages. Clin Vaccine Immunol 20:551–558

Tima HG et al (2017) Inflammatory properties and adjuvant potential of synthetic glycolipids homologous to mycolate esters of the cell wall of *Mycobacterium tuberculosis*. J Innate Immun 9:162–180

Toyonaga K et al (2014) Characterization of the receptors for mycobacterial cord factor in Guinea pig. PLoS ONE 9:e88747

Toyonaga K et al (2016) C-type Lectin receptor DCAR recognizes mycobacterial phosphatidyl-inositol mannosides to promote a Th1 response during infection. Immunity 45:1245–1257

Troegeler A et al (2017) C-type lectin receptor DCIR modulates immunity to tuberculosis by sustaining type I interferon signaling in dendritic cells. Proc Natl Acad Sci U S A 114:E540–E549

Van der Peet PL et al (2015) Corynomycolic acid-containing glycolipids signal through the pattern recognition receptor Mincle. Chem Commun (Camb) 51:5100–5103

van Dissel JT et al (2014) A novel liposomal adjuvant system, CAF01, promotes long-lived *Mycobacterium tuberculosis*-specific T-cell responses in human. Vaccine 32:7098–7107

Van Haren SD et al (2016) Age-specific adjuvant synergy: Dual TLR7/8 and Mincle activation of human newborn dendritic cells enables Th1 polarization. J Immunol 197:4413–4424
Vijayan D et al (2012) Mincle polarizes human monocyte and neutrophil responses to *Candida albicans*. Immunol Cell Biol 90:889–895
Vono M et al (2018) Overcoming the neonatal limitations of inducing germinal centers through liposome-based adjuvants including C-type lectin agonists trehalose dibehenate or curdlan. Front Immunol 9:381
Watanabe Y et al (2016) Isoliquiritigenin attenuates adipose tissue inflammation in vitro and adipose tissue fibrosis through inhibition of innate immune responses in mice. Sci Rep 6:23097
Weis WI et al (1998) The C-type lectin superfamily in the immune system. Immunol Rev 163:19–34
Wells CA et al (2008) The macrophage-inducible C-type lectin, Mincle, is an essential component of the innate immune response to *Candida albicans*. J Immunol 180:7404–7413
Wevers BA et al (2014) Fungal engagement of the C-type lectin Mincle suppresses Dectin-1-induced antifungal immunity. Cell Host Microbe 15:494–505
Williams SJ (2017) Sensing lipids with Mincle: structure and function. Front Immunol 8:1662
Wilson GJ et al (2015) The C-type lectin receptor CLECSF8/CLEC4D is a key component of anti-mycobacterial immunity. Cell Host Microbe 17:252–259
Wu XY et al (2012) Macrophage-inducible C-type lectin is associated with anti-cyclic citrullinated peptide antibodies-positive rheumatoid arthritis in men. Chin Med J (Engl) 125:3115–3119
Xie Y et al (2017) Human albumin attenuates excessive innate immunity via inhibition of microglial Mincle/Syk signaling in subarachnoid hemorrhage. Brain Behav Immun 60:346–360
Yamasaki S (2013) Signaling while eating: MCL is coupled with Mincle. Eur J Immunol 43:3167–3174
Yamasaki S et al (2008) Mincle is an ITAM-coupled activating receptor that senses damaged cells. Nat Immunol 9:1179–1188
Yamasaki S et al (2009) C-type lectin Mincle is an activating receptor for pathogenic fungus, Malassezia. Proc Natl Acad Sci U S A 106:1897–1902
Yonekawa A et al (2014) Dectin-2 is a direct receptor for mannose-capped lipoarabinomannan of mycobacteria. Immunity 41:402–413
Yu H et al (2010) *Chlamydia muridarum* T-cell antigens formulated with the adjuvant DDA/TDB induce immunity against infection that correlates with a high frequency of gamma interferon (IFN-gamma)/tumor necrosis factor alpha and IFN-gamma/interleukin-17 double-positive CD4+ T cells. Infect Immun 78:2272–2282
Yu GR et al (2018) Mincle in the innate immune response of mice fungal keratitis. Int J Ophthalmol 11:539–547
Zhang XQ et al (2014) C-type lectin receptor Dectin-3 mediates trehalose 6,6′-Dimycolate (TDM)-induced Mincle expression through CARD9/Bcl10/MALT1-dependent nuclear factor (NF)-κB activation. J Biol Chem 289:30052–30062
Zhang Q et al (2018) Integrin CD11b negatively regulates Mincle-induced signaling via the Lyn-SIRPα-SHP1 complex. Exp Mol Med 50:e439
Zheng RB et al (2017) Insights into Interactions of mycobacteria with the host innate immune system from a novel array of synthetic mycobacterial glycans. ACS Chem Biol 12:2990–3002
Zhou H et al (2016) IRAKM-Mincle axis links cell death to inflammation: pathophysiological implications for chronic alcoholic liver disease. Hepatology 64:1978–1993
Zoccola E et al (2017) Immune transcriptome reveals the Mincle C-type lectin receptor acts as a partial replacement for TLR4 in lipopolysaccharide-mediated inflammatory response in barramundi (Lates calcarifer). Mol Immunol 83:33–45

Chapter 3
CLEC5A: A Promiscuous Pattern Recognition Receptor to Microbes and Beyond

Pei-Shan Sung, Wei-Chiao Chang and Shie-Liang Hsieh

Abstract CLEC5A is a spleen tyrosine kinase (Syk)-coupled C-type lectin that is highly expressed by monocytes, macrophages, neutrophils, and dendritic cells and interacts with virions directly, via terminal fucose and mannose moieties of viral glycans. CLEC5A also binds to *N*-acetylglucosamine (GlcNAc) and *N*-acetylmuramic acid (MurNAc) disaccharides of bacterial cell walls. Compared to other C-type lectins (DC-SIGN and DC-SIGNR) and TLRs, CLEC5A binds its ligands with relatively low affinities. However, CLEC5A forms a multivalent hetero-complex with DC-SIGN and other C-type lectins upon engagement with ligands, and thereby mediates microbe-induced inflammatory responses via activation of Syk. For example, in vivo studies in mouse models have demonstrated that CLEC5A is responsible for flaviviruses-induced hemorrhagic shock and neuroinflammation, and a CLEC5A polymorphism in humans is associated with disease severity following infection with dengue virus. In addition, CLEC5A is co-activated with TLR2 by *Listeria* and *Staphylococcus*. Furthermore, CLEC5A-postive myeloid cells are responsible for Concanavilin A-induced aseptic inflammatory reactions. Thus, CLEC5A is a promiscuous pattern recognition receptor in myeloid cells and is a potential therapeutic target for attenuation of both septic and aseptic inflammatory reactions.

Keywords CLEC5A · MDL-1 · DAP12 · ITAM · Syk

P.-S. Sung · S.-L. Hsieh (✉)
Genomics Research Center, Academia Sinica, Taipei, Taiwan
e-mail: slhsieh@gate.sinica.edu.tw

P.-S. Sung
e-mail: peijon77120@gate.sinica.edu.tw

W.-C. Chang
School of Pharmacy, Taipei Medical University, Taipei, Taiwan
e-mail: wcc@tmu.edu.tw

S.-L. Hsieh
School of Medicine, Institute of Clinical Medicine, National Yang-Ming University, Taipei, Taiwan

S.-L. Hsieh (ed.), *Lectin in Host Defense Against Microbial Infections*,
Advances in Experimental Medicine and Biology 1204,
https://doi.org/10.1007/978-981-15-1580-4_3

3.1 Introduction

C-type lectins are characterized by a common structural C-type lectin domain (CTLD) that can bind glycan and non-glycan ligands in a Ca^{2+}-independent manner; CTLDs that bind glycans in Ca^{2+}-dependent manner are known as "carbohydrate recognition domains" (CRD) (Zelensky and Gready 2005). The myeloid C-type lectin CLEC5A (also known as MDL-1) (Bakker et al. 1999) is a spleen tyrosine kinase (Syk)-coupled type II membrane protein comprising a C-terminal CTLD and a short N-terminal cytoplasmic domain. Among the 15 groups of C-type lectins, CLEC5A falls into Group V (the NK cell receptor family) (Hsieh 2016), which includes CLEC7A (Dectin-1), CLEC5A, CLEC2, CLEC1, NK receptors (such as NKG2D, the NKRP1 family, the NKG2 family, CD69 and CD94), mast cell-associated functional antigen (MAFA), osteoclast inhibitory lectin (OCIL), and CD72. Similar to NKG2D, CLEC5A signals via the ITAM-containing DNAX-activating protein 12 (DAP12) when it is phosphorylated by Src upon activation (Hsieh 2016). However, the role of DAP10, a DAP12-related adaptor protein associated with NKG2D, in CLEC5A-mediated signaling remains unclear.

The human CLEC5A mRNA encodes a 165-residue polypeptide with an N-terminal signal peptide (a.a. 1–22), followed by a short intracellular cytoplasmic domain (a.a. 23–56), a transmembrane domain (a.a. 57–70) and an extracellular domain (a.a. 71–165). The transmembrane domain contains a positively charged amino acid, Lys-58, which recruits DAP10 and DAP12 to associate with CLEC5A after activation. CLEC5A is mainly expressed by myeloid cells, including monocytes, macrophages, neutrophils, and dendritic cells (Chen et al. 2008), and is further upregulated by interferon-gamma (IFN-γ)(Joyce-Shaikh et al. 2010). In addition, CLEC5A expression is under the control of the PU.1 transcription factor, which is a central regulator of myeloid cell differentiation (Batliner et al. 2011). Recent studies further demonstrate that CLEC5A expression is upregulated by the nuclear factor erythroid 2-related factor 2 (Nrf2) (Cheng et al. 2016), suggesting CLEC5A is regulated by oxidative stress.

An X-ray crystal structure has revealed that CLEC5A is a homodimeric protein when it binds to dengue virus serotype 1–4. Moreover, the CLEC5A crystal structure revealed conformational flexibility, suggesting that CLEC5A can adopt various conformations in vivo and that its conformation is ligand-dependent (Watson et al. 2011).

While NK receptors recognize stress-associated autologous antigens and are crucial for immunosurveillance, group V spleen tyrosine kinase (Syk)-coupled C-type lectins in myeloid cells recognize diverse exogenous and endogenous antigens and are involved in host defense, aseptic inflammation, platelet activation and development (Brown et al. 2018). The ligand for CLEC5A was unknown until it was shown to interact with glycan moieties on dengue virus (DV) (Chen et al. 2008), Japanese encephalitis virus (JEV) (Chen et al. 2012) and type A influenza virus (IAV) (Teng et al. 2017). In addition, CLEC5A has been found to interact with *N*-acetylglucosamine (GlcNAc) and *N*-acetylmuramic acid (MurNAc) disaccharides on gram-positive bacteria (*Listeria monocytogenes* and *Staphylococcus aureus*) (Chen et al. 2017). Moreover, CLEC5A has a critical role in the inflammatory responses associated with collagen-induced rheumatoid arthritis (Joyce-Shaikh et al. 2010) and Concanavalin A-induced liver inflammation (Cheung et al. 2011). Most recently it was demonstrated that CLEC5A interacts with exosomes released from activated platelets (Sung et al. 2019), though the ligand responsible has not been identified yet. In this review, we will discuss protective and pathogenic role of CLEC5A in viral and bacterial infections, and how CLEC5A collaborates with other C-type lectins in the recognition of various antigens.

3.2 CLEC5A Is the Pattern Recognition Receptor for Viral Glycans

It has been shown that host cells contain abundant endosomal sensors (such as TLR3, TLR7, TLR8, and TLR9) and cytoplasmic sensors (such as RIG-I, MDA5, cGAS) that detect the nucleic acids of DNA and RNA viruses (Schlee and Hartmann 2016). After endocytosis, viruses release nucleic acids that trigger receptor- and sensor-mediated signaling pathways to induce the secretion of proinflammatory cytokines and interferons. These nucleic acid receptors and sensors nicely demonstrate how virus-infected cells produce interferons and inflammatory cytokines to limit viral replication and spreading. However, it is hard to explain how some viruses (such as dengue virus, H5N1 virus, SARS-Coronavirus, Ebola viruses etc.) trigger massive cytokine release from leukocytes (known as cytokine storm), thereby causing systemic permeability change and hemorrhagic shock. For example, dengue virus (DV) infection frequently causes high fever (>40 °C), rash, thrombocytopenia, back pain, retroorbital pain and joint pain (Wilder-Smith and Schwartz 2005), but can give rise to more severe symptoms, i.e., dengue hemorrhagic fever/dengue shock syndrome, due to elevated systemic vascular permeability and activation of platelets (Wilder-Smith and Schwartz 2005). In contrast, Hepatitis B virus (HBV) infection is associated with low-grade fever and persistent viremia, leading to fibrotic changes and hepatocellular carcinoma (Ganem and Prince 2004). These observations suggest that each virus must be able to trigger a distinct pathway in addition to nucleic acid receptor-/sensor-mediated signaling to modulate host immunity, thus resulting in various clinical symptoms and outcomes.

The viral nucleocapsid is surrounded by a lipid bilayer that contains viral proteins (such as the DV E protein and HBsAg) to enable virus binding to host cells. Even though the lipid bilayer of the viral envelope is derived from host cell membranes, the envelope also contains proteins encoded by the viral genome. One of the best-studied viral envelope proteins is the E protein of dengue virus, which has two N-linked glycosylation sites at Asn-67 and Asn-153 (Pokidysheva et al. 2006), where the terminal sugars include mannose, fucose, and GluNac (Modis et al. 2003). While the glycosylation site at Asn-153 is conserved in most flaviviruses, glycosylation at Asn-67 is unique for DV serotype 2 (DV2) (Heinz and Allison 2003). It has been shown that dengue virus interacts with the C-type lectin DC-SIGN via Asn-67 (Tassaneetrithep et al. 2003), while JEV interacts with DC-SIGNR via Asn-163 (Shimojima et al. 2014). In addition, E protein is a viral structural protein and is immunogenic; thus we speculated that this glycosylated viral envelope protein could be regarded as a "virus-associated molecular pattern" (VAMP) that triggers host immune responses.

By lectin array screening, we demonstrated that the intact dengue virion binds to DC-SIGN, mannose receptor (MR) and CLEC5A specifically(Chen et al. 2008). While DC-SIGN and MR interact with the dengue virion via mannose in a Ca^{2+}-dependent manner, CLEC5A interacts with the dengue virion via fucose in a Ca^{2+}-independent manner. Even though DC-SIGN (Tassaneetrithep et al. 2003) and MR (Miller et al. 2008) have been reported to interact with dengue viruses, both receptors lack motifs capable of transducing signals upon ligand binding. In contrast, DV activates CLEC5A to recruit DAP12, thereby triggering downstream signaling in immune cells (Chen et al. 2008).

By comparing the binding of CLEC5A to JEV mutants lacking either Asn-67 or Asn-153, we found that CLEC5A interacts with flavivirus via Asn-153 (Fig. 3.1). DC-SIGN is shown to interacts with dengue virion via Asn-67 (Mondotte et al. 2007), thus dengue virion seems interacting with DC-SIGN and CLEC5A via Asn-67 and Asn-153, respectively. Furthermore, the affinity of the CLEC5A-DV interaction is much lower than that of the DC-SIGN-DV interaction, and colocalization of CLEC5A

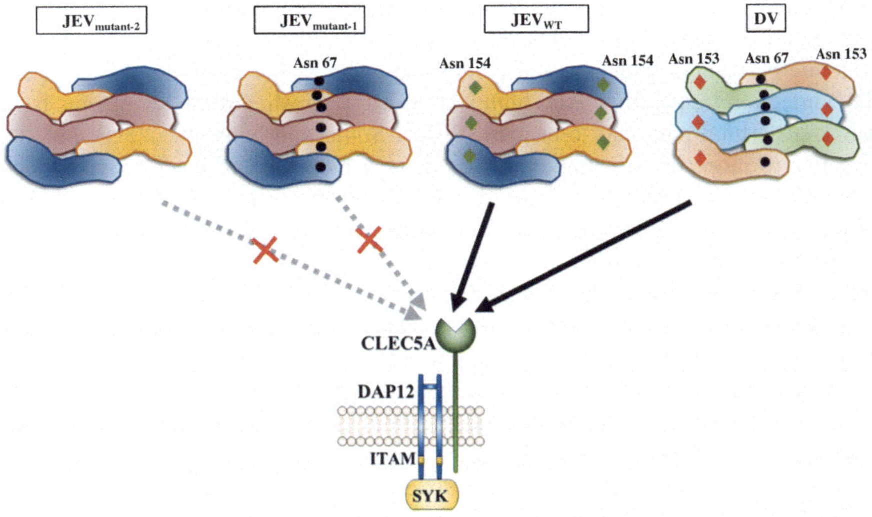

Fig. 3.1 Interaction between CLEC5A and Flaviviruses. Dengue virus (DV) contains two glycosylation sites at Asn-67 Asn-153. Japanese encephalitis virus (JEV) contains a glycosylation site at Asn-154. JEV mutant-1 contains a glycosylation site at Asn-67, while JEV mutant-1 contains no glycosylation site. Because CLEC5A interacts with both DV and wild type JEV, JEV-154 and DV-153 are crucial for the interaction between CLEC5A with JEV and DV, respectively

and DC-SIGN on macrophages was detected upon engagement with DV (Lo et al. 2016). All these observations suggest that DV form a multivalent hetero-complex with CLEC5A and DC-SIGN to activate macrophages via Syk-coupled CLEC5A. Because DC-SIGN is crucial for virus entry into dendritic cells and other DC-SIGN-positive myeloid cells (such as macrophages), DV2 is more efficient than other DV serotypes (DV1, DV3, DV4) and flaviviruses at infecting and replicating in dendritic cells and macrophages. On the other hand, CLEC5A recognizes all flaviviruses via the conserved Asn-153, supporting the notion that CLEC5A is a pan-pattern recognition receptor for all the flaviviruses. These observations suggest that viruses activate host defense mechanisms not only via nucleic acids (denoted as virus-associated molecular pattern-1, VAMP-1), but also activate innate immune receptors via their surface glycans (denoted as virus-associated molecular pattern-2, VAMP-2) (Fig. 3.2). While nonmyeloid cells are only activated by VAMP-1, macrophages and other myeloid cells are activated by both VAMP-1 and VAMP-2 via nucleic acid receptors/sensors and Syk-coupled C-type lectin, respectively; this enhances NFkB-mediated proinflammatory cytokine release to induce a severe cytokine storm. This dual activation pathway provides an explanation for why viruses capable of activating or replicating in macrophages (such as dengue virus, Ebola virus, Bunya virus, H5N1 IAV) frequently cause cytokine storms.

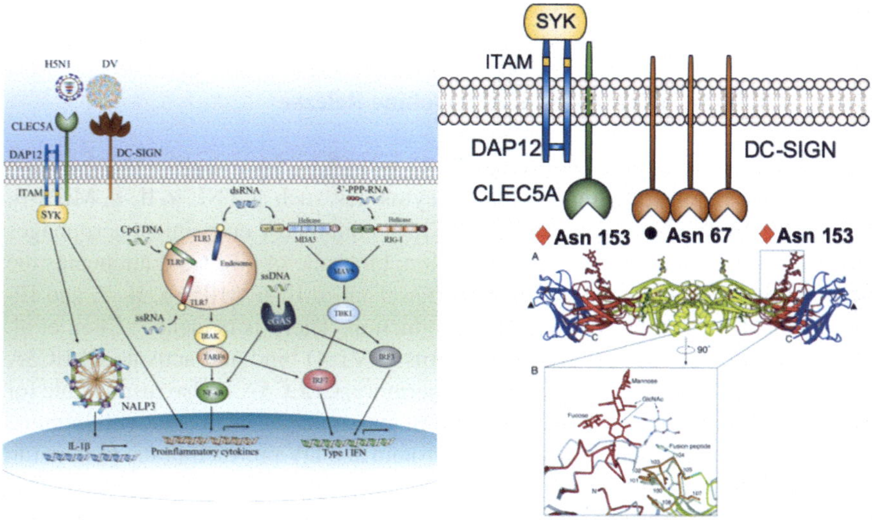

VAMP-1 and cellular sensors **VAMP-2 and lectin receptors**

Fig. 3.2 Activation of CLEC5A and nucleic acid sensors by viruses. Dengue virus infection of dendritic cells results in release of viral nucleic acids (virus-associated molecular pattern-1/VAMP-1) that interact with endosomal TLRs and cellular sensors to activate IRAK and TBK1 and induce the production of proinflammatory cytokines and interferons. In addition, dengue virus activates CLEC5A via viral glycans (virus-associated molecular pattern-2/VAMP-2) to trigger NALP3-inflammasome formation and NFkB activation. Similar pathways are associated with H5N1 IAV infection

3.3 CLEC5A in Flaviviral Infections

The genus *Flavivirus* contains a group of single-stranded, enveloped RNA viruses that cause serious diseases in humans and animals, including DV, JEV, WNV, and ZIKA virus (ZIKV). Even though the RNA structures of these four flaviviruses are similar, their clinical symptoms are very diverse. CLEC5A has been shown to be critical in the pathogenesis of dengue hemorrhagic fever, dengue shock syndrome and JEV-induced encephalitis and neuronal inflammation.

3.3.1 CLEC5A in Dengue Virus Infection

Compared to other flaviviruses (i.e., JEV, WNV, ZIKV), higher viral titers and longer viremia durations have been noted in DV infection. The clinical symptoms of DV infection include high fever (frequently >40 °C), severe headache, retro-orbital pain, back pain, and petechia. In more severe cases, severe thrombocytopenia with increased systemic vascular permeability leads to dengue hemorrhagic fever and shock syndrome. These observations suggest that systemic viremia correlates with viral potency to trigger more severe inflammatory reactions than is the case for other flaviviruses.

3.3.1.1 CLEC5A in Virus-Induced Cytokine Release

DV infection activates both endosomal TLR7 and CLEC5A to induce the secretion of interferons (IFNs) and proinflammatory cytokines, such as TNF-α, IL-6, MIP1-α, IL-8, and IP-10. While knockdown of DC-SIGN inhibits DV entry into macrophages with mild impairment on cytokine production, blockade of CLEC5A attenuates the release of proinflammatory cytokines, such as TNF-α, IL-6, MIP1-α, IL-8, and IP-10 dramatically, without affecting IFN production. There is evidence (Chen et al. 2008) that all four serotypes of dengue virus (DV1-4) bind and activate CLEC5A to secrete proinflammatory cytokines. Furthermore, CLEC5A is also responsible for enhanced inflammatory cytokine release from macrophages incubated with immune complexes containing antibodies to DV E protein and DV pre-membrane protein (an in vitro system to mimic antibody-dependent enhancement of DV infection). This observation demonstrates that DV can activate dual receptor pathways; i.e., endosomal TLR7 and other nucleic acid sensors via viral RNA (VAMP-1), and CLEC5A via glycans and proteins on the viral envelope (VAMP-2). Thus, it should be possible to attenuate virus-induced inflammatory reactions without suppressing IFN-mediated anti-viral immunity. This was confirmed in vivo using an anti-CLEC5A mAb to treat DV-infected mice. Compared to isotype control, anti-CLEC5A mAb inhibited DV-induced plasma leakage, as well as subcutaneous and vital-organ hemorrhaging, thereby reducing mortality by about 50% in STAT1-deficient mice. Thus, blockade of CLEC5A offers a novel strategy for alleviating tissue damage and increasing survival rate in patients suffering from dengue hemorrhagic fever and shock syndrome (Chen et al. 2008).

3.3.1.2 CLEC5A-Mediated NALP3 Inflammasome Activation

Fever is caused by endogenous pyrogens produced by stimulated leukocytes (or other cell types) upon pathogen invasion. The most abundant endogenous pyrogens are TNF-α and IL-1β derived from activated macrophages (Netea et al. 2000; Dinarello 2004), both of which can regulate local and systemic inflammation by activating lymphocytes and promoting leukocyte infiltration. In contrast to TNF-α, IL-1β production relies on the activation of the NACHT, LRR and PYD domain-containing protein 3 (NALP3) inflammasome and caspase-I (Schroder and Tschopp 2010). Thus, it is crucial to understand how DV activates inflammasomes to release IL-1β, which contributes significantly to high fever and inflammatory reactions in dengue patients.

Although both M-CSF and GM-CSF contribute to macrophage differentiation, during infection this is primarily regulated by GM-CSF, which is upregulated during inflammatory responses (Wu et al. 2013) and can cause massive expansion of the macrophage population in the spleen leading splenomegaly (Krakowski et al. 2002). In addition, CLEC5A expression is much higher in GM-CSF-derived macrophages (GM-Mϕ), compared to M-CSF-derived macrophages (M-Mϕ) (Gonzalez-Dominguez et al. 2015). We found that DV activates CLEC5A to induce the activation of the NALP3 inflammasome and caspase-1, leading to abundant release of IL-1β from GM-Mϕ (Wu et al. 2013). Moreover, blockade of CLEC5A by an antagonistic mAb inhibits DV-induced NALP3 inflammasome activation and IL-1β secretion. While serum IL-1β is elevated after DV infection in $stat1^{-/-}$ mice, it is almost undetectable in $stat1^{-/-}clec5a^{-/-}$ mice. These observations help to explain how DV binding to CLEC5A contributes to the severe inflammatory reaction during DV infection (Wu et al. 2013).

3.3.1.3 CLEC5A in DV-Induced Osteoclast Activation

In addition to high fever, the clinical symptoms of dengue infection include severe headache and retro-orbital pain, arthralgia, myalgia, anorexia and abdominal discomfort. Therefore, dengue fever is also known as "break-bone" disease. However, the underlying mechanism for the pain experienced by dengue patients is still unclear. Because increased osteoclastic bone resorption is associated with pain, we asked whether DV infects osteoclasts and increases their osteolytic activity. We found that osteoclasts are highly susceptible to DV infection and that they produce high levels of TNF-α and IL-6 following infection. In contrast, osteoclasts are resistant to WNV and JEV infection, and do not produce detectable proinflammatory cytokines in response to these viruses. Furthermore, DV activates NFATc1 to upregulate osteolytic activity via CLEC5A, and DV infection causes bone tissue inflammation and disturbance of bone homeostasis in *stat1*$^{-/-}$ mice. While CLEC5A-induced proinflammatory cytokine release is via DAP12-mediated signaling, CLEC5A-induced osteolytic activity requires the formation of a CLEC5A/DAP12/DAP10 tri-molecular complex (Inui et al. 2009). Nevertheless, blockade of CLEC5A is able to suppress DV-induced bone inflammation and osteolytic activity in vivo (Huang et al. 2016), confirming the critical role of CLEC5A in osteoclast activation. It would be interesting to explore whether blockade of CLEC5A is able to reduce pain in dengue patients in the future.

3.3.2 CLEC5A in JEV-Induced Neuronal Inflammation

Even though JEV is structurally similar to DV its tropism and the clinical symptoms it induces are distinct. JEV is less effective at infecting peripheral macrophages and inducing proinflammatory cytokines and, while DV is detectable in sera following infection, JEV is almost undetectable in patients' sera. However, JEV infects neuronal cells efficiently, and JEV-induced neuronal inflammation associated with microglia activation. In contrast to the transient nature of DV-induced hemorrhagic fever/shock syndrome, JEV-infected victims experience permanent neuropsychiatric sequelae, including persistent motor defects and severe cognitive and language impairments (Mackenzie et al. 2004). However, it is unclear whether neuronal death after JEV infection is due to a direct cytotoxic effect of JEV on infected neuronal cells, or is via inflammatory mediators released from microglia cells.

We found that supernatants from JEV-infected mouse glial cells have potent neurotoxic effects. Incubation of macrophages and microglia with JEV induces proinflammatory cytokine and chemokine release, where addition of an anti-CLEC5A mAb inhibits JEV-induced proinflammatory cytokine release, and attenuates the neurotoxic effects of supernatants from JEV-infected mixed glial cells in vitro. Although blockade of CLEC5A cannot inhibit JEV infection of neurons and astrocytes, anti-CLEC5A mAb inhibits JEV-induced proinflammatory cytokine release from microglia and prevents bystander damage to neuronal cells. These observations suggest that JEV activates CLEC5A to induce the release of neurotoxic substances from activated microglia. Moreover, injection of anti-CLEC5A mAb into JEV-infected mice maintains the integrity of the blood–brain-barrier, attenuates infiltration of $CD45^{+}Ly6G^{+}$ and $CD45^{+}Ly6C^{+}$ inflammatory cells into the brain, and protects mice from JEV-induced lethality; surviving mice develop protective humoral and cellular immunity against JEV infection. Thus, it is evident that CLEC5A plays a critical role in the pathogenesis of JEV-induced neurotoxicity, where blockade of CLEC5A protects against JEV-induced brain damage (Chen et al. 2012).

3.4 CLEC5A in Type A Influenza Virus (IAV) Infection

Influenza virus is the most common pathogen associated with human respiratory infections, and causes high morbidity and mortality in susceptible hosts who are unable to tolerate the negative consequences of the immune response during the acute stage or reduce pathogen burden in later stages. Innate immune receptors and sensors involved in recognition of influenza viruses include Toll-like receptors TLR3, TLR7, TLR8, retinoic acid-inducing gene (RIG-I), NOD-like receptor family members and NALP3. However, there are no published data to evidence a direct interaction between the NALP3 inflammasome and influenza virus.

As seen in DV infection, C-type lectins expressed on myeloid cells, including DC-SIGN (CLEC4L), DC-SIGNR (CLEC4M), macrophage mannose receptor (MMR) and macrophage galactose-type lectin (MGL), have been shown to mediate influenza virus internalization in a sialic acid-independent manner (Londrigan et al. 2011; Ng et al. 2014; Reading et al. 2000; Upham et al. 2010). However, it is still unclear whether influenza virus can bind and trigger signaling via myeloid C-type lectins to further modulate the host immune response.

We employed a lectin Fc array to identify pattern recognition receptors for H5N1 and H1N1 influenza viruses, using pseudotyped lentiviral particles with surface expression of influenza HA proteins derived from H5N1 or H1N1 isolates. We found that both H1N1 and H5N1 bind to DC-SIGN (CLEC4L) via mannose moieties associated with the influenza virus membrane protein (Londrigan et al. 2011). Moreover, DC-SIGNR (CLEC4M), Mincle (CLEC4E) and CLEC5A also interact with H1N1 and H5N1, where the highest affinity interaction is between CLEC5A and H5N1 (Teng et al. 2017). We further found that CLEC5A plays a critical role in H5N1 IAV-induced release of proinflammatory cytokine (TNF-α), chemokines (IP-10, MCP-1, MIP-1α) and IFN-α (Londrigan et al. 2011). Similar observations were made when human macrophages were incubated with H1N1 and H5N7 IAVs.

The cytokine profile observed for *Clec5a*$^{-/-}$ mouse bone marrow-derived macrophages (mBMDM) was slightly different from that seen for human macrophages blockaded with anti-CLEC5A mAb, when these cells were infected with IAV. While the production of TNF-α and IP-10 was downregulated in *Clec5a*$^{-/-}$ mBMDM, IFN-α/β production was upregulated. Since mice lack orthologues of DC-SIGN and DC-SIGNR, which are involved in IAV infection of human macrophages, the differential responses of mouse and human macrophages to IAV infection can be attributed to the presence or absence of DC-SIGN/DC-SIGNR. Nevertheless, *Clec5a*$^{-/-}$ mice were protected from the sublethal challenge of H5N1 virus by reducing the level of proinflammatory cytokine in the lungs, suggesting activation of CLEC5A plays a pathogenic role in IAV infection.

3.5 CLEC5A in Gram-Positive Bacterial Infection

While TLR2 and TLR4 recognize PAMPs expressed on gram-positive and gram-negative bacteria, respectively (Medzhitov 2001; Akira 2001), TLR2-deficient mice are not always more susceptible than wild type to Gram-positive infection, e.g., in the case of *Listeria* infection (Edelson and Unanue 2002). These observations suggest that other receptors/sensors may play more important roles than TLR2 in host defense against some gram-positive bacterial infection.

3.5.1 CLEC5A Is Critical for Bacteria-Induced NET Formation and Proinflammatory Cytokine Production

Neutrophils clear invading bacteria by phagocytosis and production of reactive oxygen species (ROS) to kill engulfed bacteria. In addition, neutrophils produce neutrophil extracellular traps (NETs) to ensnare a variety of microbes (Brinkmann et al. 2004). However, the key receptors that trigger bacteria-induced NET formation are not clearly defined. It has been shown that free radical production is required for NET formation; however, TLR-mediated signaling does not contribute to ROS production (Gantner et al. 2003). In contrast, Syk-mediated signaling leads to ROS production in the cytosol (Mocsai et al. 2010), thus we asked whether CLEC5A contributes to bacteria-induced NET formation.

We demonstrated that CLEC5A is required for *Listeria*- and *Staphylococcus*-induced NOD-like receptor (NLR) 4 inflammasome activation and IL-1β production in macrophages and also to restrict bacteria-spreading and NET formation in vivo (Chen et al. 2017). While TLR2 binds to teichoic acids, CLEC5A binds to *N*-acetylglucosamine (GlcNAc) and *N*-acetylmuramic acid (MurNAc) disaccharides of the *Listeria* cell wall. Thus, engagement of *Listeria* by neutrophils is associated with co-activation of TLR2 and CLEC5A to induce ROS and proinflammatory cytokine production. Moreover, $Clec5a^{-/-}$ mice are more susceptible to infection than $Tlr2^{-/-}$ mice after intravenous inoculation with *Listeria monocytogenes*; $Clec5a^{-/-}Tlr2^{-/-}$ mice are extremely susceptible to infection with all mice dying within 5 days. These observations suggest co-activation of CLEC5A and TLR2 is critical to host defense against Listeria and Staphylococcus infections. Moreover, CLEC5A deficiency is associated with impaired production of TNF-α and IL-1β, and reduced the numbers of IL-17A-producing γδ T cells, which are critical for immunity to *L. monocytogenes* (Romagnoli et al. 2016) as well as contributing to autoimmune diseases (Papotto et al. 2018; Akitsu and Iwakura 2018). Therefore, CLEC5A/TLR2 activation is not only central to bacteria-induced inflammatory responses, but may also play a key role in the pathogenesis of autoimmunity.

3.6 Multivalent Binding Between Microbes and C-Type Lectins

Using genetic approaches, we have clearly demonstrated critical roles for CLEC5A in viral and bacterial infections as mentioned above. Immunofluorescent staining revealed that CLEC5A colocalizes with mannose receptor (MR) and DC-SIGN in the presence of DV infection, where both MR and DC-SIGN display higher binding affinities for DV compared to CLEC5A (Lo et al. 2016). This observation suggests that CLEC5A and other C-type lectins form a hetero-multivalent receptor complex to engage with DV (Fig. 3.3a). This model is in accord with our observation that dengue virions bind to DC-SIGN/CLEC5A via N-linked glycans associated with Asn-67 (mannose) and Asn-153 (fucose) on DV E protein (VAMP-2), thus activating CLEC5A to trigger downstream signaling. In addition, dengue virus enters macrophages via DC-SIGN and activates TLR7 by interaction with viral RNA (VAMP-1). Co-activation of CLEC5A and endosomal TLR7 leads to enhanced proinflammatory cytokine release from DV-infected macrophages (Chen et al. 2008). These observations suggest that viruses are captured by high-affinity receptors such as DC-SIGN, DC-SIGN and DC-SIGNR, and form a hetero-multivalent receptor

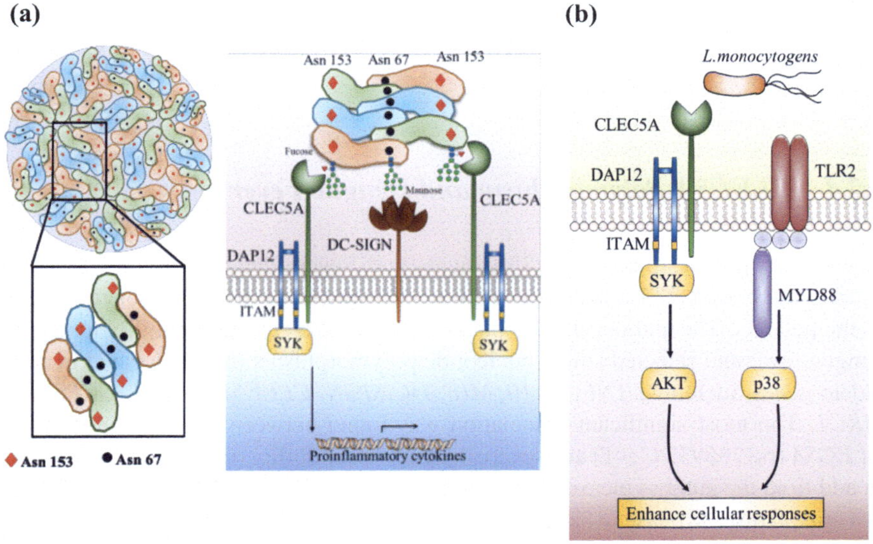

Fig. 3.3 **a** Formation of a hetero-multivalent receptor complex upon engagement with DV. Acrophages express abundant DC-SIGN and CLEC5A, which capture dengue virions via mannose and fucose moieties associated with glycans on Asn-67 and Asn-153 of DV E protein to form a hetero-multivalent receptor complex. **b** Co-activation of CLEC5A and TLR2 by Listeria. Neutrophils and macrophage express abundant TLR2 and CLEC5A, which interact with teichoic acid and GlcNAc/MurNAc disaccharides on bacterial cell walls and are co-activated leading to production of free radicals and NET formation by neutrophils, as well as inducing proinflammatory cytokine and chemokine expression by macrophages

complex with the low-affinity receptor CLEC5A via fucose on the viral envelope. This model is consistent with our observations in flaviviral and H5N1 viral infections and also predicts that CLEC5A on macrophages and dendritic cells will be activated by other microbes that harbor mannose and fucose moieties on their surfaces.

In addition to C-type lectins, we have demonstrated that TLR2 and CLEC5A bind teichoic acid and GlcNAc-MurNAc disaccharides of Listeria cell wall, thereby forming CLEC5A-TLR2 heterocomplex upon engagement with *Listeria monocytogenes* (Chen et al. 2017) (Fig. 3.3b). Because microbial surfaces are rich in various glycans with terminal mannose and fucoses, CLEC5A may also be co-activated with other TLRs during microbial infections. Because activation of CLEC5A induces the production of chemokines and free radicals by macrophages and neutrophils, blockade of CLEC5A may represent a means of attenuating inflammation due to excessive NET formation.

3.7 CLEC5A Polymorphism and Disease Susceptibility

Associations between *CLEC5A* polymorphisms and disease susceptibility have been investigated in previous clinical studies. Furthermore, in vitro studies have been conducted to assess the relationship between *CLEC5A* genetic variants and the expression levels of *CLEC5A* and various immunological mediators (Xavier-Carvalho et al. 2013, 2017).

3.7.1 CLEC5A Polymorphism and Dengue Fever

Xavier-Carvalho et al. recruited 88 patients with severe dengue virus infection in Brazil, along with 335 healthy controls who shared the same living environment as the patients. The authors studied the association between susceptibility to severe dengue fever and selected single nucleotide polymorphisms (SNPs) located in candidate genes, including *TNF*, *IL-10*, *MIF*, *DC-SIGN*, *CLEC5A*, *NOD2*, *CCR5* and *MRC1*. The most significant association was found between the TT genotype of *CLEC5A* (rs1285933 C > T) and severe dengue susceptibility (OR = 2.25; $p = 0.03$). In addition, the authors measured the serum levels of TNF, IL-10, IL-13 and IFN-γ at a critical disease phase in patients with different rs1285933 genotypes. Those with CT or TT genotypes exhibited higher levels of serum TNF than CC individuals, suggesting that *CLEC5A* variants might increase the risk of disease susceptibility and serum TNF level in dengue patients. In a subsequent analysis, the control cohort was divided into groups according to their IgG status, where IgG^+ implied recent infection with mild to no symptoms and without hospital admission; here, only borderline significance was reported for the association between rs1285933 genotype and dengue severity (Xavier-Carvalho et al. 2013, 2017). To validate their findings and to decipher the role of *CLEC5A* genetic variants in dengue viral infection, Xavier-Carvalho et al. conducted a follow-up study. They recruited 213 hospitalized dengue patients and investigated the association of rs1285933 with disease severity. Consistent with their previous study, T carriers at rs1285933 were associated with a more severe manifestation in dengue fever. Furthermore, to investigate the role of *CLEC5A* after DV infection, the authors collected monocytes from healthy individuals and treated the cells with DV. The DV-infected monocytes showed increased *CLEC5A* mRNA and protein expression after infection. Among the groups, monocytes from CC genotype individuals showed higher CLEC5A protein expression 24 h after infection. Furthermore, a positive correlation was found between TNF concentration in supernatants and CLEC5A protein 48 h after infection. However, there was no difference in dengue viral load or TNF expression among the different genotypes. Moreover, the authors observed only borderline significance for the association between *CLEC5A* mRNA levels in patients with mild dengue infection and the CC genotype; TNF mRNA expression showed no variation among mild dengue patients with different genotypes (Xavier-Carvalho et al. 2017).

There is evidence that dengue patients with different ethnicities exhibit different levels of clinical symptoms. For example, migrant Chinese have a higher risk of severe dengue fever than local Chinese in Singapore (Xu et al. 2018). Furthermore, African ethnicity has been identified as a protective factor for severe symptoms in Cuban dengue patients (Sierra et al. 2017). Using data from the 1000 Genomes project, we screened the frequency of rs1285933 alleles of *CLEC5A* in the African, European, American, South Asian and East Asian populations. As shown in Fig. 3.4, the frequency of the T allele, which is associated with increased risk of severe dengue fever, has the lowest frequency in Africans (AFR, 34.9%), followed by the Europeans (EUR, 46.8%), Americans (AMR, 58.1%), South Asians (SAS, 60.9%) and East Asians (EAS, 77.2%). The results indicate that the highly polymorphisms CLEC5A

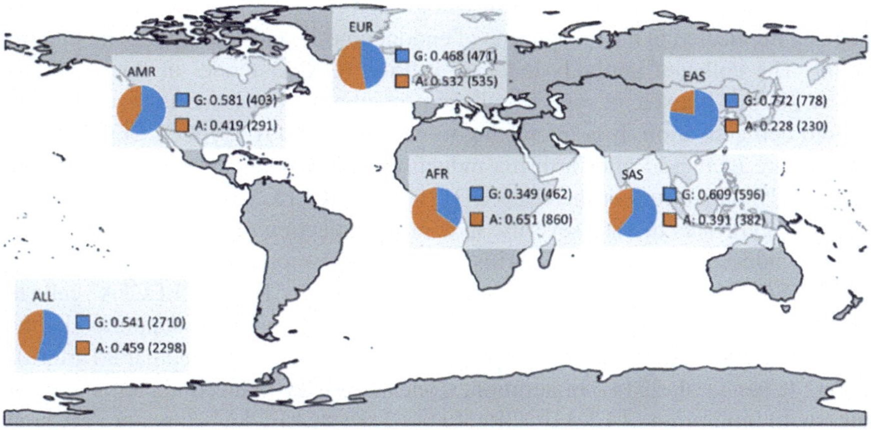

Fig. 3.4 Worldwide allele frequency at rs1285933 in CLEC5A in different ethnicities (shown as frequency (numbers)). ALL: all populations; EUR: Europeans; AMR: Americans; SAS: south Asians; EAS: east Asians

may associate with human migration distance as well as susceptibility of infectious diseases such as dengue fever.

3.7.2 CLEC5A Polymorphism and Kawasaki Disease

Kawasaki disease (KD) is an acute systemic childhood vasculitis without clear etiology. The clinical symptoms include fever, enlargement of lymph nodes, rash in the genital area and sore throat. The most severe complications of KD are aseptic meningitis and aortic aneurysm (Romagnoli et al. 2016). An association study based on 381 KD patients and 664 normal controls was conducted to identify SNPs associated with KD. In that study, four tagging single nucleotide polymorphisms (tSNPs) (i.e., rs1285968, rs11770855, rs1285935, rs1285933) were selected for genotyping. However, no clear association was found between the four tSNPs and susceptibility of coronary artery lesion formation or response to intravenous immunoglobulin treatment (Yang et al. 2012). Since only the SNP with a minor allele frequency of more than 10% were selected for testing, in this study, the effects of the rare causal SNPs are very likely to be underestimated. To understand the genetic functions of CLEC5A in the pathogenesis of KD, a large-scale DNA sequencing to CLEC5A is needed.

3.8 The Role of CLEC5A in Aseptic Inflammatory Reactions

CLEC5A is not only involved in host recognition to viruses and bacteria, but has also been implicated in the recognition of endogenous ligands during aseptic inflammation. The collagen antibody-induced arthritis (CAIA) model in mice is initiated by passive transfer of type II collagen-specific antibody and is dependent on myeloid cell activation. It is interesting to note that $CD11b^+Ly6G^+CLEC5A^+$ cell numbers are increased after arthritis induction, and that injection of an agonistic anti-CLEC5A mAb further increases $CD11b^+Ly6G^+CLEC5A^+$ cell infiltration into joint and aggravates disease severity. In contrast, injection of CLEC5A.Fc inhibits collagen-induced autoimmune arthritis. These observations suggest that activation of CLEC5A can enhance the pathogenic effect of $CD11b^+Ly6G^+CLEC5A^+$ cells in autoimmune arthritis (Joyce-Shaikh et al. 2010). This argument is supported by the observation that CLEC5A is highly upregulated in monocytes of rheumatoid arthritis patients (Chen et al. 2014). In addition, Concanavalin A-induced hepatitis is associated with infiltration of $CD11b^+GR1^+Ly6G^+Ly6C^+CLEC5A^+$ immature myeloid cells into the liver, and activation of this cell population by an agonistic mAb activates eNOS leading to high levels of NO and TNF-α, thereby resulting in shock syndrome in this mouse model (Cheung et al. 2011).

Recent studies have demonstrated that DV infection activates nuclear factor Nrf2, which further upregulates CLEC5A and TNF-α expression (Cheng et al. 2016). These observations suggest that CLEC5A expression is upregulated under oxidative stress. Thus, overactivation of CLEC5A$^+$ immature myeloid cells by endogenous ligands under stressful conditions can lead to systemic inflammatory response syndrome (SIRS), and blockade of CLEC5A may represent a novel strategy for prevention of septic and aseptic shock syndrome.

3.9 Summary

CLEC5A is a myeloid Syk-coupled C-type lectin, which preferentially binds to fucose and mannose. As microbial surfaces are rich in glycans, microbes are captured by lectins, including CLEC5A. Thus, CLEC5A can form multivalent heterocomplex with other C-type lectins or TLRs, resulting in receptor activation, upon engagement with microbial and nonmicrobial antigens. Moreover, CLEC5A is involved in the pathogenesis of aseptic inflammation, suggesting that it may also recognize various fucosylated endogenous danger signals. Blockade of CLEC5A attenuates inflammatory reactions, but without downregulating antiviral immunity (since it does not influence TLR-mediated IFN production), thereby protecting the host from virus-induced systemic inflammatory reactions. In addition to infectious diseases, CLEC5A$^+$ cells mediate aseptic inflammation, e.g., in the contexts of collagen-induced arthritis and cigarette smoke-induced chronic obstructive pulmonary disease (COPD); blockade of CLEC5A inhibits the onset of arthritis and COPD. Thus, CLEC5A is a potential therapeutic target for various inflammatory diseases, and blockade of CLEC5A and associated TLRs may provide a means of protecting against both septic and aseptic inflammatory diseases.

Acknowledgements This work was supported by Academia Sinica (106-2101-01-11-01, 107-2101-01-18-03, AS-TP-106-L11, AS-TP-106-L11-1), and Ministry of Science and Technology (MOST 107-2321-B-001-015, MOST 106-2320-B-001-023-MY3), and Summit and Thematic Research Projects (Academia Sinica), and VGH, TSGH, AS Joint Research Program (VTA107-V2-1-2).

References

Akira S (2001) Toll-like receptors and innate immunity. Adv Immunol 78:1–56

Akitsu A, Iwakura Y (2018) Interleukin-17-producing gammadelta T (gammadelta17) cells in inflammatory diseases. Immunology 155(4):418–426

Bakker AB et al (1999) Myeloid DAP12-associating lectin (MDL)-1 is a cell surface receptor involved in the activation of myeloid cells. Proc Natl Acad Sci U S A 96(17):9792–9796

Batliner J et al (2011) CLEC5A (MDL-1) is a novel PU.1 transcriptional target during myeloid differentiation. Mol Immunol 48(4):714–719

Brinkmann V et al (2004) Neutrophil extracellular traps kill bacteria. Science 303(5663):1532–1535
Brown GD, Willment JA, Whitehead L (2018) C-type lectins in immunity and homeostasis. Nat Rev Immunol 18(6):374–389
Chen ST et al (2008) CLEC5A is critical for dengue-virus-induced lethal disease. Nature 453(7195):672–676
Chen ST et al (2012) CLEC5A regulates Japanese encephalitis virus-induced neuroinflammation and lethality. PLoS Pathog 8(4):e1002655
Chen DY et al (2014) A potential role of myeloid DAP12-associating lectin (MDL)-1 in the regulation of inflammation in rheumatoid arthritis patients. PLoS ONE 9(1):e86105
Chen ST et al (2017) CLEC5A is a critical receptor in innate immunity against Listeria infection. Nat Commun 8(1):299
Cheng YL et al (2016) Activation of Nrf2 by the dengue virus causes an increase in CLEC5A, which enhances TNF-alpha production by mononuclear phagocytes. Sci Rep 6:32000
Cheung R et al (2011) Activation of MDL-1 (CLEC5A) on immature myeloid cells triggers lethal shock in mice. J Clin Invest 121(11):4446–4461
Dinarello CA (2004) Infection, fever, and exogenous and endogenous pyrogens: some concepts have changed. J Endotoxin Res 10(4):201–222
Edelson BT, Unanue ER (2002) MyD88-dependent but Toll-like receptor 2-independent innate immunity to Listeria: no role for either in macrophage listericidal activity. J Immunol 169(7):3869–3875
Ganem D, Prince AM (2004) Hepatitis B virus infection–natural history and clinical consequences. N Engl J Med 350(11):1118–1129
Gantner BN et al (2003) Collaborative induction of inflammatory responses by dectin-1 and Toll-like receptor 2. J Exp Med 197(9):1107–1117
Gonzalez-Dominguez E et al (2015) CD163L1 and CLEC5A discriminate subsets of human resident and inflammatory macrophages in vivo. J Leukoc Biol 98(4):453–466
Heinz FX, Allison SL (2003) Flavivirus structure and membrane fusion. Adv Virus Res 59:63–97
Hsieh SL (2016) The DAP12-assoicated myeloid C-type lectin. In: C-type lectin receptors in immunity, pp 35–49 (Chapter 3)
Huang YL et al (2016) CLEC5A is critical for dengue virus-induced osteoclast activation and bone homeostasis. J Mol Med (Berl) 94(9):1025–1037
Inui M et al (2009) Signal adaptor DAP10 associates with MDL-1 and triggers osteoclastogenesis in cooperation with DAP12. Proc Natl Acad Sci U S A 106(12):4816–4821
Joyce-Shaikh B et al (2010) Myeloid DAP12-associating lectin (MDL)-1 regulates synovial inflammation and bone erosion associated with autoimmune arthritis. J Exp Med 207(3):579–589
Krakowski M et al (2002) Granulocyte macrophage-colony stimulating factor (GM-CSF) recruits immune cells to the pancreas and delays STZ-induced diabetes. J Pathol 196(1):103–112
Lo YL et al (2016) Dengue virus infection is through a cooperative interaction between a mannose receptor and CLEC5A on macrophage as a multivalent hetero-complex. PLoS ONE 11(11):e0166474
Londrigan SL et al (2011) N-linked glycosylation facilitates sialic acid-independent attachment and entry of influenza A viruses into cells expressing DC-SIGN or L-SIGN. J Virol 85(6):2990–3000
Mackenzie JS, Gubler DJ, Petersen LR (2004) Emerging flaviviruses: the spread and resurgence of Japanese encephalitis, West Nile and dengue viruses. Nat Med 10(12 Suppl):S98–S109
Medzhitov R (2001) Toll-like receptors and innate immunity. Nat Rev Immunol 1(2):135–145
Miller JL et al (2008) The mannose receptor mediates dengue virus infection of macrophages. PLoS Pathog 4(2):e17
Mocsai A, Ruland J, Tybulewicz VL (2010) The SYK tyrosine kinase: a crucial player in diverse biological functions. Nat Rev Immunol 10(6):387–402
Modis Y et al (2003) A ligand-binding pocket in the dengue virus envelope glycoprotein. Proc Natl Acad Sci U S A 100(12):6986–6991
Mondotte JA et al (2007) Essential role of dengue virus envelope protein N glycosylation at asparagine-67 during viral propagation. J Virol 81(13):7136–7148

Netea MG, Kullberg BJ, Van der Meer JW (2000) Circulating cytokines as mediators of fever. Clin Infect Dis 31(Suppl 5):S178–S184

Ng WC et al (2014) The macrophage galactose-type lectin can function as an attachment and entry receptor for influenza virus. J Virol 88(3):1659–1672

Papotto PH et al (2018) Innately versatile: gammadelta17 T cells in inflammatory and autoimmune diseases. J Autoimmun 87:26–37

Pokidysheva E et al (2006) Cryo-EM reconstruction of dengue virus in complex with the carbohydrate recognition domain of DC-SIGN. Cell 124(3):485–493

Reading PC, Miller JL, Anders EM (2000) Involvement of the mannose receptor in infection of macrophages by influenza virus. J Virol 74(11):5190–5197

Romagnoli PA et al (2016) IL-17A-producing resident memory $\gamma\delta$ T cells orchestrate the innate immune response to secondary oral Listeria monocytogenes infection. Proc Natl Acad Sci U S A 113(30):8502–8507

Schlee M, Hartmann G (2016) Discriminating self from non-self in nucleic acid sensing. Nat Rev Immunol 16(9):566–580

Schroder K, Tschopp J (2010) The inflammasomes. Cell 140(6):821–832

Shimojima M et al (2014) Distinct usage of three C-type lectins by Japanese encephalitis virus: DC-SIGN, DC-SIGNR, and LSECtin. Arch Virol 159(8):2023–2031

Sierra B et al (2017) OSBPL10, RXRA and lipid metabolism confer African-ancestry protection against dengue haemorrhagic fever in admixed Cubans. PLoS Pathog 13(2):e1006220

Sung PS, Huang TF, Hsieh SL (2019) Extracellular vesicles from CLEC2-activated platelets enhance dengue virus-induced lethality via CLEC5A/TLR2. Nat Commun 10(1):2402

Tassaneetrithep B et al (2003) DC-SIGN (CD209) mediates dengue virus infection of human dendritic cells. J Exp Med 197(7):823–829

Teng O et al (2017) CLEC5A-mediated enhancement of the inflammatory response in myeloid cells contributes to influenza virus pathogenicity in vivo. J Virol 91(1)

Upham JP et al (2010) Macrophage receptors for influenza A virus: role of the macrophage galactose-type lectin and mannose receptor in viral entry. J Virol 84(8):3730–3737

Watson AA et al (2011) Structural flexibility of the macrophage dengue virus receptor CLEC5A: implications for ligand binding and signaling. J Biol Chem 286(27):24208–24218

Wilder-Smith A, Schwartz E (2005) Dengue in travelers. N Engl J Med 353(9):924–932

Wu MF et al (2013a) CLEC5A is critical for dengue virus-induced inflammasome activation in human macrophages. Blood 121(1):95–106

Wu MF, Chen ST, Hsieh SL (2013b) Distinct regulation of dengue virus-induced inflammasome activation in human macrophage subsets. J Biomed Sci 20:36

Xavier-Carvalho C et al (2013) Single nucleotide polymorphisms in candidate genes and dengue severity in children: a case-control, functional and meta-analysis study. Infect Genet Evol 20:197–205

Xavier-Carvalho C et al (2017) Association of rs1285933 single nucleotide polymorphism in CLEC5A gene with dengue severity and its functional effects. Hum Immunol 78(10):649–656

Xu C et al (2018) Differences in clinical features and dengue severity between local and migrant Chinese with dengue infection in Singapore. PLoS ONE 13(8):e0201441

Yang YL et al (2012) Lack of association between CLEC5A gene single-nucleotide polymorphisms and Kawasaki disease in Taiwanese children. J Biomed Biotechnol 2012:398628

Zelensky AN, Gready JE (2005) The C-type lectin-like domain superfamily. FEBS J 272(24):6179–6217

Chapter 4
Collectins: Innate Immune Pattern Recognition Molecules

Valarmathy Murugaiah, Anthony G. Tsolaki and Uday Kishore

Abstract Collectins are collagen-containing C-type (calcium-dependent) lectins which are important pathogen pattern recognising innate immune molecules. Their primary structure is characterised by an N-terminal, triple-helical collagenous region made up of Gly-X-Y repeats, an a-helical coiled-coil trimerising neck region, and a C-terminal C-type lectin or carbohydrate recognition domain (CRD). Further oligomerisation of this primary structure can give rise to more complex and multimeric structures that can be seen under electron microscope. Collectins can be found in serum as well as in a range of tissues at the mucosal surfaces. Mannanbinding lectin can activate the complement system while other members of the collectin family are extremely versatile in recognising a diverse range of pathogens via their CRDs and bring about effector functions designed at the clearance of invading pathogens. These mechanisms include opsonisation, enhancement of phagocytosis, triggering superoxidative burst and nitric oxide production. Collectins can also potentiate the adaptive immune response via antigen presenting cells such as macrophages and dendritic cells through modulation of cytokines and chemokines, thus they can act as a link between innate and adaptive immunity. This chapter describes the structure-function relationships of collectins, their diverse functions, and their interaction with viruses, bacteria, fungi and parasites.

Keywords Collectins · Pathogens · Innate immunity · Phagocytes · Receptors

4.1 Introduction

Collectins (collagen-containing C-type lectins) are soluble mammalian C-type lectins, which represent an important group of pattern-recognition molecules and serve multiple functions in the innate immune system. The term "collectin" was first

V. Murugaiah · A. G. Tsolaki · U. Kishore (✉)
College of Health and Life Sciences, Brunel University London, London UB8 3PH, UK
e-mail: ukishore@hotmail.com; uday.kishore@brunel.ac.uk

A. G. Tsolaki
e-mail: anthony.tsolaki@brunel.ac.uk

S.-L. Hsieh (ed.), *Lectin in Host Defense Against Microbial Infections*,
Advances in Experimental Medicine and Biology 1204,
https://doi.org/10.1007/978-981-15-1580-4_4

used by Malhotra et al. (1992). They are known to mediate pathogen recognition through calcium-dependent carbohydrate recognition domains (CRDs). The following nine collectins have been identified to date: mannan-binding lectin (MBL), three bovine serum collectins, conglutinin, CL-43 and CL-46, lung surfactant proteins SP-A and SP-D, and more recently discovered collectins including, collectin kidney 1 (CL-K1, also called CL-11), collectin liver 1 (CL-L1, also called CL-10) and collectin placenta 1 (CL-P1 also called CL-12). The overall functions of collectins include microbial aggregation and neutralisation, opsonisation, complement activation, and modulation of inflammatory responses.

4.2 Structure of Collectins

Collectins are oligomers of trimeric subunits, For most collectins, the subunits are homotrimers (made up of three identical polypeptides) but heterotrimers can be found for SP-A, which is made up of highly homologous SPA-1 and SPA-2 polypeptides. Hetrotrimers can also form in the case of CL-10 and CL-11 (Fig. 4.1). The subunit of each collectin is composed of (i) a short N-terminal (7-28 amino acid residues) cysteine-rich domain, involved in multimerisation (by disulphide bridging); (ii) a collagen-like domain composed of Gly-X-Y triplets repeats, where X and Y represent any amino acids; (iii) a short segment which can form coiled-coil helices, and (iv)

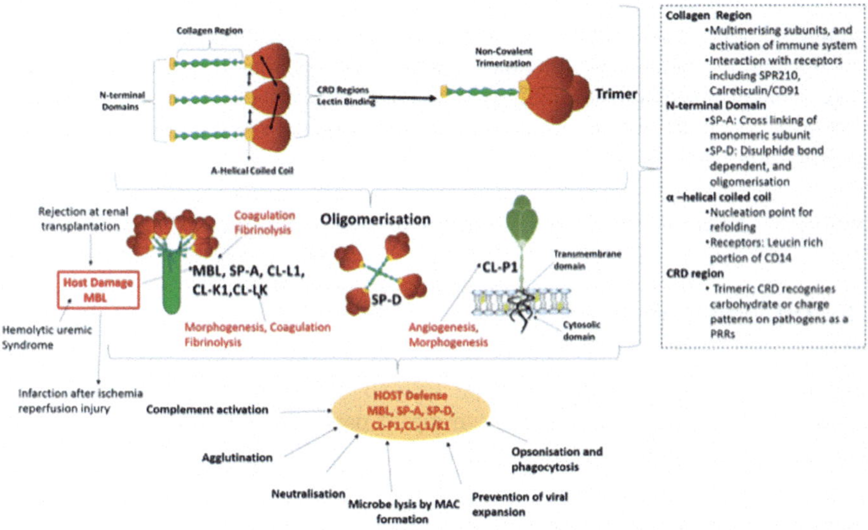

Fig. 4.1 Molecular structural representation and biological functions of human collectins. Collectins are shown as monomeric subunits, followed by trimeric subunits, composed of an N-terminal domain, collagen-like region, α-helical coiled-coil neck region and C-terminal carbohydrate recognition domain (CRD). Biological functions of each domain are also briefly described

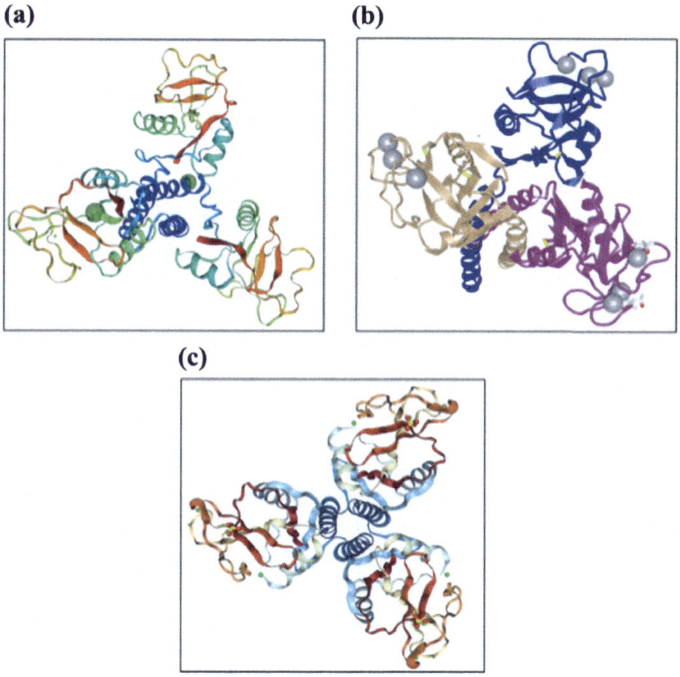

Fig. 4.2 Three-dimensional structures of trimeric human SP-A (**a**), SP-D (**b**), and MBL (**c**). Representations of the trimeric "head" of collectins. These structures represent the 'neck', and the CRDs of three polypeptides which make up the trimeric subunit. The helix interacts with a neighbouring carbohydrate recognition domains (Kishore et al. 2006; Skjoedt et al 2012)

the C-terminal globular C-type lectin domain, also called the CRD (carbohydrate recognition domain) (Uemura et al. 2006) (Fig. 4.1).

The triple-helical collagen region provides significant rigidity and stability to the molecule (Colley and Baenziger 1987). Another structural feature of the collagen-like domain of collectins is that it can be O-glycosylated (Colley and Baenziger 1987). Both MBL and SP-A show an interruption of the Gly-X-Y triplet repeats, which introduces a bend in the otherwise straight triple helix. This enables the fully assembled multi-subunit structure to angle away from the central core, producing a structure resembling a bouquet of flowers (Fig. 4.2) (Voss et al. 1991). Several distinct functions of the collagen domain of collectins have been reported. The collagen domains of SP-A and MBL are involved in receptor-mediated properties. A GEKGEP specific motif found within the collagen domain of MBL is suggested to bind C1q receptor (Arora et al. 2001), and mediates the enhancement of phagocytosis through C1qR (Arora et al. 2001). A similar motif is within the collagen domain of SP-A (White et al. 1985), which is also involved in the interaction with C1q receptor (Malhotra et al. 1992; Malhotra et al. 1990), and mediates phagocytosis of *Staphylococcus aureus* by monocytes (Geertsma et al. 1994). Furthermore, the collagen domain of MBL is shown to bind MBL-associated serum proteases, MASP1, 2 and 3, which mediate complement activation via the lectin pathway (Thiel et al. 1997; Tan et al. 1996). Additionally, the positively charged collagen region found in the membrane bound CL-P1 is involved in the uptake of oxidised LDL particles (Ohtani et al. 2001).

The cysteine residues found within the N-terminal domain (7-28 amino acids) form disulphide bonds between monomers, thereby, stabilising trimeric subunits as well as a larger multimers. It was believed that at least two cysteine residues are required at the N-terminal domain for the formation of multimers of trimeric subunits (Brown-Augsburger et al. 1996; McCormack et al. 1999; McCormack et al. 1997a, b). However, in the case of CL-43, it is secreted as a single trimeric subunit, despite having two cysteine residues (Rothmann et al. 1997; Lim et al. 1994a, b). Therefore, other factors contribute to oligomerisation of trimeric subunits, in addition to the of N-terminal cysteine residues.

The C-terminal region contains a coiled-coil trimerizing neck region (residues 112-130 in human MBL) (Fig. 4.1), and the CRD (residues 134-245 in human MBL) which folds up into an independent globular carbohydrate–binding structure for each polypeptide chain. Each subunit is held together covalently through disulphide bonds, or non-covalently structured into oligomers of up to six subunits. C-type CRDs are connected to the collagen-like domain through the 'neck' region (24-28 amino acid residues) (Hoppe and Reid 1994). Furthermore, the neck region is involved in aligning the collagen chains.

4.2.1 Ligand Specificity of Collectins

A broad carbohydrate specificity is required by collectins in order to recognise and bind a large repertoire of (pathogen-associated molecular patterns) PAMPs. Such broad specificity is achieved by an open and flexible trough -like binding pocket found within the CRDs. The selection of ligands by this site depends on the positioning of vicinal hydroxyl groups of sugars, which form coordination bonds with a ligated calcium ion, hydrogen bonds and a polar Van der Waals contact (Ng et al. 1996). Ligand specificity of collectins is divided into two main sub-classes (mannose-binding or galactose-binding type), which is based on a three amino acid residue motif found in the Ca^{++} ion binding site. The sequence 185-Glu-Pro-Asn is associated with binding of mannose-like sugars, while the sequence 185-Gln-Pro-Asp is associated with binding galactose-like sugars. The molecular differences based on which CRDs discriminate between mannose and galactose-type ligands depend on the orientation of C3 and C4 vicinal hydroxyl groups presented on monosaccharides. Mannose-specific CRDs bind ligands in which hydroxyl groups at the C3 and C4 positions are in an equatorial orientation (mannose, glucose, glucosamine), while in galactose these vicinal hydroxyls are in an axial orientation (Drickamer and Taylor 2015). Inhibition studies using monosaccharides have shown that most likely, all the above described collectins, except CL-P1, prefer mannose ligands over galactose (Ohtani et al. 2001; Holmskov et al. 1994).

However, a wider range of binding specificity has been reported for MBL and lung surfactant proteins SP-A and SP-D, as these collectins are also capable of binding to nucleic acids (Nadesalingam et al. 2003), phospholipids (Sano et al. 1999), as well as non-glucosylated proteins.

Fucose, a hexose deoxy sugar is bound by mannose-specific CRDs in a different manner as it has equatorial hydroxyl groups placed on its C2 and C3 position of the sugar ring, not the C3 and C4 (Weis et al. 1991a, b; Ng et al. 1996; Iobst and Drickamer 1994). Computational docking studies have demonstrated that αD-glucose docks into the CRD of SP-D via vicinal equatorial hydroxyl groups on the 2- and 3- position of its sugar ring (Allen et al. 2001a, b). Although MBL affinity is reported to be very low for monosaccharide galactose, MBL crystallographic studies demonstrate that galactose is ligated in the MBL binding region via coordination bonds with hydroxyl groups placed at C1 and C2 position of the sugar ring (Ng et al. 1996). In addition to galactose and mannose, binding of collectins to a range of sugars has also been studied (Holmskov et al. 1994); they exhibit preferences for certain sugar residues over others. For instance, despite SP-D being structurally similar to conglutinin, it displays a greater affinity for maltose, a glucose disaccharide, which is a weak ligand for conglutinin. SP-D is suggested to have a lower affinity for GlcNAc, which is the best ligand for conglutinin. Moreover, binding of CL-43 to sugars is closely related to MBL, although the structure of CL-43 is closer to SP-D and conglutinin (Lu et al. 2002).

The sugar-binding specificity of CL-11/CL-K1 has been investigated (Venkatraman Girija et al. 2015). It has a larger recognition interface than MBL, and recognises predominantly mannose-rich structures, interacting with two sugar residues at a glycan terminal, rather than a single sugar.

4.3 Biosynthesis and Localisation of Collectins

Human MBL is synthesised by hepatocytes and secreted into the blood stream (Sastry et al. 1991; Ezekowitz et al. 1988; Hansen et al. 2000). Initially, MBL was isolated from the liver of the rabbit, rat and chicken, where expression levels were detected in the soluble cytosol, rather than on the cell surface. Two forms of MBL (MBL-A and MBL-C) were detected in rodents (Hansen et al. 2000; Drickamer et al. 1986), rabbits (Kawasaki et al. 1978; Kozutsumi et al. 1980) and rhesus monkeys (Mogues et al. 1996). However, only one form of MBL is present in humans and chimpanzees (Mogues et al. 1996). Although the liver is the main production site of MBL-A and MBL-C in mice, mRNA expression of MBL was also detected in various tissues (Table 4.1) (Shushimita et al. 2015). Substantial expression levels of MBL-A and MBL-C were reported in kidney and intestine (Table 4.1). Detection of MBL proteins in the small intestine suggests that MBL may have similar roles to secretory IgA (Reichhardt et al. 2012).

Table 4.1 Origin of collectins and their tissue distribution

Collectins	Tissues of origin	Tissues of presentation	Remarks
MBL	Liver and small intestine	Serum	Two different variants of MBL (A and C) have been identified in animals, while only one variant is found in humans and chimpanzee.
Conglutinin, CL-43 and CL-46	Bovine liver	Serum	These bovine collectins plays an important role in the first line of defense against rumen microbes without eliciting general inflammatory response
SP-A and SP-D	Clara cells, intestinal mucosa, thymus, prostrate gland, Eustachian tube, paranasal sinuses, middle ear, synovium	Alveolar space, mucosal surfaces, semen	Extrapulmonary expression of SP-A is limited to a few organs, while SP-D expression has been detected in many non-pulmonary mucosal tissues
CL-P1	Placenta, and vascular endothelial cells	Endothelial cells	CL-P1 is the only membrane bound collectin with an intracellular domain. It is suggested to play many roles which differ from those of soluble collectins
CL-K1		Serum	Different from all other collectins, but seems to have functions phylogenetically similar to CL-L1

(continued)

The collectins SP-A and SP-D are primarily detected in the alveolar space of the lungs, and synthesised by alveolar type-II cells (Table 4.1) (Voorhout et al. 1992, Nayak et al. 2012), and nonciliated bronchial epithelial cells, also known as Clara cells (Voorhout et al. 1992; Crouch et al. 1992). Although the lung is the main

Table 4.1 (continued)

Collectins	Tissues of origin	Tissues of presentation	Remarks
CL-L1	Liver and hepatocyte	Ubiquitous	Immunoblot analyses using human liver demonstrated that CL-L1 was distributed to the cytoplasm. It is also involved in embryonic development

site of SP-A and SP-D synthesis, presence of SP-D has also been reported at extra-pulmonary sites. SP-D expression has been shown immunohistochemically in human trachea, brain, heart, kidneys, testis, salivary gland, placenta, prostate, small intestine, and pancreas (Table 4.1). A low expression level has been detected in spleen, uterus, adrenal gland and mammary glands (Fisher and Mason 1995; Madsen et al. 2000; Herías et al. 2007). Furthermore, immunoreactivity of SP-D has also been shown in the epithelial cells of both small and large ducts of the parotid gland, lacrimal and sweat glands, epithelial cells of intra-hepatic bile ducts and gall bladder, as well as esophagus, exocrine pancreatic ducts, and in the urinary tract (Madsen et al. 2000; Bräuer et al. 2007). In the case of SP-A, low levels are detected in small intestines from human and rat (Table 4.1) (Lin et al. 2001, van Iwaarden et al. 1990). In addition to its presence in the murine uterus, very low SP-A expression is found in human prostate, amniotic fluid, thymus and salivary gland (Madsen et al. 2003). SP-A and SP-D have also been localised in the fetal membranes, and choriodecidual layer of the late pregnancy uterus (Miyamura et al. 1994). As a result of pulmonary microbial infection, the protein levels of both SP-A and SP-D have been reported to increase in the alveolar compartment (Atochina et al. 2001). Thus, the level of SP-D increases in response to allergen-induced eosinophilia (Kasper et al. 2002), suggesting that both SP-A and SP-D may function as acute phase reactants within the lungs. Furthermore, hypoxia results in an increased concentration of both SP-A and SP-D in the alveolar compartment (White et al. 2001).

Conglutinin, CL-46 and CL-43 are serum collectins identified in bovidae and synthesised in the liver (Hansen et al. 2002). These collectins provide a first line of defense against microbial pathogens. CL-L1 mRNA was detected in the liver, and studies using Northern blot analysis have suggested that low levels occur in the placenta. Although most collectins are secreted, CL-L1 was found in the cytosol of hepatocytes, which may suggest its interaction with intracellular ligands (Ohtani et al. 1999). The presence of CL-P1 was reported in vascular endothelial cells (Table 4.1); CL-P1 is suggested to be membrane bound, and it contains an intracellular domain (Ohtani et al. 2001). Expression of MBL, SP-A and SP-D at the mucosal surfaces suggest the innate immune roles of these collectins against invading pathogens. During

Helicobacter pylori infection, an increased level of SP-D has been detected, suggesting the possible role of SP-D in the mucosal defense outside the lungs (Murray et al. 2002), eg. gastrointestinal tract.

4.4 Role of Collectins in Microbial Infection

Collectins are important soluble pattern-recognition receptors (PRRs) of the humoral arm of the innate immune response. Collectins are able to recognise and bind to a wide variety of microbes and are involved in their clearance and forming a central link to adaptive immunity against microbial infections. In this section, we will discuss the well-known collectins: MBL, SP-A and SP-D, as well as newly discovered collectins: liver collectin (CL-L1), kidney collectin (CL-K1), and placenta collectin (CL-P1). We will also briefly discuss bovine collectins, conglutinin, CL-43 and CL-46. Microbes can be cleared by collectins via a number of mechanisms such as aggregation, opsonisation, phagocytosis, microbial growth inhibition, complement activation, as well as modulation of adaptive immunity.

4.5 Interaction of Collectins with Bacteria

4.5.1 SP-A and SP-D

Pulmonary surfactant is composed of 90% phospholipids and 10% proteins (made up of surfactant proteins, SP-A, SP-B, SP-C and SP-D. Whilst, SP-B and SP-C are hydrophobic and essential for the physiology of the alveolar surfaces, SP-A and SP-D are hydrophilic and contribute to lung immunity. An early study showed that pulmonary surfactant enhanced the killing of *Staphylococcus aureus* by alveolar macrophages (AM), in vitro (LaForce et al. 1973). Both Gram-negative and Gram-positive bacteria are recognised by SP-A and SP-D, enhancing their phagocytosis by AMs (Fig. 4.3) (Pikaar et al. 1995). For Gram-negative bacteria, SP-A and SP-D both bind to lipopolysaccharide (LPS) but differ in preferential targets on the molecule. SP-A binds to the lipid A moiety of rough LPS (which lacks the O-antigen and shortened oligosaccharides) (Van Iwaarden et al. 1994), and enhances phagocytosis of bacteria by AM (Kalina et al. 1995), but not to smooth LPS (which contains the O-antigen) (Van Iwaarden et al. 1994). In contrast, SP-D binds strongly to smooth LPS from *Escherichia coli* and *Salmonella* species but does not recognise the lipid A moiety or oligosaccharide deficient LPS (Kuan et al. 1992). This indicates that SP-D preferentially targets the core terminal saccharides in the bacterial ligand, whilst SP-A prefers lipid A. SP-D has also able been shown to bind to rough LPS via its trimeric carbohydrate recognition domain (CRD), (targeting shortened oligosaccharides) and agglutinating *E. coli* (Kuan et al. 1992), and rough LPS from *Klebsiella*

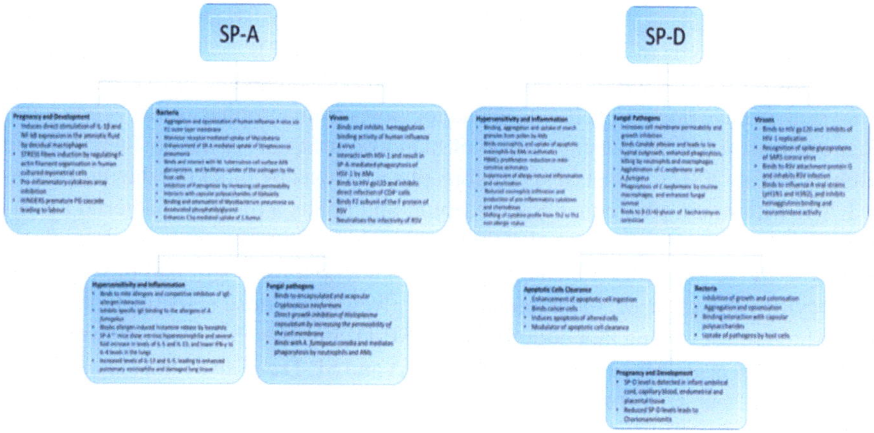

Fig. 4.3 Multiple functions of SP-A and SP-D in human health and disease

pneumoniae and *Pseudomonas aeruginosa* (Lim et al. 1994a, b; Kishore et al. 1996). In addition to LPS, SP-A is able to bind to capsular polysaccharides of *Klebsiella* species, enhancing their phagocytosis by AM (Kabha et al. 1997). However, bacterial peptidoglycan is not a ligand for SP-A (Murakami et al. 2002).

SP-A and SP-D directly inhibit the growth of several Gram-negative bacteria by increasing the membrane permeability of the bacterial cell wall (Fig. 4.3) (Wu et al. 2003). SP-A and SP-D also inhibit biosynthetic functions in strains of *E. coli*, *K. pneumoniae* and *Enterobacter aerogenes* (Wu et al. 2003). Similarly, SP-A inhibits the growth of *P. aeruginosa* by increasing membrane permeability (Van Iwaarden et al. 1994), but the bacterium can resist through quorum-sensing and the secretion of a flagellum-mediated exoprotease that degrades SP-A (Kuang et al. 2011a). Furthermore, SP-A downregulates TNF-α secretion via toll-like receptor 2/NF-κB mediated pathway, indicating its role in modulating inflammatory responses against bacterial ligands (Murakami et al. 2002). SP-A can bind to the outer membrane protein (OMP) of *Haemophilus influenzae* type A and to a lesser extent, type B (McNeely and Coonrod, 1994). SP-A can also aggregate and opsonise *H. influenzae* type A, facilitating killing by AM (McNeely and Coonrod 1994). Similarly, SP-A binds to the capsular polysaccharide of some strains of *K. pneumoniae*, agglutinating the bacteria and increase phagocytosis by macrophages (Kabha et al. 1997), and treatment with SP-A plus SP-B^{N} (N-terminal saponin domain of SP-B) significantly reduced bacterial infection and enhanced neutrophil recruitment (Coya et al. 2015). SP-A has a bacteriostatic effect on *Mycoplasma pneumoniae* via binding to di-saturated phosphatidylglycerols on the bacterial membrane (Piboonpocanun et al. 2005). SP-A can interact with *Mycobacterium tuberculosis* putative adhesin Apa glycoprotein on its surface (Ragas et al. 2007). SP-D can also bind to Gram-positive bacterial ligands such as lipoteichoic acid and peptidoglycan via its CRD (van de Wetering et al. 2001) and to lipoarabinomannan (LAM) from *M. tuberculosis* and *Mycobacterium avium* (Ferguson et al. 1999; Kudo et al. 2004). SP-D is also able to interact with cell membrane lipids of *M. pneumoniae* (Chiba et al. 2002).

It is intriguing that although both SP-A and SP-D bind and agglutinate *M. tuberculosis*, they have opposing effects on phagocytosis by macrophages. SP-A enhances phagocytosis via increased expression of mannose receptor on the host cell surface (Beharka et al. 2002), whilst SP-D inhibits phagocytosis by blocking the interaction of LAM with macrophage mannose receptor, and not as a result of bacterial agglutination by SP-D (Ferguson et al. 1999, 2002). In a mouse model of tuberculosis infection, SP-A$^{-/-}$, SP-D$^{-/-}$, and SP-A/D$^{-/-}$ knockout mice still had the ability to phagocytose and clear M. tuberculosis when given a low-dose aerosol challenge of the pathogen, suggesting that both SP-A and SP-D could be redundant in this animal model (Lemos et al. 2011). Similarly, both SP-A and SP-D can also bind to *Legionella pneumophila*, but seem to inhibit intracellular bacterial growth in the macrophage (Sawada et al. 2010).

SP-A and SP-D can also directly facilitate phagocytosis without the need for microbial binding, by up-regulating the expression of cell surface phagocytic receptors in macrophages, such as mannose receptor (Beharka et al. 2002; Kudo et al. 2004). In SP-A$^{-/-}$ knockout mice, expression of mannose receptor is down-regulated, showing that SP-A is important in regulating the expression of this receptor (Beharka et al. 2002). Similarly, SP-A is able to enhance phagocytosis of *Streptococcus pneumoniae* by AM, independent of its binding to the bacterium, via the increased expression of scavenger receptor A (SR-A) (Kuronuma et al. 2004). Interestingly, the vast majority of clinical strains of the opportunist *Pseudomonas aeruginosa* secrete an elastase that degrades SP-A and facilitates evasion of opsonisation by the collectin during phagocytosis (Kuang et al. 2011b).

SP-A and SP-D can play important roles in modulating the intracellular environment after phagocytosis by stimulating reactive oxygen and nitrogen intermediates facilitating the killing of intracellular pathogens. This is of particular note in mycobacteria, which are specialist intracellular bacteria. SP-A enhances the killing of intracellular *Mycobacterium bovis* BCG by increasing nitric oxide (NO) production, in addition to enhancing the release of inflammatory mediators such as TNF-α (Weikert et al. 2000). In contrast, in IFN-γ primed AM, SP-A decreases NO production in response to intracellular infection with *M. tuberculosis* and *M. avium* by inhibiting TNF-α secretion and nuclear factor-kappa B (NF-κB) activation (Pasula et al. 1999; Hussain et al. 2003). SP-A can also enhance the intracellular killing of *Mycoplasma pulmonis* via a NO dependent mechanism (Hickman-Davis et al. 1998).

Bacteria-derived cell-wall molecules such as LPS and peptidoglycan are potent stimulators of inflammation and can also interact with pattern-recognition receptors (PRRs) such as CD14 or toll-like receptors (TLR), via pathogen-associated molecular patterns (PAMPs), and activate downstream intracellular signalling. SP-A and SP-D can also directly bind to PRRs (e.g. TLR and CD14) and thus can modulate the inflammatory response. SP-A and SP-D can alter LPS interactions with CD14 via different mechanisms (SP-A via neck domain; SP-D via CRD) (Sano et al. 2000). Furthermore, via direct interaction with CD14, SP-A inhibits production of TNF-α induced by smooth LPS, but not rough LPS in U937 macrophages (Sano et al.

1999). In SP-A$^{-/-}$ knockout mice, TNF-α induced by smooth LPS, significantly increased, compared to wild-type mice (Borron et al. 2000), whilst SP-A has also been shown to inhibit TNF-α induction by peptidoglycan via direct binding to TLR-2 (Murakami et al. 2002). Thus, SP-A significantly decreases peptidoglycan or smooth LPS-induced pro-inflammatory responses (via NF-κB activation). SP-A has no effect or increases the inflammatory response induced by rough LPS. In tuberculosis, SP-A has pleiotropic effects being able to promote inflammation in the presence of infection and suppresses inflammation in uninfected macrophages, probably protecting uninfected lung areas from the deleterious effects of inflammation (Gold et al. 2004).

In humans, SP-A exists in two isoforms, SP-A1 and SP-A2, which are encoded by distinct genes. Fully assembled SP-A protein contains both gene products. A number of studies have described polymorphisms in these genes and the SP-D gene which may have a role in susceptibility to microbial infection, particularly tuberculosis. Polymorphisms within and flanking the SP-A1 and SP-A2 genes have been described which indicate protection or susceptibility toward pulmonary TB in the populations studied in Mexico, Ethiopia, India and China (Floros et al. 2000; Madan et al. 2002; Malik et al. 2006; Vaid et al. 2006; Yang et al. 2014). Two SP-A1 alleles (SFTPA1 307A, SFTPA1 776T) and two SP-A2 alleles (SFTPA2 355C and SFTPA2 751C) were associated with tuberculosis susceptibility in Ethiopia (Malik et al. 2006). The SFTPA2 751A/C polymorphism and the haplotype $1A^3$ in SP-A2, which both affect the amino acids in CRD region of SP-A, may alter binding to *M. tuberculosis* and thus were found to be strongly linked with tuberculosis susceptibility (Malik et al. 2006). Another study also found two polymorphisms (SP-A2 G1649C and SP-A2 A1660G) in the introns of SP-A1 that were associated with tuberculosis in an Indian population, but none in the SP-A1 gene (Madan et al. 2002). In a Chinese population, the polymorphism 1649G in the SP-A2 gene was strongly associated with tuberculosis, mirroring the findings in the Ethiopian and Indian populations (Yang et al. 2014). The SP-A2 1649G leads to a transversion (proline to alanine), affecting the triple helical structure of SP-A (Yang et al. 2014). In SP-D, the polymorphism, G459A, is significantly associated with tuberculosis susceptibility in an Indian population, but the molecular basis for susceptibility is not understood (Vaid et al. 2006). These observations illustrate the complexities of host-pathogen interactions in bacterial infection mediated by these collectins.

4.5.2 MBL

MBL is the recognition subcomponent of the lectin pathway of the complement system and is present mostly in the serum. The structure of MBL is similar to that of SP-A, and in the presence of Ca^{2+}, it has been observed to target terminal sugars (e.g. D-mannose, L-fucose, and *N*-acetyl-D-glucosamine), on the surface of a number of Gram-positive and Gram-negative bacterial species (Ip et al. 2009; Lugo-Villarino et al. 2011). The binding of MBL to microbial surfaces can activate complement through MBL-associated serine proteases (MASPs), resulting in enhanced microbial clearance via opsonisation (C3 and C4 deposition) and complement-mediated lysis. However, MBL also has complement-independent activity such as inhibition of bacterial adhesion (Jack et al. 2005) and opsonisation to enhance bacterial uptake (Kuhlman et al. 1989; Polotsky et al. 1997; Jack et al. 2005). Strong in vitro binding of MBL to *S. aureus*, *Streptococcus pyogenes*, *Listeria monocytogenes* and non-encapsulated *Neisseria meningitidis* has been described (Levitz et al. 1993; van Emmerik et al. 1994; Neth et al. 2000). Moderate levels of MBL binding were observed in *E. coli*, *Haemophilus influenzae* and *Klebsiella* species, whilst no binding has been observed for *P. aeruginosa*, *Enterococcus* species and *Streptococcus pneumoniae* (Levitz et al. 1993; van Emmerik et al. 1994; Neth et al. 2000). Bacterial pathogens have involved strategies to interfere with MBL binding and functions for survival, via the synthesis of a polysaccharide capsule and sialylation of LPS ligands on the bacterial surface which reduces the binding of MBL (Jack et al. 2005; Krarup et al. 2005). This effectively masks or alters the bacterial ligands for MBL interaction. A number of studies have characterised the bacterial ligands for MBL. MBL is able to bind to peptidoglycan and lipoteichoic acid from *S. aureus* (Polotsky et al. 1996; Nadesalingam et al. 2005a, b), LAM from *M. avium* (Polotsky et al. 1997), and mannosylated lipoarabinomannan (ManLAM) from a number of mycobacteria *(M. tuberculosis*, *M. bovis*, *M. kansasii*, *M. gordonae* and *M. smegmatis*) (Bartlomiejczyk et al. 2014). There is also a report of MBL binding to the antigen 85 (Ag85) complex of M. tuberculosis (Swierzko et al. 2016). Neisseria (*M. meningitidis* and *M. gonorrhoeae*) are Gram-negative diplococci that have shorter versions of LPS on their surface called lipooligosaccharides (LOS) that are commonly terminated in sialic (neuraminic) acid, instead of the O-antigen. Neisseria bacteria are able to decrease binding of MBL to their surface by the sialylation on LOS (Jack et al. 1998; Devyatyarova-Johnson et al. 2000; Jack et al. 2001; Gulati et al. 2002). *M. meningitidis* can also interfere with MBL binding through encapsulation (van Emmerik et al. 1994), whilst *M. gonorrhoeae* is not able to form capsules. Encapsulation seems to be less robust at decreasing MBL binding than sialyation of LOS (Jack et al. 1998). Bound MBL can activate complement and the ability of Neisseria species to cascade complement all the way to C9 (membrane attack complex (MAC)) is crucial for protection against infection, since they are otherwise poorly phagocytosed by neutrophils and macrophages when opsonised by C3 (Ross and Densen 1984). MBL bound to the surface of Neisseria is able to increase bacterial killing via increased complement activation (Jack et al. 1998, 2001; Gulati et al. 2002), and similar observations of bactericidal activity have been reported for *E. coli* and Salmonella species (Kawasaki et al. 1989; Ihara et al. 1991). For most other bacteria, complement activation to the C3 deposition stage is enough for protection via increased phagocytosis through opsonisation by complement products on the bacterial cell surface. MBL can increase C3b deposition on *S. aureus* (Neth et al. 2002), but this does not appear to result in increased complement activation (Cunnion et al. 2001). MBL targets wall teichoic acid in *S. aureus* and this interaction is particularly important in infants that have not developed adaptive immunity, leading to bacterial clearance via MBL-mediated complement activation (Kurokawa et al. 2016).

In addition to its complement-mediated activities, MBL is also has various intrinsic effects, being able to act as an opsonin independently and other direct effects. MBL enhances uptake and intracellular killing of Salmonella by neutrophils and monocytes (Kuhlman et al. 1989), but this may also involve interaction with fibronectin (Ghiran et al. 2000). Recently, MBL has also been shown to have a direct inhibitory effect on flagellar activity in pathogenic Salmonella bacteria, impairing their motility (Xu et al. 2016). MBL can also increase uptake of mycobacteria by macrophages (Polotsky et al. 1997) and *N. meningitidis* by neutrophils, monocytes and macrophages (Jack et al. 2001), but this uptake by neutrophils may not result in intracellular killing (Drogari-Apiranthitou et al. 1997). MBL also appears to improve the efficiency of internalisation of bacteria bound to the macrophage plasma membrane (Neth et al. 2002). MBL co-interacts with TLR2 in sensing *S. aureus* and thus influencing the subsequent inflammatory response (Nauta et al. 2003; Ip et al. 2008).

MBL deficiency increases susceptibility to microbial infection even though the majority of MBL-deficient individuals are usually healthy (Eisen and Minchinton 2003). The concentration of MBL in the plasma varies considerably in humans (0–10, 000 ng/ml) due to polymorphisms in the MBL gene (Steffensen et al. 2000). MBL deficiency is commonly observed in around 25% of Caucasians (having low levels (<500 ng/ml)), which renders them susceptible to infection (Valdimarsson et al. 2004). MBL-deficient mice are susceptible to *S. aureus* infection (Shi et al. 2004), whilst MBL deficiency increases susceptibility to postburn infection with *P. aeruginosa* (Moller-Kristensen et al. 2006). A large cohort study has also found a strong association between MBL deficiency and meningococcal infection, and pneumococcal pneumonia, in patients undergoing chemotherapy (Gaynor et al. 1995; Kronborg et al. 2002; Roy et al. 2002). In contrast, normal or increased levels of MBL are linked to frequent infection with *M. tuberculosis* and *M. leprae* (Garred et al. 1994, 1997b), probably through complement-mediated phagocytosis of the pathogen. Up to 30% of healthy individuals have polymorphisms linked to MBL deficiency and these, together with serum levels, have been associated with susceptibility to tuberculosis and other inflammatory diseases in some ethnic populations (Takahashi and Ezekowitz 2005; Thiel et al. 2006; Goyal et al. 2016).

4.5.3 CL-L1, CL-K1, CL-P1 and the Bovine-Unique Collectins, Conglutinin, CL-43 and CL-46

Of the three more recently discovered collectins (CL-L1, CL-K1, CL-P1), CL-L1 and CL-P1 have been shown to have bacterial interactions. CL-K binds to *E. coli*, *K. pneumoniae*, *P. aeruginosa* and *M. tuberculosis* (Keshi et al. 2006; Hansen et al. 2010; Troegeler et al. 2015), whilst CL-P1 can bind to *E. coli* and *S. aureus* (Ohtani et al. 1999; Jang et al. 2009). Both CL-L1 and CL-K1 can activate the complement lectin pathway as can the heteromeric form CL-LK, which interacts with the MASPs (Henriksen et al. 2013). CL-P1 can activate the alternative and classical pathways via its interaction with C-reactive protein (CRP) (Roy et al. 2016). There is limited data on the activity of CL-LK in vivo and in vitro, but due to average serum concentrations being below that of MBL (0.3 μg/ml vs. 1.5 μg/ml), pathogen recognition and clearance through complement activation is likely to have a minor role to play for these collectins. It is not clear whether these collectins can act directly as opsonins in a complement-independent manner. CL-L1 can bind D-mannose, *N*-acetylglucosamine, D-galactose, L-fucose and D-fructose in a Ca^{2+} dependent manner (Ohtani et al. 1999; Axelgaard et al. 2013). Similarly, CL-K1 can also bind L-fucose, D-mannose and *N*-acetylmannosamine (Ohtani et al. 1999; Hansen et al. 2010). Furthermore, CL-LK was recently demonstrated to be a PRR for *M. tuberculosis*, being able to primarily target mannose-capped lipoarabinomannan (ManLAM), in a Ca^{2+} dependent manner, on the surface of the mycobacterium, but not to *M. smegmatis* due to the lack of mannose caps on its LAM (Troegeler et al. 2015). Mice deficient in CL-K1 did not show altered susceptibility to *M. tuberculosis* infection and CL-LK opsonised *M. tuberculosis* did not result in altered phagocytosis or intracellular survival of the pathogen in human macrophages (Troegeler et al. 2015). Interestingly, the levels of CL-LK in serum of tuberculosis patients is reduced, compared to controls, correlating inversely to the immune response to *M. tuberculosis* and suggesting that it may be useful as a biomarker for the disease (Troegeler et al. 2015).

Conglutinin was the first mammalian collectin to be discovered and is found in *Bovidae*, together with other lesser known collectins (CL-43 and CL-46) (Hansen and Holmskov 2002). Conglutinin is similar in overall structure to SP-D and is able to bind to microbial surfaces in the presence of Ca^{2+} (Hansen and Holmskov 2002). Conglutinin is secreted by the liver and found predominantly in bovine serum at an average concentration of 12 μg/ml (Holmskov et al. 1998). Conglutinin has been shown to have anti-microbial properties. Low serum levels of conglutinin have been associated with acute infections (e.g. pneumonia and metritis) and predisposition to respiratory infection (Ingram and Mitchell 1971; Holmskov et al. 1998). Conglutinin is able to bind many microbes, including Gram-negative bacteria such as *E. coli* and *Salmonella typhimurium* (Friis-Christiansen et al. 1990; Friis et al. 1991), LPS and peptidoglycan (Wang et al. 1995) and Gram-positive bacteria such as mycobacteria (Dec et al. 2012; Mehmood et al. 2019). Conglutinin is uniquely able to bind to iC3b, via the mannose sugars on the α-chain of iC3b (Laursen et al. 1994). Conglutinin is able to bind and agglutinate iC3b-coated erythrocytes (Lachmann and Muller-Eberhard 1968; Laursen et al. 1994), and as well as *E. coli*, increasing the respiratory burst of phagocytes (Friis et al. 1991). Conglutinin has also been shown to be protective against bacterial infection in vivo, being able to increase the survival of mice experimentally infected with highly virulent strains of *S. typhimurium* (Friis-Christiansen et al. 1990). A recombinant truncated form of conglutinin, composed of the α-helical neck region and the CRD of conglutinin (Wang et al. 1995), was recently shown to bind to able to bind to the vaccine strain *Mycobacterium bovis* BCG (probably via LAM), and act as an anti-opsonin both in the presence and absence of complement deposition. Thus, Conglutinin can interfere with the uptake of the bacterium by THP-1 macrophages and alter their inflammatory response (Mehmood et al. 2019). This suggests that conglutinin interfers with uptake of mycobacteria by macrophages via two important mechanisms: (1) blocking interaction of mycobacterial LAM with mannose receptor, and (2) blocking iC3b interaction with complement receptors CR3 and CR4 (Mehmood et al. 2019). These data potentially have important implications for bovine tuberculosis.

CL-43 and CL-46 are also bovine-unique collectins, but their role in the physiology and innate immunity against bacteria has not been fully studied. There is one report of CL-43 binding to *E. coli* strain K12, enhancing attachment to phagocytes (Hansen and Holmskov 2002).

4.6 Interaction of Collectins with Viruses

4.6.1 SP-A and SP-D

There are numerous studies that describe direct interaction of SP-A and SP-D with a range of viruses, enhancing their phagocytosis, as well as neutralising viral infection of host cells (Fig. 4.3). Experiments on SP-A$^{-/-}$ and SP-D$^{-/-}$ knockout mice infected with influenza A virus (IAV) suggest that both collectins are protective, but this is dependent on viral strain-specific factors, such as the nature of glycosylation in HA and NA (LeVine et al. 2001, 2002; Hawgood et al. 2004). Also, mice lacking both SP-A and SP-D, have an IAV infection phenotype almost identical to SP-D$^{-/-}$ mice (Hawgood et al. 2004). Moreover, SP-D, but not SP-A, enhanced the clearance of IAV infection in the mouse lung (LeVine et al. 2001). Thus, these studies suggest that SP-D plays a more significant role than SP-A in the host innate immune response to infection with IAV.

SP-A binds to IAV, neutralises the virus and inhibits the release of viral particles from infected cells, by targeting mannose residues of viral surface haemagglutinin (HA) or neuraminidase (NA) (Malhotra et al. 1994; Benne et al. 1995). SP-D strongly inhibits hemagglutination activity of IAV, resulting in viral aggregation and neutralisation (Hartshorn et al. 1994). SP-D is also able to inhibit NA activity, with inhibition being stopped in the presence of D-mannose (Reading et al. 1997). SP-D has a stronger inhibitory effect on NA compared to SP-A (Tecle et al. 2007). SP-D binds to mannosylated, N-linked sugars on viral HA and NA via its CRD, resulting in potent anti-IAV infectivity (Hartshorn et al. 1994, 2000). SP-D was able to inhibit virus-induced HA activity, block the enzymatic activity of viral NA, and neutralise the ability of seasonal H1N1 strains of IAV to infect human respiratory epithelial cells (Job et al. 2010). However, in the same study, some pandemic H1N1 were found to be resistant to SP-D inhibition that correlated with the degree of *N*-glycosylation in the globular head of HA (Job et al. 2010). It has been shown that porcine SP-D has an increased ability to inhibit, not just seasonal IAV strains, but also a number of pandemic and avian strains (van Eijk et al. 2003; Hillaire et al. 2012). This is important as pigs are a source of IAV pandemic strains (H1N1) that can be transmitted to humans, so studying porcine SP-D could provide further insights into this host reservoir.

A recombinant truncated form of SP-A (rfhSP-A) made up of α-helical neck and CRD, promotes IAV infection, replication, upregulation of viral factors (M1) in lung epithelial A549 cells and enhances the pro-inflammatory response (Al-Qahtani et al. 2019). This contrasts with full-length SP-A which inhibits IAV infection and dampens the pro-inflammatory response, demonstrating that the full-length SP-A molecule is required for IAV protection (Al-Qahtani et al. 2019). However, in a similar study, a recombinant truncated form of SP-D (rfhSP-D) was shown to inhibit IAV entry, down-regulate viral factors (M1) and down-regulate the pro-inflammatory response (Al-Ahdal et al. 2018). These opposing effects of rfhSP-A and rfhSP-D provide further insight into IAV pathogenesis and the possible utility of rfhSP-D as a therapeutic molecule. In bronchoalveolar lavage (BAL), SP-D enhances IAV uptake and virus-induced respiratory burst by neutrophils (White et al. 2005), but other collectins (SP-A), mucins and gp-340 dampen SP-D's effect, and thus, significantly reduce the ability of SP-D to promote neutrophil oxidative response (White et al. 2005). Therefore, the net effect of BAL is to increase neutrophil uptake of IAV while reducing the respiratory burst response to virus (White et al. 2005).

SP-A is also able to bind to herpes simplex virus type 1 (HSV-1) via viral N-linked sugars and enhance phagocytosis of the virus by macrophages (van Iwaarden et al. 1991; Van Iwaarden et al. 1992a, b). The mechanism of binding of SP-A to HSV-1 is similar to binding to IAV, involving interaction with the sialylated carbohydrate on the collectin's CRD. SP-A also has an opsonin activity, increasing uptake of HSV-1 by AM (van Iwaarden et al. 1991). Similarly, SP-A binds to cytomegalovirus and enhances viral entry into rat lung cells (Weyer et al. 2000). It is unknown whether SP-D has any activity against other Herpesviridae. SP-A is able to bind to respiratory syncytial virus (RSV) targeting the F2 subunit of the viral F antigen and is able neutralise the virus (Ghildyal et al. 1999; Sano et al. 1999, 2000). Children with severe RSV infection have reduced levels of SP-A and SP-D in BAL samples compared to healthy controls (Kerr and Paton 1999). In SP-A$^{-/-}$ knockout mice, RSV infection was more severe than in SP-A$^{+/+}$ mice and the addition of exogenous SP-A to SP-A$^{-/-}$ mice reduced viral load and inflammation, and enhanced RSV clearance (LeVine et al. 1999). SP-D can bind to RSV protein G and is able to neutralise RSV infectivity in vitro (Hickling et al. 1999). Interestingly, RSV itself can alter SP-A expression in human pulmonary epithelial cells, upon infection by interfering with protein translation (Bruce et al. 2009). SP-A binds to Human Immunodeficiency Virus 1 (HIV-1) via the viral envelope gp120 glycoprotein and inhibits direct infection of CD4^{+} T cells (Gaiha et al. 2008). Yet, in dendritic cells (DC), SP-A increases HIV uptake, through enhanced binding to gp120 and facilitates transfer of HIV from DC to CD4^{+} T cells (Gaiha et al. 2008). SP-D is also able to bind to HIV gp120 and inhibit viral infectivity (Meschi et al. 2005), whilst rfhSP-D was also able to bind to gp120 and prevent infection of Jurkat T cells, U937 monocytic cells and PBMC, and significantly suppress the HIV-1 induced cytokine storm in these cells (Pandit et al. 2014). Interestingly, a direct protein–protein interaction between rfhSP-D and

DC-SIGN (dendritic cell-specific intercellular adhesion molecule-3-grabbing non-integrin) modulates the capture of HIV-1 and its transfer to $CD4^+$ T cells, revealing a novel and distinct anti-viral mechanism against HIV-1 by SP-D (Dodagatta-Marri et al. 2017). This same rfhSP-D has also been recently shown to restrict the transfer of HIV across the vaginal epithelial barrier, by altering the gene expression signature of the epithelium (Pandit et al. 2019). These recent studies demonstrate the therapeutic potential of rfhSP-D against HIV infection.

Elevated levels of serum SP-D have been reported in severe acute respiratory syndrome (SARS) coronavirus infected patients (Wu et al. 2009). SP-D is able to bind to the glycosylated spike protein (S-protein) on the SARS coronavirus (Leth-Larsen et al. 2007). Both SP-A and SP-D bind to coronavirus strain HCoV-229E, and inhibit viral infection of human bronchial epithelial (16HBE) cells. Whilst SP-A only modestly reduced infection in AM, whereas SP-D had no effect (Funk et al. 2012). Human and porcine SP-D can interact with Ebola virus glycoprotein and enhance viral infection in pulmonary cells, suggesting that SP-D may enhance viral spread (Favier et al. 2018). SP-A has been shown to enhance clearance of pulmonary adenovirus infection and inhibit lung inflammation (Harrod et al. 1999). Bovine SP-D is also able to bind to bovine rotaviruses via the VP7 glycoprotein and neutralise infectivity (Reading et al. 1998). SP-D binds to the A27 protein of vaccinia virus. SP-$D^{-/-}$ knockout mice challenged with vaccinia virus resulted in increased mortality, compared to SP-$D^{+/+}$ mice, suggesting that SP-D has a protective role against vaccinia infection (Perino et al. 2013).

4.6.2 MBL

MBL is able to interact with a number of viral pathogens and its effect is generally protective, although there are examples of negative as well as positive outcomes for infection as a result of MBL-mediated binding (Fig. 4.4). Several studies have shown that MBL is a potent inhibitor of IAV infection (Hartley et al. 1992; Hartshorn et al. 1993b; Reading et al. 1995, 1997). Moreover, MBL also has the added ability to deposit complement on IAV-infected cells (Reading et al. 1995). There are also elevated levels of MBL in the lung during IAV infection, suggesting that it may be important for protection against IAV pathogenesis (Reading et al. 1997; Fidler et al. 2009). MBL can inhibit viral hemagglutination and directly neutralise IAV in either a complement-dependent or independent manner (Hartshorn et al. 1993b; Anders et al. 1994; Kase et al. 1999). MBL binds to IAV HA and NA, and without involving complement, neutralises the virus (Kase et al. 1999). However, some IAV strains are resistant to the effects of MBL which is dependent on the degree of glycosylation on the viral HA globular domain (Reading et al. 1997; Job et al. 2010; Tokunaga et al. 2011). Furthermore, $MBL^{-/-}$ mice were more susceptible to infection from highly glycosylated viral strains of IAV than wild-type mice (Chang et al. 2010). However, pandemic strain H1N1 and avian influenza A H9N2 produced more severe disease (enhanced production of pro-inflammatory response) in wild-type mice compared

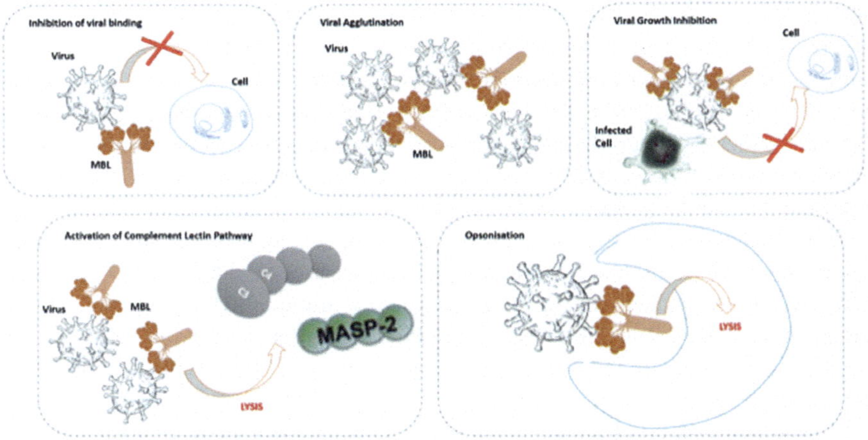

Fig. 4.4 Anti-viral activity of mannose-binding lectin (MBL). MBL binds to viruses, including influenza virus, acting as an opsonin (not through direct neutralisation), eliminating viral particles by phagocytosis. Binding of MBL to carbohydrate groups found on the surface of viral particles triggers the lectin activation pathway of complement leading to lysis

to $MBL^{-/-}$ mice, suggesting that MBL may have a deleterious effect in some IAV infections (Ling et al. 2012).

MBL is able to neutralise HIV-1 in vitro by binding to the N-linked mannose glycans of viral gp120, and binding to HIV-1 infected $CD4^+$ T cells and monocytes and inhibiting reverse transcriptase activity (Ezekowitz et al. 1989; Teodorof et al. 2014). Another study has also shown MBL can also bind to viral gp41 as well as gp120 (Saifuddin et al. 2000), whilst MBL also activates complement on gp120 binding (Haurum et al. 1993). Studies have shown a tentative link between low levels of MBL and increased risk of HIV-1 transmission or progression to AIDS, but this remains contentious (Garred et al. 1997a; Takahashi and Ezekowitz 2005; Ballegaard et al. 2014). There has also been a report of a positive correlation between the rate of AIDS progression and MBL plasma concentration (Mangano et al. 2008). However, other studies have found no correlation between MBL levels and AIDS disease progression (Nielsen et al. 1995; McBride et al. 1998). In general, SP-D is better able to inhibit HIV-1 than MBL, but as is the case for MBL, SP-D's inhibitory activity against HIV-1 is lower than what has been observed for IAV (Meschi et al. 2005). MBL has also been shown to contribute to HIV-1 pathogenesis, where MBL mediates enhancement of HIV-1 dissemination to the brain by soluble gp120, which is taken up by the CXCR4 receptor on neurones, and then intracellularly trafficked by MBL, thus resulting in the apoptosis of neuronal cells (Bachis et al. 2006; Teodorof et al. 2014).

Epidemiological studies have revealed association of MBL with hepatitis B virus (HBV) and hepatitis C virus (HCV) infection and disease severity, based on genetic polymorphisms (Thomas et al. 1996; Matsushita et al. 1998; Yuen et al. 1999; Sasaki et al. 2000; Hakozaki et al. 2002). However, one study found no link between MBL polymorphisms and HBV infection (Hohler et al. 1998). MBL is able to bind to HCV envelope glycoproteins, E1 and E2, and is able to activate complement (via MASP-2), resulting in the neutralisation of HCV particles (Brown et al. 2010). MBL probably binds to N-linked glycosylated forms or HBV surface antigen (HBsAg) (Brown et al. 2007), but it is unknown whether this interaction neutralises the infectivity of the virus.

MBL is also able to bind to Ebola virus via its envelope glycoprotein (GP), which contains high mannose glycan sites, and is able to inhibit the binding of Ebola (and Marburg) viruses to DC-SIGN, blocking attachment to host cells and also neutralise the virus through complement activation (Ji et al. 2005). Furthermore, soluble GP is a key component of Ebola viral pathogenesis and MBL was found to be able to negate GP activity and the virally induced cytokine storm (Escudero-Perez et al. 2014), and thus MBL could be involved in protection against increased vascular permeability which is a characteristic of Ebola haemorrhagic disease. Nevertheless, high dose MBL therapy in a mouse model, where mice we given recombinant human MBL at levels greater than seven times above average human levels, survived otherwise fatal Ebola viral infection and became resistant to reinfection (Michelow et al. 2011).

There is limited or circumstantial data on the interaction of MBL with a number of other viral pathogens. In mice, MBL appears to modulate the immune responses to HSV-2 (Gadjeva et al. 2004), MBL deficiency seems to be linked with recurrent infections (Gadjeva et al. 2004; Seppanen et al. 2009). MBL also binds to flaviviruses such as Dengue and West Nile virus and is able to neutralise infection through complement-dependent and independent mechanisms (Avirutnan et al. 2011; Fuchs et al. 2011). Genetic polymorphism affecting MBL serum levels may also contribute to the pathogenesis and disease severity of Dengue fever (Avirutnan et al. 2011).

4.6.3 CL-L1, CL-K1, CL-P1 and the Bovine-Unique Collectins, Conglutinin, CL-43 and CL-46

For collectins CL-L1, CL-K1, CL-P1, there is limited data on their interaction with viruses. Only CL-K1 has been shown to bind IAV and partially decrease the infectivity of the virus (Hansen et al. 2010; Selman and Hansen 2012). The binding of CL-L1 and CL-P1 to viruses has not been reported.

Like SP-D, conglutinin binds to IAV resulting in the inhibition of hemagglutination and infectivity of the virus (Hartshorn et al. 1993a). Conglutinin binds via its CRD to the high mannose sites on the viral HA. IAV treated with conglutinin also boosted neutrophil respiratory burst (Hartshorn et al. 1993a). Conglutinin, CL-43 and bovine SP-D have been reported to bind the bovine rotavirus Nebraska calf diarrhoea virus, targeting its VP7 glycoprotein (Reading et al. 1998). Binding resulted in hemagglutination and neuralisation of rotavirus, with CL-43 showing the highest activity against the virus; it is the first report of collectin activity against a non-enveloped virus (Reading et al. 1998). However, conglutinin has a higher inhibitory activity against IAV (strain HKx31) than bovine SP-D or CL-43 (Reading et al. 1998). Conglutinin binds to HSV-2 and enhances infection in mice (Fischer et al. 1994). It is also able to bind to HIV-1 gp160 and inhibit interaction of the virus with the CD4 receptor (Andersen et al. 1991). Interestingly, a collectin-like protein analogous, to bovine conglutinin, was purified from human serum (called conglutinin-like protein) and this was demonstrated to bind to HIV-1 gp120 via its CRD and inhibit viral infectivity (Ushijima et al. 1992).

4.7 Interaction of Collectins with Fungi

Collectins are able to recognise and bind to a number of fungi, both primary and opportunistic fungal pathogens, at various stages in their life cycle. Collectins can exhibit direct growth inhibition and enhance phagocytosis of fungi; in some cases, they can contribute to the fungal pathogenesis.

4.7.1 SP-A and SP-D

Both SP-A and SP-D can bind to the conidia of *Aspergillus fumigatus*, via its β-(1-6)-glucan carbohydrate structures on the fungal cell surface in a Ca^{2+} dependent manner (Fig. 4.3) (Madan et al. 1997a; Allen et al. 2001a, b). SP-A and SP-D can cause inhibition of conidia infectivity via agglutination, enhancement of phagocytosis and intracellular killing of *A. fumigatus* conidia by neutrophils and AM (Madan et al. 1997a). The fungal ligands of SP-A are 2 N-linked glycosylated glycoproteins (gp45 and gp55) isolated from culture filtrate and are also used for ELISA diagnosis of allergic aspergillosis (Madan et al. 1997b). Fungal melanin was recently determined to be the primary ligand for SP-D on the *A. fumigatus* conidia cell surface, and is able to facilitate fungal phagocytosis and modulate the anti-fungal immune response (Wong et al. 2018).

Utilising a mouse model of invasive pulmonary aspergillosis (IPA), SP-D, but not SP-A, was found to be protective against a normally fatal challenge of *A. fumigatus* conidia (Madan et al. 2001a, b). In this study, IPA mice-treated intranasally with purified human SP-D or rfhSP-D showed 60 and 80% survival respectively (Madan et al. 2001a, b). The basis of this therapeutic protection by SP-D and rfhSP-D was observed to be enhanced phagocytosis of conidia by macrophages and neutrophils, fungistatic effects on the growth of conidia and a dampening of pathogenic Th2 cytokines (IL-4 and IL-5), whilst enhancing protective Th1 cytokines (TNF-α and IFN-γ) (Singh et al. 2009). $SP\text{-}D^{-/-}$ knockout mice are more susceptible to IPA (Madan et al. 2010). However, $SP\text{-}A^{-/-}$ knockout mice demonstrate resistance to IPA, suggesting that SP-A may be involved in the pathogenesis of IPA (Madan et al. 2010). Both SP-A and SP-D also have a direct effect on *Histoplasma capsulatum*, inhibiting its growth by increasing the permeability of the fungal membrane (McCormack et al.

2003). However, no aggregation of *H. capsulatum* was observed by SP-A or SP-D, and neither collectin altered the phagocytosis of the fungus or inhibited the growth of macrophage-infected *H. capsulatum* (McCormack et al. 2003).

SP-A is also able to bind to *Cryptococcus neoformans*, both in its encapsulated and non- encapsulated yeast form, but this does not result in increased phagocytosis of the acapsular form (Walenkamp et al. 1999). SP-A binding was Ca^{2+}-dependent and was inhibited by glucose and mannose, but not galactose (Walenkamp et al. 1999). Intranasal infection with *C. neoformans* gave the same survival outcome in SP-A$^{-/-}$ knockout mice and wild-type mice, suggesting that the fungus is resistant to SP-A mediated host defence mechanisms (Giles et al. 2007). A subsequent study found that SP-D increases the phagocytosis of hypocapsular *C. neoformans* by murine macrophages and that this facilitated fungal survival (Geunes-Boyer et al. 2009). Other studies have also shown that SP-D can agglutinate *C. neoformans* and *A. fumigatus* (Schelenz et al. 1995; Madan et al. 1997a). Furthermore, SP-D can bind to both encapsulated and acapsular *C. neoformans* and can aggregate acapsular *C. neoformans* in particular (van de Wetering et al. 2004a). The cryptococcal capsular components glucuronoxylomannan (GXM) and mannoprotein 1 (MP1) are the ligands for SP-D (van de Wetering et al. 2004a). SP-D is able to facilitate *C. neoformans* infection further by protecting the fungus against oxidative stress allowing for disease progression in the mouse model of infection (Geunes-Boyer et al. 2012).

SP-D is also able to bind *Blastomyces dermatitidis*, via β-glucan on its surface, and subsequently block interactions with β-glucan-receptors on AM (Lekkala et al. 2006). This study also showed a reduction in TNF-α, dampening the host inflammatory response and thus may facilitate disease progression (Lekkala et al. 2006). SP-D and SP-A can also bind to *Coccidioides posadasii* via its surface antigens. In a mouse model of infection, *C. posadasii* infection is able to suppress the expression of pulmonary SP-A and SP-D, possibly facilitating fungal disease progression and dissemination (Awasthi et al. 2004). SP-D can also bind to *Candida albicans* via its surface antigens and agglutinate the pathogen and directly inhibiting its growth without the requirement of macrophage dependent phagocytosis (van Rozendaal et al. 2000). Similarly, SP-A is able to bind to *C. albicans* and interfere with attachment to AM, inhibiting phagocytosis (Rosseau et al. 1997). SP-A is also able to dampen the pro-inflammatory response elicited by *C. albicans* by human AM and monocytes, which may be important in regulating excessive inflammation in the lung during Candida infection (Rosseau et al. 1999). In *Saccharomyces cerevisiae*, SP-D is observed to bind to its surface, but not SP-A, whilst the fungal ligand for SP-D is yeast β-(1-6)-glucan (Allen et al. 2001a, b).

The opportunistic fungus, Pneumocystis is able to infect a number of mammals with each species of the fungus displaying strict host specificity. For example, *P. carinii* and *P. wakefieldiae* infect rats, *P. murina* infects mice, *P. oryctolagi* infects rabbits, and *P. jirovecii* infects humans. SP-A and SP-D are able to recognise and bind Pneumocystis species via the major surface glycoprotein (MSG; also known as gpA) of the fungus (O'Riordan et al. 1995; McCormack et al. 1997a, b). MSG contains an N-linked carbohydrate chain made up of glucose, mannose, and *N*-acetylglucosamine and is involved in attachment of the fungus to alveolar epithelium in Pneumocystis pneumonia (Zimmerman et al. 1992; Vuk-Pavlovic et al. 2001). Cruciform dodecamers and other large oligomers of SP-D have a higher affinity of binding to *P. carinii* than do smaller oligomeric versions of SP-D (Vuk-Pavlovic et al. 2001). SP-D is also able to recognise *Pneumocystis* cysts via surface β-glucans (Vuk-Pavlovic et al. 2001). However, SP-D binding does not appear to increase the phagocytosis of the fungus (McCormack et al. 1997a, b; Vuk-Pavlovic et al. 2001). Despite this, SP-D does aggregate *P. carinii* in large complexes that may restrict phagocytosis by macrophages and may allow for persistence of the fungus within the host lungs (Vuk-Pavlovic et al. 2001). Pneumocystis pneumonia does alter the expression of SP-A in the lungs (Atochina et al. 2001), with a threefold increase in the levels of SP-A and SP-D (Phelps et al. 1996; Aliouat et al. 1998; Qu et al. 2001), but decreases total phospholipid content (Atochina et al. 2001).

Human SP-A enhances attachment of *P. carinii* to rat AM in vitro (Williams et al. 1996). SP-A also reduces phagocytosis of *P. carinii* in human AM in vitro (Koziel et al. 1998). These data suggest that increased levels of SP-A during *Pneumocystis pneumonia* (Phelps et al. 1996) may contribute to the pathogenesis through binding to the fungus and interfering with its AM recognition (Koziel et al. 1998). Immunosuppressed SP-A$^{-/-}$ mice also have increased susceptibility to *P. carinii* infection (Linke et al. 2001), whilst removal of immunosuppression resulted in efficient clearance of the infection (Linke et al. 2006), showing that SP-A does not enhance *P. carinii* clearance, but does modulate the host immune response during the resolution of infection. SP-D modulates interaction of *P. carinii* with AM (Limper et al. 1995) and also aggregates *P. carinii*, impairing phagocytosis by AM (Yong et al. 2003). In SP-D$^{-/-}$ mice, there was delayed clearance of *P. carinii* infection, increased inflammation and altered nitric oxide response (Atochina et al. 2004). Similarly, in immunosuppressed mice, SP-D was found to enhance *P. carinii* infection (Vuk-Pavlovic et al. 2006).

4.7.2 MBL

MBL has been reported to interact with various primary and opportunistic fungal pathogens. Low serum levels of MBL have been linked to increased likelihood of fungal disease (Mullighan et al. 2002; Granell et al. 2006). MBL is able to bind to *A. fumigatus* (Neth et al. 2000), *B. dermatitidis* (Koneti et al. 2008), *C. albicans* (Kitz et al. 1992; Neth et al. 2000; Ip and Lau et al. 2004; van Asbeck et al. 2008), *Candida parapsilosis* (van Asbeck et al. 2008), and *C. neoformans* (Chaka et al. 1997; van Asbeck et al. 2008). The ligands for MBL binding to *C. albicans* and *C. neoformans* are mannan and mannoprotein, respectively (Chaka et al. 1997; Ip and Lau, 2004), whilst 1,3-β-glucan and mannose are the MBL ligands on *B. dermatitidis* and *A. fumigatus*, respectively (Neth et al. 2000; Koneti et al. 2008).

MBL is able to bind *A. fumigatus* conidia showing aggregation, enhancing phagocytosis, and complement deposition (Kaur et al. 2007). However, MBL binding of conidia did not always result in the killing of *A. fumigatus* by phagocytes (Madan et al. 2005a, b; Kaur et al. 2007). Moreover, MBL may be less important in this context, since it is mainly a serum protein and may not be in significant levels in the lung. Nevertheless, genetic polymorphisms in the MBL gene have been shown to be associated with severe aspergillosis (Crosdale et al. 2001; Vaid et al. 2007). Similarly, MBL deficiency is a risk factor for aspergillosis in immunocompromised patients, cancer patients and transplant recipients. In the mouse model of infection, MBL deficiency does not necessarily affect the survival of mice infected with *A. fumigatus* conidia, due to redundancy since mice having two copies of the MBL gene (Mbl1 and Mbl2), encoding MBL-A and MBL-C proteins in mouse serum (Hogaboam et al. 2004). However, treatment with recombinant MBL does enhance survival in IPA mice (Kaur et al. 2007). Thus, the role of MBL in *A. fumigatus* infection may also depend on the route of infection and the level of immunosuppression of the host.

MBL interaction with *B. dermatitidis* has only been studied in the mouse system. Both MBL mouse proteins (MBL-A and MBL-C) bind to *B. dermatitidis* yeast cells (Koneti et al. 2008). Inhibition of macrophage response to *B. dermatitidis* is also mediated by MBL, binding to 1,3-β-glucan ligand on *B. dermatitidis*, and thus inhibiting 1,3-β-glucan interaction with Dectin-1 receptor on macrophages and also decreasing TNF-α production (Brown et al. 2002; Kimberg and Brown 2008). Moreover, macrophage production of G-CSF, IFN-γ, MCP-1, and RANTES were significantly inhibited by MBL in response to *B. dermatitidis*, but not IL-6 (Brummer et al. 2007).

MBL can bind to *C. albicans* yeast and pseudohyphae and to *C. parapsilosis* yeast cells (Denton and Disalvo 1964; Sugar and Picard 1988; Brummer et al. 2005; van Asbeck et al. 2008). MBL is able to aggregate *C. albicans* resulting in its growth inhibition and complement deposition of C4b and C3b on its surface via MASPs (Ip and Lau 2004; van Asbeck et al. 2008). Similar levels of MBL-mediated complement deposition were also observed for *C. parapsilosis* (van Asbeck et al. 2008). However, the binding of MBL to *C. albicans* may inhibit its phagocytosis by macrophages or dendritic cells (Zimmerman et al. 1992; Schelenz et al. 1995; Chaka et al. 1997; Vuk-Pavlovic et al. 2001; Ip and Lau 2004; van de Wetering et al. 2004a). MBL seems to inhibit *Candida*-induced macrophage responses in THP-1 cells through TLR-2 and TLR-4, suggesting that *C. albicans* modifies TLR signalling pathways in the macrophage (Wang et al. 2013). However, in the case of neutrophils, MBL enhances the phagocytosis of both *C. albicans* and *C. parapsilosis* yeast cells (van Asbeck et al. 2008). MBL greatly facilitates complement-mediated uptake of *C. albicans* via CR1 receptor in neutrophils and this results in the stimulation of reactive oxygen species by intracellular Dectin-1, which recognises the phagocytosed fungal β-1,3 glucan (Li et al. 2012). The binding of MBL with *C. albicans* yeast also increases TNF-α production by monocytes in vitro (Ghezzi et al. 1998) and in vivo (Lillegard et al. 2006). Double knockout (MBL-A and MBL-C) mice were found to be more susceptible to systemic infection with *C. albicans* compared to wild-type mice (Held et al. 2008). Vaginal candidiasis is an important mycosis in women. MBL protein is present in vaginal secretions (Pellis et al. 2005); MBL levels seem to increase in vulvovaginal candidiasis. However, MBL levels were found to be lower in women with recurrent vulvovaginal candidiasis, because of polymorphisms in their MBL gene (Babula et al. 2003; Liu et al. 2006; Giraldo et al. 2007; Donders et al. 2008; Milanese et al. 2008). The precise role of MBL in candidiasis remains to be fully explored.

MBL can bind to acapsular *C. neoformans* yeast cells (Chaka et al. 1997), but this does not cause aggregation (Eisen et al. 2008). However, MBL binding to acapsular *C. neoformans* did facilitate complement deposition and enhancement of fungal phagocytosis by neutrophils (van Asbeck et al. 2008). Furthermore, TNF-α production was induced in peripheral blood mononuclear cells by *C. neoformans* mannoprotein and this effect was enhanced by MBL (Chaka et al. 1997). It is unknown whether MBL binds to *H. capsulatum* or *P. carinii*. It is unlikely that MBL binds to *H. capsulatum*, since the cell wall contains 1,3-α-glucan (Rappleye et al. 2007); however, in *P. carinii*, the cell surface of cyst forms does contain β-1,3-glucan (Williams et al. 1996), which may bind MBL. In Coccidioides species, it is also unknown whether MBL interaction occurs, but patients with active coccidioidomycosis have been shown to have low serum MBL levels, compared to healthy individuals previously infected with Coccidioides, and that low levels of MBL were associated with polymorphisms in their MBL gene (Corredor et al. 1999).

4.7.3 CL-L1, CL-K1, CL-P1 and the Bovine-Unique Collectins, Conglutinin, CL-43 and CL-46

Very few studies have investigated the interaction of these minor collectins with fungal species. CL-K1 can bind *C. albicans* (Selman and Hansen 2012) and cellular extracts (mannan) of *S. cerevisiae* (Keshi et al. 2006; Selman and Hansen 2012). CL-P1 has also been reported to bind to *S. cerevisiae* and mediate phagocytosis of yeast-derived zymosan, suggesting that CL-P1 mediates phagocytosis for fungi in the vascular endothelium (Ohtani et al. 1999; Jang et al. 2009). Interestingly, CP-P1 also partially binds to *A. fumigatus*, via its CRD, and in association with properdin, can activate the complement alternative pathway, resulting in C3b deposition and formation of the membrane attack complex (Ma et al. 2015). This shows a novel mechanism of triggering the alternative pathway of complement (Ma et al. 2015). There are no reports of CL–L1 interaction with fungi.

There are also limited reports of the binding of bovine-unique collectins to fungi. CL-43 is able to bind to acapsular *C. neoformans* in vitro in a Ca^{2+}-dependent manner (Schelenz et al. 1995), and immobilised yeast mannan (Holmskov et al. 1996). Conglutinin is able to bind to glycoproteins and polysaccharides derived from *S. cerevisiae* (*N*-acetylglucosamine, mannose, mannan) (Loveless et al. 1989; Lim and Holmskov 1996).

4.8 Interaction of Collectins with Protozoal and Helminth Pathogens

An area of collectin that is yet to be fully explored is the interaction of collectins with protozoal and helminth pathogens, which are responsible for some of the most important global infections. There are limited studies and these are mostly based on genetic polymorphisms in collectin genes that are associated with predisposition or severity of these diseases. There is a limited number of functional studies on the role of collectins in protozoal and helminth infections.

Increases in levels of SP-D were observed in serum, renal and cerebral tissues in mice experimentally infected with *Plasmodium berghei*, compared to control mice (Cahayani et al. 2016). Low MBL serum levels and genetic polymorphisms in the MBL gene have been associated with more severe malaria, particularly in children (Luty et al. 1998; Holmberg et al. 2008). MBL can bind to *P. falciparum* protein extracts, but it does not appear to inhibit the parasite directly (Klabunde et al. 2002). MBL does not opsonise *P. falciparum*, but it can bind to *P. falciparum*-infected erythrocytes, recognising the 78-kDa glucose-regulated stress glycoprotein of the parasite (Garred et al. 2003). MBL binding to *P. falciparum* merozoite adhesins have also been reported, having the ability to activate complement (Korir et al. 2014).

The complement lectin pathway can be activated by Trypanosoma and Leishmania (Cestari et al. 2013). MBL binds to glycosylated antigens on *Trypanosoma cruzi*, on the surface of metacyclic trypomastigotes, resulting in complement activation (Cestari Idos et al. 2009). In a mouse model of *T. cruzi* infection, MBL influences host resistance and pathology (Rothfuchs et al. 2012). In some strains of *T. cruzi*, MBL mediates resistance to complement lysis of the parasite and enhances invasion of host cells (Evans-Osses et al. 2014).

MBL also binds to major cell surface glycoconjugates (lipophosphoglycans) on Leishmania parasites, triggering lectin pathway activation and promastigote lysis (Green et al. 1994; Ambrosio and De Messias-Reason 2005). Certain genotypes of the MBL2 gene were also predictive for the risk for developing visceral leishmaniasis and other clinical complications in infections with *Leishmania chagasi* (Alonso et al. 2007). Similarly, there was a strong correlation found between serum levels of MBL and the probability of developing visceral leishmaniasis (Santos et al. 2001). Monocytes challenged with MBL-opsonised *L. chagasi* promastigotes secreted higher levels of TNF-α and IL-6 than controls, suggesting that MBL may play an important role in pathogenesis (Santos et al. 2001).

In helminth infections, MBL binds to the surface glycoproteins of *Schistosoma mansoni* cercariae and adult worms and is able activate the lectin pathway (Klabunde et al. 2000). Curiously, no differences in serum MBL levels were observed between patients infected with Schistosoma and in healthy controls (Klabunde et al. 2000). Another study has shown that high MBL serum levels are associated with protection in schistosomiasis (Antony et al. 2013). Interestingly, high levels of MBL and CL-K1 were inversely correlated with urogenital infections with *S. haematobium* (Antony et al. 2015b). Although CL-K1 has not been shown to bind directly to the parasitic worm, it was observed to be a risk factor for urinary schistosomiasis (Antony et al. 2015a). Furthermore, concomitantly elevated IL-6 levels were also observed in urinary schistosomiasis cases compared to controls that correlated with MBL levels (Antony et al. 2015b). Similar findings linking IL-6 and MBL have also been described in patients with visceral leishmaniasis (Santos et al. 2001; Antony et al. 2015b).

SP-D has also been shown to bind to fucosylated glycoconjugates (α-1–3 linked fucose) on the surface of *S. mansoni* larval stages, although the significance of this interaction remains unclear (van de Wetering et al. 2004b, c). However, a recent study has suggested that SP-D is essential for protection against helminth infection, using the experimental model nematode *Nippostrongylus brasiliensis* (Thawer et al. 2016). *N. brasiliensis* infection of SP-$D^{-/-}$ knockout mice resulted in severe susceptibility to parasitic disease, whilst treatment with rfhSP-D enhanced parasite clearance and anti-parasitic immune responses (Thawer et al. 2016). SP-D was determined to bind to *N. brasiliensis* larvae via its CRD, and to enhance their killing by AM (Thawer et al. 2016).

4.9 Collectins and Allergy

A considerable number of in vitro and in vivo studies have focused on the immunomodulatory functions of collectins and their contribution to the host defense system. Through activation of complement, and production of pro-inflammatory cytokines, MBL makes a major impact on the generation and regulation of the immune-mediated inflammatory response. Allergen-mediated activation of the complement lectin pathway has been demonstrated (Varga et al. 2003). Allergen extracts (parietaria (PA) and house dust (HD) mite) were shown to bind purified MBL, and trigger the lectin complement pathway. Differences in plasma MBL levels may affect the degree of complement activation in different individuals, thus, susceptibility to allergic diseases. Significantly elevated serum MBL levels were observed in pediatric mild-asthma patients, suggesting the possible role of MBL in the pathogenesis of asthma by contributing to airway inflammation, or increasing the risk of asthma development (Uguz et al. 2005). Enhanced levels of serum MBL also correlate with an increased peripheral blood eosinophils in these individuals. It is also suggested that oxidative stress increases the MBL synthesis, and triggers complement activity. MBL can initiate complement activation following oxidative stress in asthma (Collard et al. 2000; Nadeem et al. 2003; Uguz et al. 2005), and aggravate inflammation. Significantly increased MBL levels and MBL pathway was also detected in patients with bronchial asthma, rhinitis and allergic bronchopulmonary aspergillosis (ABPA) (Kaur et al. 2005).

A higher level of plasma MBL is likely to contribute to activation of lectin pathway, and an increased severity, including enhanced blood eosinophil counts. In addition, production of MBL in the liver is suggested to increase by up to three fold in response to environmental stimuli. Therefore, higher levels of plasma MBL in allergic patients, compared to the non-allergic patients, may result from elevated hepatic synthesis caused by allergen exposure. Furthermore, the circulating level of mouse MBL-A was also measured in *Aspergillus fumigatus* allergen-sensitised and non-sensitised mice. Increased level of mMBL-A was observed following allergic sensitisation, suggesting that challenging these mice with allergen may contribute to a higher level of MBL in sensitised mice as well as allergic patients (Kaur et al. 2005). Earlier in vivo studies using mouse MBLs have reported the likely role of MBL-A as a mediator of inflammation (Santos et al. 2001; Takahashi et al. 2002). Moreover, a substantial decline in the airway hyperresposiveness to *A. fumigatus conidia* was seen in MBL-A–deficient mice (MBL-A$^{-/-}$) when compared to MBL-A$^{+/+}$ control mice, which suggest the possible role of MBL-A and its ability to trigger progression of airway hyper-responsiveness (Hogaboam et al. 2004).

Since levels of plasma MBL are genetically determined, it is of interest to study the genetic polymorphisms in MBL in relation to allergic susceptibility. In order to address the correlation between polymorphisms in the MBL gene and the progression of atopic diseases, Nagy et al. found a contribution of variant MBL alleles to the susceptibility to acute or chronic Chlamydia pneumoniae infection in asthmatic children (Nagy et al. 2003). Another study that focused on the genetic association of MBL related single nucleotide polymorphisms (SNPs) with allergic patients (Kaur et al. 2006), reported the identification of G1011A, an intronic SNP found in the MBL gene, and presence of 1011A allele of SNP G1011A to be associated with an enhanced level of plasma MBL., SNP G1011A has also been suggested to play a role in regulating MBL expression. Additional polymorphisms were found at positions 550 (H/L variants) and 221 (X/Y variants) in the promoter region of the MBL gene, which associated with high MBL levels in the plasma. 1011A allele was also associated with bronchial asthmatic patients with allergic rhinitis and ABPA, which positively correlated with allergic markers, including high peripheral blood eosinophil counts, and reduced levels of forced expiratory volume at timed interval of 1 s (FEV1) in these patients. However, no structural SNPs have been observed within the MBL gene in these allergic patients.

As carbohydrate recognition immune molecules, both SP-A and SP-D have been shown to interact with gp55 and gp45 of *A. fumigatus* in a calcium and carbohydrate specific dependent manner (Madan et al. 1997b). Both these collagenous molecules inhibit specific IgE binding to these glycoproteins, and block allergen triggered histamine release from human basophils isolated from Derp- and *A. fumigatus*-sensitised patients (Madan et al. 1997a, b). Dodecameric forms of human SP-D mediate binding, aggregation, and phagocytosis of starch granules, containing grass pollen allergens from *Dactylis glomerata* and *Phleum pratense* via alveolar macrophages (Erpenbeck et al. 2005). SP-D can suppress proliferation of PBMCs isolated from children with Derp–sensitive asthma (Wang et al. 1998), and suppress secretions of IL-2 by PBMCs (Borron et al. 1998). Suppressive effects of SP-A on the production and release of IL-8 by eosinophils were also reported, which is stimulated by ionomycin in a concentration-dependent manner (Cheng et al. 1998). Since IgE cross-linking, release of histamine and PMBCs proliferation are crucial immunological factors contributing to the development of asthmatic symptoms, both SP-A and SP-D are crucial immune modulators in resisting allergenic challenge, as well as suppressing substantial hypersensitivity reactions in the lungs (Kishore et al. 2002).

Intranasal administration of SP-D or rfhSP-D caused reduced levels of peripheral and pulmonary eosinophilia, and the effect persisted up to 16 days in the ABPA mice. These observations therefore indicate the potential of SP-D as a therapeutic agent (Kishore et al. 2002; Madan et al.2001a; 2005a, b). In addition, protective role of rhfSP-D has also reported in murine model of Derp allergens-induced pulmonary hypersensitivity (Singh et al. 2003). Shifting of Th2 to a Th1 following SP-D treatments appeared to be crucial to the protective mechanism, since, IFN-γ gamma is suggested to inhibit differentiation of Th2 in response to IL-4 (Elser et al. 2002). Additionally, production of nitric oxide was significantly inhibited when Derp mice derived alveolar macrophages are pre-incubated with rfhSP-D, and resulted in low levels of TNF-α in the rfhSP-D treated Derp mice (Liu et al. 2005). Culturing alveolar macrophages with allergen and SP-D has induced an increased production of IL-10, IL-12, and IFN- γ, indicating a positive correlation between macrophages and SP-D triggered inhibition of airway inflammation and airway hyper-responsiveness (AHR) (Takeda et al. 2003).

A study by Madan et al. has focused on the susceptibility of SP-A$^{-/-}$ and SP-D$^{-/-}$ mice to challenge with *A. fumigatus* allergen compared to wild-type mice (Madan et al. 2005a, b).

Intrinsic hypereosinophilia and seven fold increase in IL-5 and IL-13 levels were seen in both SP-A$^{-/-}$ and SP-D$^{-/-}$ mice. However, lower levels of IFN- γ to IL-4 ratio in the lungs were observed, suggesting the possible Th2 basis of immune response. Treating these mice with exogenous intranasal SP-A and SP-D resulted in reversal of Th2 polarisation. SP-D$^{-/-}$ mice was reported to be more susceptible to *A. fumigatus* allergen-induced pulmonary hypersensitivity when compared to wild-type mice. However, resistant to sensitisation was seen with SP-A$^{-/-}$ mice. Intranasal administration of SP-D or rfhSP-D led to rescue of the sensitised SP-D$^{-/-}$ mice, while SP-A$^{-/-}$ mice demonstrated an enhanced levels of IL-5 and IL-13, causing greater pulmonary eosinophilia. Genetic polymorphisms in the collagen region of SP-A2 (SP-A2 G1649C and SP-A2 A1660G) may also increase susceptibility to allergic bronchopulmonary aspergillosis (ABPA) (Saxena et al. 2003).

4.10 Collectin (and C1q) Receptors

The collectins are structurally related to the complement protein C1q (having a collagenous region and similar overall tertiary structure. A common receptor for SP-A, MBL and C1q was described in 1990 (Malhotra et al. 1990) (Fig. 4.5), as collagen region binding cC1qR. This was subsequently identified as Calreticulin (~56kDa). Two other candidate receptors were subsequently proposed:

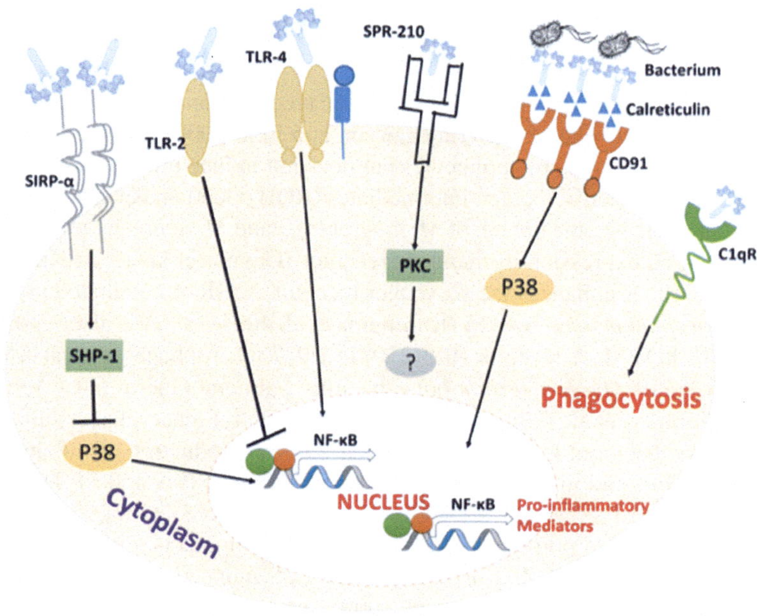

Fig. 4.5 Collectin receptors on immune cells. Collectins have been shown to bind a number of receptors and putative receptors, which lead to immunomodulatory responses. Binding of collectins to Toll-like receptor 2 (TLR-2), TLR-4, SP-A receptor 210 (SP-R210), CD91-calreticulin, and signal inhibitory regulatory protein-α (SIRP-α) alters production of pro-inflammatory mediators. For example, SP-A and SP-D binds to SIRP-α via their collagenous tails, and stimulates pro-inflammatory chemokine production via calreticulin/CD91 interaction. Furthermore, bacterium bound collectins induces the conformational changes of calreticulin/CD91 interaction, which then activates P38-mitogen-activated protein kinase (MAPK) signalling pathway, leading to transcriptional activation of NF-κB (nuclear factor kappa-light-chain-enhancer of activated B cells), and expression of pro-inflammatory cykines, including tumour necrosis factor alpha (TNF-α). SIRP-α is abundantly expressed in macrophages, and ligation of SIRP-α by lung collectins are crucial in preventing damage to the airways caused by the production of pro-inflammatory responses. Thereby, phosphorylated cytoplasmic region of SIRP-α recruits SHP-1 (Src homology region 2 domain-containing phosphatase-1), which in turn dephosphorylates protein substrates involved in mediating physiological effects. Thus, the interaction between SIRP-α and SHP-1 negatively regulates P38-MAP kinase signalling, and stimulates NF-κB activity, and cells become resistant to TNF-mediated effects, such as apoptosis

(1) C1qRp (CD93) (C1q receptor associated with phagocytosis stimulated by C1q, MBL or SP-A): but this has subsequently been shown not to bind any of these ligands. It may be an adhesion receptor (McGreal et al. 2002; Norsworthy et al. 2004).
(2) CR1, the complement C3b receptor, does interact with C1q and MBL, but functional aspects are not yet widely explored (Jacquet et al. 2018).
(3) Calreticulin remains the main candidate as a receptor/adapter involved in phagocytosis mediated by C1q and collectins (Ogden et al. 2001; Vandivier et al. 2002). Calreticulin bound to the cell surface CD91 mediates uptake of apoptotic cells

to which C1q, MBL, SP-A and SP-D are bound. It also mediates uptake of microorganisms targeted by the collectins.

SP-A and SP-D can interact with phagocytic receptors and are able to influence receptor-mediated uptake of bacteria (Lawson and Reid 2000). SP-A enhances the phagocytosis of *S. aureus* by monocytes but does not induce intracellular killing or the production of reactive oxygen intermediates (ROI) (Geertsma et al. 1994). SP-A is also able to enhance the uptake of *M. tuberculosis* and *M. avium* by macrophages via the increased expression of mannose receptor (Gaynor et al. 1995; Kudo et al. 2004), whilst SP-A enhances the scavenger receptor A (SR-A)-mediated uptake of *Streptococcus pneumoniae* by AM (Kuronuma et al. 2004). SP-A is also reported to bind to a 210-kDa SP-A receptor (SPR210) in U937 macrophages and rat AM and mediate uptake of *Mycobacterium bovis* bacillus Calmette-Guérin (BCG) via this receptor (Chroneos et al. 1996; Weikert et al. 1997); in rat macrophages, this led to enhanced mycobacterial killing and an increase in the production of inflammatory mediators, TNF-α and nitric oxide (Weikert et al. 2000). SP-A and SP-D can also bind to the major LPS receptor, CD14 that is present on alveolar macrophages. SP-A binds to the peptide portion of CD14, whilst SP-D interacts with the glycan of the receptor (Sano et al. 2000). SP-A modulates LPS-induced cellular responses by direct interaction with CD14 (Sano et al. 1999), serving as an important mechanism for the recognition and clearance of this endotoxin. As noted above, SP-A and SP-D can interact with the versatile protein calreticulin (Malhotra et al. 1990). When SP-A and SP-D bind to surface target ligands such as LPS or apoptotic cells, multiple collagen regions are presented on the surface, which interact with the Calreticulin -CD91 receptor complex (Gardai et al. 2003). This can then lead to the promotion of phagocytosis and initiation of cell signalling pathways leading to the production of pro-inflammatory cytokines and priming of adaptive immunity (Ogden et al. 2001; Vandivier et al. 2002; Gardai et al. 2003). In contrast, SP-A and SP-D can also suppress inflammatory responses by binding to signal regulating protein α (SIRP-α) on macrophages and epithelia via their CRD region, and this leads to a signalling pathway that blocks pro-inflammatory mediator production (Fig. 4.5) (Table 4.2) (Gardai et al. 2003). Therefore, the orientation of binding of SP-A and SP-D (collagen or lectin (CRD)) domains to host receptors Calreticulin-CD91 or SIRP-α, respectively, have opposing effects and illustrate a dual function of these collectins, which could be, (1) protection of the naïve lung via maintenance of immune homeostasis and (2) protection via the triggering of inflammation to clear pathogens, allergens, necrotic and apoptotic cells. Another receptor, gp-340 has also been found to bind to SP-D and SP-A (Holmskov et al. 1997a; Tino and Wright 1999), but any microbial interaction has been described for Influenza virus (Hartshorn et al. 2006a, b) and not bacteria, whilst binding of SP-A to gp-340 stimulates macrophage chemotaxis (Tino and Wright 1999).

Calreticulin (cC1qR) is a collectin binding protein of 56 kDa, and is known to bind C1q, MBL, SP-A, conglutinin and CL-43 (Fig. 4.5) (Table 4.2) (Malhotra et al. 1993). This interaction is independent of calcium ions, and ionic in nature. It is mediated by the collagen domain 'c' of the collectins (cC1qR). The C1q binding site has

Table 4.2 Binding proteins and Receptors for lung collectins

Binding proteins/receptors	Cells expressing the receptors	Protein domain	Proposed functions
Calreticulin/CD91	Ubiquitous	Collagen region of SP-A and SP-D	Phagocytosis and production of pro-inflammatory cytokines
C1qRp/CD93	Monocytes, macrophages, neutrophils, platelets, endothelial cells and microglia	Reported as a receptor, but may not bind C1q or collectins	Initially proposed to have roles in phagocytosis
gp340	Soluble opsonin	CRD region of SP-A and SP-D	Stimulation and migration of alveolar macrophages
CD14	Alveolar macrophages	Neck region of SP-A which interacts with the peptide portion containing the leucine-rich repeats of CD14. CRDs of SP-D that binds and interact with carbohydrate moiety of CD14	CD14-LPS modulation and release of pro-inflammatory cytokines and chemokines
SIRP α/CD172	Antigen presenting cells, endothelial cells, myeloid cells, and neurons	CRD of SP-A and SP-D	Inhibition of pro-inflammatory cytokine and chemokine production
SPR210	Type II cells, alveolar macrophages, and bone marrow derived macrophages	SP-A- 36 residues collagen region composed of RGD motif	Regulation of phospholipid secretion and cytokine production, phagocytosis, and inhibition of T-cell proliferation

been mapped to the S-domain of calreticulin (Stuart et al. 1996). Conglutinin and CL-43 also bind C1qR and calreticulin (Dec and Wernicki 2006). SP-A receptor 210 (SP-R210) is a 210 kDa oligomeric molecule that earlier has been purified from the macrophage cell line U937 by affinity chromatography (Fig. 4.5 (Table 4.2) (Chroneos et al. 1996). SP-R210 is another candidate receptor for SP-A, and also found in type II cells and alveolar macrophages. The direct interaction between SP-R210 and SP-A occurs through the collagen region of SP-A. Antibodies raised against SP-R210 inhibit SP-A binding to alveolar type II cells and alveolar macrophages, thus, prevent SP-A-mediated uptake of *Mycobacterium bovis*, and block SP-A-bacillus Calmette-Guerin complexes to phagocytes. Furthermore, SP-R210 inhibited SP-A-mediated phospholipid secretion by alveolar type II cells (Weikert et al. 1997). Additionally, a study by Yang et al. reported SP-R210 as a cell surface myosin 18A (Yang et al. 2005), and expressed in multiple isoforms. Gp340 is a 340 kDa SP-D binding glycoprotein purified from lung bronchoalvelar lavage of alveolar proteinosis patients (Holmskov et al. 1997), and belongs to the scavenger receptor superfamily, consisting of multiple scavenger receptor type B domains (Holmskov et al. 1999). Furthermore, gp340 has been shown to be identical to salivary agglutinin, and its binding interaction with *Streptococcus mutans* and *Helicobacter pylori* has been reported (Ligtenberg et al. 2001). The direct binding of SP-D to gp340 occurs in a calcium dependent manner, and is inhibited by EDTA. The interaction is not affected by the presence of maltose, suggesting that the binding is a protein-protein interaction via the CRD region of SP-D rather than a lectin-carbohydrate interaction. Similar binding pattern was observed between gp340 and rfhSP-D, composed of trimeric neck region and CRD (Holmskov et al. 1997). Gp340 exists in a soluble form, and acts as an adaptor for SP-A and SP-D, but how gp340 is anchored in the membrane and the mechanism of binding to cell surface has not been elucidated fully yet (Table 4.2). SP-D can bind to human neutrophil defensins (HNPs) via its neck and CRD region (Hartshorn et al. 2006a, b). The interaction between SP-D and HNPs can trigger anti-viral activity in the BAL fluid. HD6, human β-defensins, and human neutrophil peptide (HNP)-4 bind SP-D with weaker affinity, while HNP-1-3 bind SP-D with high affinity and trigger inhibition of SP-D mediated anti-viral activity (Doss et al. 2009). Additionally, SP-D binds human decorin from amniotic fluid in a calcium dependent manner via sulphated *N*-acetyl galactosamine moiety of decorin (Nadesalingam et al. 2003). rfhSP-D interacts with dermatan sulphate moiety of decorin, and core protein of decorin interacts with SP-D via the collagen-like region.

The interaction between lung collectins (SP-A and SP-D) and native as well as recombinant CD14 has been reported (Table 4.2) (Sano et al. 1999). CD14 is a soluble receptor for LPS, and the neck domain of SP-D has been shown to bind the leucine-rich peptide portion of CD14, whilst the carbohydrate moiety of CD14 interacts with the lectin domain of SP-D (Sano et al. 1999). Thus, both these surfactant proteins appear to modulate the cellular response to smooth and rough LPS by interaction with CD14. (Sano et al. 1999). Furthermore, the association of SP-A and SP-D with toll-like receptors (TLR), and or the TLR-associated molecule, could be one of the mechanisms by which they function as modulators of inflammation an inflammatory mediators (Sano et al. 1999). Studies have reported direct involvement of SP-A in TLR2 signalling, and inhibition of downstream gene activation (Murakami et al. 2002). SP-A interacts directly with TLR4 and myeloid differentiation factor 2 (MD-2), which are known critical signalling receptors for LPS (Billod et al. 2016). Thus, binding of SP-A with extracellular domain of TLR4 and MD-2 was revealed in a calcium-dependent manner, involving the CRD region. Additionally, SP-A has attenuated cell surface binding of smooth LPS, and induced NF-κB activation in cells expressing TLR4 and MD-2 (Murakami et al. 2002). SP-D does not have significant effect on TLR4 expression, but it down-regulates the TLR4-mediated signalling against LPS (Henning et al. 2008). Thus, SP-A's ability to dampen TLR2 and

TLR4 signalling is associated with decrease in the phosphorylation of IkappaBalpha, a key regulator of NF-κB activity, and nuclear translocation of p65, resulting in reduced TNF-α secretion in response to TLR ligands. Furthermore, the same study also reported diminished phosphorylation of Akt, an essential regulator of NF-κB and potentially MAPKs. Therefore, there is a critical role for SP-A in modulating lung inflammatory reactions by regulating macrophage-mediated TLR4 activity.

As noted above, calreticulin was identified as a common receptor for C1q, MBL and SP-A (Malhotra et al. 1990). Calreticulin is mainly an intracellular protein but it is present on cell surfaces bound to CD91 and, thus, acts as an adaptor or co-receptor while binding to collagenous region of these collectins (Fig. 4.5) (Table 4.2) (Ogden et al. 2001; Vandivier et al. 2002). Uptake of apoptotic cells by phagocytes mediated by C1q or MBL binding to calreticulin-CD91 complex was revealed (Vandivier et al. 2002). HLA class I heavy chain (Arosa et al. 1999), or CD59 have been shown to act as a calreticulin-binding proteins on cells which do not express CD91, allowing C1q or MBL coated particles to adhere to the cells. Although a number of collectin receptors have been identified, there is still a need to fully elucidate how collectins stimulate phagocytes, and mediate phagocytosis, as well as other signalling transduction pathways. More studies on the structural aspects of receptors are needed, especially how these receptors are anchored in the membrane of immune and non-immune cells and, which co-receptors and signalling pathways are involved.

In conclusion, collectins appear to play important roles in controlling lung allergy, inflammation and hypersensitivity, in addition to dealing with a wide variety of pathogens at pulmonary and extra-pulmonary sites. They act against pulmonary allergens through their ability to resist allergen challenge by interfering with allergen triggered IgE interaction, degranulation of mast/basophils, cellular infiltration, and polarisation of helper Th response. Their roles have been also implicated in altering profiles of pro-inflammatory cytokines and chemokines as a result of infection, and allergen challenge. Further research is needed to characterise specific collectin receptors that are crucial for collectin functions other than phagocytosis.

References

Al-Ahdal MN, Murugaiah V, Varghese PM, Abozaid SM, Saba I, Al-Qahtani AA, Pathan AA, Kouser L, Nal B, Kishore U (2018) Entry inhibition and modulation of pro-inflammatory immune response against influenza A virus by a recombinant truncated surfactant protein D. Front Immunol 9:1586

Aliouat EM, Escamilla R, Cariven C, Vieu C, Mullet C, Dei-Cas E, Prevost MC (1998) Surfactant changes during experimental pneumocystosis are related to Pneumocystis development. Eur Respir J 11:542–547

Allen MJ, Laederach A, Reilly PJ, Mason RJ (2001a) Polysaccharide recognition by surfactant protein D: novel interactions of a C-type lectin with nonterminal glucosyl residues. Biochemistry 40(26):7789–7798

Allen MJ, Voelker DR, Mason RJ (2001b) Interactions of surfactant proteins A and D with *Saccharomyces cerevisiae* and *Aspergillus fumigatus*. Infect Immun 69:2037–2044

Alonso DP, Ferreira AF, Ribolla PE, De Miranda Santos IK, Do Socorro Pires E Cruz M, Aecio De Carvalho F, Abatepaulo AR, Lamounier Costa D, Werneck GL, Farias TJ, Soares MJ, Costa CH (2007) Genotypes of the mannan-binding lectin gene and susceptibility to visceral leishmaniasis and clinical complications. J Infect Dis 195:1212–1217

Al-Qahtani AA, Murugaiah V, Bashir HA, Pathan AA, Abozaid SM, Makarov E, Nal B, Kishore U, Al-Ahdal MN (2019) Full-length human Surfactant protein A inhibits Influenza A virus infection of A549 lung epithelial cells: a recombinant form containing neck and lectin domains promotes infectivity. Immunobiology (in Press)

Ambrosio AR, De Messias-Reason IJ (2005) Leishmania (Viannia) braziliensis: interaction of mannose-binding lectin with surface glycoconjugates and complement activation. An antibody-independent defence mechanism. Parasite Immunol 27:333–340

Anders EM, Hartley CA, Reading PC, Ezekowitz RA (1994) Complement-dependent neutralization of influenza virus by a serum mannose-binding lectin. J Gen Virol 75(Pt 3):615–622

Andersen O, Sorensen AM, Svehag SE, Fenouillet E (1991) Conglutinin binds the HIV-1 envelope glycoprotein gp 160 and inhibits its interaction with cell membrane CD4. Scand J Immunol 33:81–88

Antony JS, Ojurongbe O, Van Tong H, Ouf EA, Engleitner T, Akindele AA, Sina-Agbaje OR, Adeyeba AO, Kremsner PG, Velavan TP (2013) Mannose-binding lectin and susceptibility to schistosomiasis. J Infect Dis 207:1675–1683

Antony JS, Ojurongbe O, Kremsner PG, Velavan TP (2015a) Lectin complement protein Collectin 11 (CL-K1) and susceptibility to urinary schistosomiasis. PLoS Negl Trop Dis 9:e0003647

Antony JS, Ojurongbe O, Meyer CG, Thangaraj K, Mishra A, Kremsner PG, Velavan TP (2015b) Correlation of Interleukin-6 levels and lectins during Schistosoma haematobium infection. Cytokine 76:152–155

Arora M, Munoz E, Tenner AJ (2001) Identification of a site on mannan-binding lectin critical for enhancement of phagocytosis. J Biol Chem 276(46):43087–43094

Arosa FA, de Jesus O, Porto G, Carmo AM, de Sousa M (1999) Calreticulin is expressed on the cell surface of activated human peripheral blood T lymphocytes in association with major histocompatibility complex class I molecules. J Biol Chem 274(24):16917–16922

Atochina EN, Beck JM, Scanlon ST, Preston AM, Beers MF (2001) Pneumocystis carinii pneumonia alters expression and distribution of lung collectins SP-A and SP-D. J Lab Clin Med 137(6):429–439

Atochina EN, Gow AJ, Beck JM, Haczku A, Inch A, Kadire H, Tomer Y, Davis C, Preston AM, Poulain F, Hawgood S, Beers MF (2004) Delayed clearance of pneumocystis carinii infection, increased inflammation, and altered nitric oxide metabolism in lungs of surfactant protein-D knockout mice. J Infect Dis 189:1528–1539

Avirutnan P, Hauhart RE, Marovich MA, Garred P, Atkinson JP, Diamond MS (2011) Complement-mediated neutralization of dengue virus requires mannose-binding lectin. MBio 2

Awasthi S, Magee DM, Coalson JJ (2004) Coccidioides posadasii infection alters the expression of pulmonary surfactant proteins (SP)-A and SP-D. Respir Res 5:28

Axelgaard E, Jensen L, Dyrlund TF, Nielsen HJ, Enghild JJ, Thiel S, Jensenius JC (2013) Investigations on collectin liver 1. J Biol Chem 288:23407–23420

Babula O, Lazdane G, Kroica J, Ledger WJ, Witkin SS (2003) Relation between recurrent vulvovaginal candidiasis, vaginal concentrations of mannose-binding lectin, and a mannose-binding lectin gene polymorphism in Latvian women. Clin Infect Dis 37:733–737

Bachis A, Aden SA, Nosheny RL, Andrews PM, Mocchetti I (2006) Axonal transport of human immunodeficiency virus type 1 envelope protein glycoprotein 120 is found in association with neuronal apoptosis. J Neurosci 26:6771–6780

Ballegaard V, Haugaard AK, Garred P, Nielsen SD, Munthe-Fog L (2014) The lectin pathway of complement: advantage or disadvantage in HIV pathogenesis? Clin Immunol 154:13–25

Bartlomiejczyk MA, Swierzko AS, Brzostek A, Dziadek J, Cedzynski M (2014) Interaction of lectin pathway of complement-activating pattern recognition molecules with mycobacteria. Clin Exp Immunol 178:310–319

Beharka AA, Gaynor CD, Kang BK, Voelker DR, Mccormack FX, Schlesinger LS (2002) Pulmonary surfactant protein A up-regulates activity of the mannose receptor, a pattern recognition receptor expressed on human macrophages. J Immunol 169:3565–3573

Benne CA, Kraaijeveld CA, Van Strijp JA, Brouwer E, Harmsen M, Verhoef J, Van Golde LM, Van Iwaarden JF (1995) Interactions of surfactant protein A with influenza A viruses: binding and neutralization. J Infect Dis 171:335–341

Billod JM, Lacetera A, Guzmán-Caldentey J, Martín-Santamaría S (2016) Computational approaches to toll-like receptor 4 modulation. Molecules 21(8)

Borron PJ, Crouch EC, Lewis JF, Wright JR, Possmayer F, Fraher LJ (1998) Recombinant rat surfactant-associated protein D inhibits human T lymphocyte proliferation and IL-2 production. J Immunol 161(9):4599–4603

Borron P, Mcintosh JC, Korfhagen TR, Whitsett JA, Taylor J, Wright JR (2000) Surfactant-associated protein A inhibits LPS-induced cytokine and nitric oxide production in vivo. Am J Physiol Lung Cell Mol Physiol 278:L840–L847

Bräuer L, Kindler C, Jäger K, Sel S, Nölle B, Pleyer U, Ochs M, Paulsen FP (2007) Detection of surfactant proteins A and D in human tear fluid and the human lacrimal system. Invest Ophthalmol Vis Sci 48(9):3945–3953

Brown GD, Taylor PR, Reid DM, Willment JA, Williams DL, Martinez-Pomares L, Wong SY, Gordon S (2002) Dectin-1 is a major beta-glucan receptor on macrophages. J Exp Med 196:407–412

Brown KS, Ryder SD, Irving WL, Sim RB, Hickling TP (2007) Mannan binding lectin and viral hepatitis. Immunol Lett 108:34–44

Brown KS, Keogh MJ, Owsianka AM, Adair R, Patel AH, Arnold JN, Ball JK, Sim RB, Tarr AW, Hickling TP (2010) Specific interaction of hepatitis C virus glycoproteins with mannan binding lectin inhibits virus entry. Protein Cell 1:664–674

Brown-Augsburger P, Hartshorn K, Chang D, Rust K, Fliszar C, Welgus HG, Crouch EC (1996) Site-directed mutagenesis of Cys-15 and Cys-20 of pulmonary surfactant protein D. Expression of a trimeric protein with altered anti-viral properties. J Biol Chem 271(23):13724–13730

Bruce SR, Atkins CL, Colasurdo GN, Alcorn JL (2009) Respiratory syncytial virus infection alters surfactant protein A expression in human pulmonary epithelial cells by reducing translation efficiency. Am J Physiol Lung Cell Mol Physiol 297:L559–L567

Brudner M, Karpel M, Lear C, Chen L, Yantosca LM, Scully C, Sarraju A, Sokolovska A, Zariffard MR, Eisen DP, Mungall BA, Kotton DN, Omari A, Huang IC, Farzan M, Takahashi K, Stuart L, Stahl GL, Ezekowitz AB, Spear GT, Olinger GG, Schmidt EV, Michelow IC (2013) Lectin-dependent enhancement of Ebola virus infection via soluble and transmembrane C-type lectin receptors. PLoS ONE 8:e60838

Brummer E, Kethineni N, Stevens DA (2005) Immunological basis for susceptibility and resistance to pulmonary blastomycosis in mouse strains. Cytokine 32:12–19

Brummer E, Capilla J, Bythadka L, Stevens DA (2007) Production of IL-6, in contrast to other cytokines and chemokines, in macrophage innate immune responses: effect of serum and fungal (Blastomyces) challenge. Cytokine 39:163–170

Cahayani WA, Norahmawati E, Budiarti N, Fitri LE (2016) Increased CD11b and hypoxia-inducible factors-1alpha expressions in the lung tissue and surfactant protein-D levels in serum are related with acute lung injury in severe malaria of C57BL/6 mice. Iran J Parasitol 11:303–315

Cestari Idos S, Krarup A, Sim RB, Inal JM, Ramirez MI (2009) Role of early lectin pathway activation in the complement-mediated killing of *Trypanosoma cruzi*. Mol Immunol 47:426–437

Cestari I, Evans-Osses I, Schlapbach LJ, De Messias-Reason I, Ramirez MI (2013) Mechanisms of complement lectin pathway activation and resistance by trypanosomatid parasites. Mol Immunol 53:328–334

Chaka W, Verheul AF, Vaishnav VV, Cherniak R, Scharringa J, Verhoef J, Snippe H, Hoepelman AI (1997) Induction of TNF-alpha in human peripheral blood mononuclear cells by the mannoprotein of *Cryptococcus neoformans* involves human mannose binding protein. J Immunol 159:2979–2985

Chang WC, White MR, Moyo P, Mcclear S, Thiel S, Hartshorn KL, Takahashi K (2010) Lack of the pattern recognition molecule mannose-binding lectin increases susceptibility to influenza A virus infection. BMC Immunol 11:64

Cheng G, Ueda T, Nakajima H, Nakajima A, Kinjyo S, Motojima S, Fukuda T (1998) Suppressive effects of SP-A on ionomycin-induced IL-8 production and release by eosinophils. Int Arch Allergy Immunol 117(Suppl 1):59–62

Chiba H, Pattanajitvilai S, Evans AJ, Harbeck RJ, Voelker DR (2002) Human surfactant protein D (SP-D) binds *Mycoplasma pneumoniae* by high affinity interactions with lipids. J Biol Chem 277:20379–20385

Chroneos ZC, Abdolrasulnia R, Whitsett JA, Rice WR, Shepherd VL (1996) Purification of a cell-surface receptor for surfactant protein A. J Biol Chem 271(27):16375–16383

Collard CD, Väkevä A, Morrissey MA, Agah A, Rollins SA, Reenstra WR, Buras JA, Meri S, Stahl GL (2000) Complement activation after oxidative stress: role of the lectin complement pathway. Am J Pathol 156(5):1549–1556

Colley KJ, Baenziger JU (1987) Identification of the post-translational modifications of the core-specific lectin. The core-specific lectin contains hydroxyproline, hydroxylysine, and glucosylgalactosylhydroxylysine residues. J Biol Chem 262(21):10290–10295

Corredor GG, Castano JH, Peralta LA, Diez S, Arango M, Mcewen J, Restrepo A (1999) Isolation of Paracoccidioides brasiliensis from the nine-banded armadillo *Dasypus novemcinctus*, in an endemic area for paracoccidioidomycosis in Colombia. Rev Iberoam Micol 16:216–220

Coya JM, Akinbi HT, Saenz A, Yang L, Weaver TE, Casals C (2015) Natural anti-infective pulmonary proteins: in vivo cooperative action of surfactant protein SP-A and the lung antimicrobial peptide SP-BN. J Immunol 195:1628–1636

Crosdale DJ, Poulton KV, Ollier WE, Thomson W, Denning DW (2001) Mannose-binding lectin gene polymorphisms as a susceptibility factor for chronic necrotizing pulmonary aspergillosis. J Infect Dis 184:653–656

Crouch E, Parghi D, Kuan SF, Persson A (1992) Surfactant protein D: subcellular localization in nonciliated bronchiolar epithelial cells. Am J Physiol 263(1 Pt 1):L60–L66

Cunnion KM, Lee JC, Frank MM (2001) Capsule production and growth phase influence binding of complement to *Staphylococcus aureus*. Infect Immun 69:6796–6803

Dec M, Wernicki A (2006) Conglutinin, CL-43 and CL-46–three bovine collectins. Pol J Vet Sci 9(4):265–275

Dec M, Wernicki A, Puchalski A, Urban-Chmiel R, Radej S (2012) Effect of conglutinin on phagocytic activity of bovine granulocytes. Pol J Vet Sci 15:455–462

Denton JF, Disalvo AF (1964) Isolation of blastomyces dermatitidis from natural sites at Augusta, Georgia. Am J Trop Med Hyg 13:716–722

Devyatyarova-Johnson M, Rees IH, Robertson BD, Turner MW, Klein NJ, Jack DL (2000) The lipopolysaccharide structures of *Salmonella enterica* serovar Typhimurium and *Neisseria gonorrhoeae* determine the attachment of human mannose-binding lectin to intact organisms. Infect Immun 68:3894–3899

Dodagatta-Marri E, Mitchell DA, Pandit H, Sonawani A, Murugaiah V, Idicula-Thomas S, Nal B, Al-Mozaini MM, Kaur A, Madan T, Kishore U (2017) Protein-protein interaction between surfactant protein D and DC-SIGN via C-type lectin domain can suppress HIV-1 transfer. Front Immunol 8:834

Donders GG, Babula O, Bellen G, Linhares IM, Witkin SS (2008) Mannose-binding lectin gene polymorphism and resistance to therapy in women with recurrent vulvovaginal candidiasis. Br J Obs Gyn 115:1225–1231

Doss M, White MR, Tecle T, Gantz D, Crouch EC, Jung G, Ruchala P, Waring AJ, Lehrer RI, Hartshorn KL (2009) Interactions of alpha-, beta-, and theta-defensins with influenza A virus and surfactant protein D. J Immunol 182(12):7878–7887.

Drickamer K, Taylor ME (2015) Recent insights into structures and functions of C-type lectins in the immune system. Curr Opin Struct Biol 34:26–34.

Drickamer K, Dordal MS, Reynolds L (1986) Mannose-binding proteins isolated from rat liver contain carbohydrate-recognition domains linked to collagenous tails. Complete primary structures and homology with pulmonary surfactant apoprotein. J Biol Chem 261(15):6878–6887

Drogari-Apiranthitou M, Fijen CA, Thiel S, Platonov A, Jensen L, Dankert J, Kuijper EJ (1997) The effect of mannan-binding lectin on opsonophagocytosis of *Neisseria meningitidis*. Immunopharmacology 38:93–99

Eisen DP, Minchinton RM (2003) Impact of mannose-binding lectin on susceptibility to infectious diseases. Clin Infect Dis 37:1496–1505

Eisen DP, Dean MM, O'sullivan MV, Heatley S, Minchinton RM (2008) Mannose-binding lectin deficiency does not appear to predispose to cryptococcosis in non-immunocompromised patients. Med Mycol 46:371–375

Elser B, Lohoff M, Kock S, Giaisi M, Kirchhoff S, Krammer PH, Li-Weber M (2002) IFN-gamma represses IL-4 expression via IRF-1 and IRF-2. Immunity 17(6):703–712

Erpenbeck VJ, Malherbe DC, Sommer S, Schmiedl A, Steinhilber W, Ghio AJ, Krug N, Wright JR, Hohlfeld JM (2005) Surfactant protein D increases phagocytosis and aggregation of pollen-allergen starch granules. Am J Physiol Lung Cell Mol Physiol 288(4):L692–L698

Escudero-Perez B, Volchkova VA, Dolnik O, Lawrence P, Volchkov VE (2014) Shed GP of Ebola virus triggers immune activation and increased vascular permeability. PLoS Pathog 10:e1004509

Evans-Osses I, Mojoli A, Beltrame MH, Da Costa DE, Darocha WD, Velavan TP, De Messias-Reason I, Ramirez MI (2014) Differential ability to resist to complement lysis and invade host cells mediated by MBL in R4 and 860 strains of *Trypanosoma cruzi*. FEBS Lett 588:956–961

Ezekowitz RA, Day LE, Herman GA (1988) A human mannose-binding protein is an acute-phase reactant that shares sequence homology with other vertebrate lectins. J Exp Med. 167(3):1034–1046.

Ezekowitz RA, Kuhlman M, Groopman JE, Byrn RA (1989) A human serum mannose-binding protein inhibits *in vitro* infection by the human immunodeficiency virus. J Exp Med 169:185–196

Favier AL, Reynard O, Gout E, Van Eijk M, Haagsman HP, Crouch E, Volchkov V, Peyrefitte C, Thielens NM (2018) Involvement of surfactant protein D in ebola virus infection enhancement via glycoprotein interaction. Viruses 11

Ferguson JS, Voelker DR, Mccormack FX, Schlesinger LS (1999) Surfactant protein D binds to *Mycobacterium tuberculosis* bacilli and lipoarabinomannan via carbohydrate-lectin interactions resulting in reduced phagocytosis of the bacteria by macrophages. J Immunol 163:312–321

Ferguson JS, Voelker DR, Ufnar JA, Dawson AJ, Schlesinger LS (2002) Surfactant protein D inhibition of human macrophage uptake of *Mycobacterium tuberculosis* is independent of bacterial agglutination. J Immunol 168:1309–1314

Fidler KJ, Hilliard TN, Bush A, Johnson M, Geddes DM, Turner MW, Alton EW, Klein NJ, Davies JC (2009) Mannose-binding lectin is present in the infected airway: a possible pulmonary defence mechanism. Thorax 64:150–155

Fisher JH, Mason R (1995) Expression of pulmonary surfactant protein D in rat gastric mucosa. Am J Respir Cell Mol Biol 12(1):13–18

Fischer PB, Ellermann-Eriksen S, Thiel S, Jensenius JC, Mogensen SC (1994) Mannan-binding protein and bovine conglutinin mediate enhancement of herpes simplex virus type 2 infection in mice. Scand J Immunol 39:439–445

Floros J, Lin HM, Garcia A, Salazar MA, Guo X, Diangelo S, Montano M, Luo J, Pardo A, Selman M (2000) Surfactant protein genetic marker alleles identify a subgroup of tuberculosis in a Mexican population. J Infect Dis 182:1473–1478

Friis P, Svehag SE, Andersen O, Gahrn-Hansen B, Leslie RG (1991) Conglutinin exhibits a complement-dependent enhancement of the respiratory burst of phagocytes stimulated by *E. coli*. Immunology 74:680–684

Friis-Christiansen P, Thiel S, Svehag SE, Dessau R, Svendsen P, Andersen O, Laursen SB, Jensenius JC (1990) *In vivo* and *in vitro* antibacterial activity of conglutinin, a mammalian plasma lectin. Scand J Immunol 31:453–460

Fuchs A, Pinto AK, Schwaeble WJ, Diamond MS (2011) The lectin pathway of complement activation contributes to protection from West Nile virus infection. Virology 412:101–109

Funk CJ, Wang J, Ito Y, Travanty EA, Voelker DR, Holmes KV, Mason RJ (2012) Infection of human alveolar macrophages by human coronavirus strain 229E. J Gen Virol 93:494–503

Gadjeva M, Paludan SR, Thiel S, Slavov V, Ruseva M, Eriksson K, Lowhagen GB, Shi L, Takahashi K, Ezekowitz A, Jensenius JC (2004) Mannan-binding lectin modulates the response to HSV-2 infection. Clin Exp Immunol 138:304–311

Gaiha GD, Dong T, Palaniyar N, Mitchell DA, Reid KB, Clark HW (2008) Surfactant protein A binds to HIV and inhibits direct infection of $CD4^+$ cells, but enhances dendritic cell-mediated viral transfer. J Immunol 181:601–609

Gardai SJ, Xiao YQ, Dickinson M, Nick JA, Voelker DR, Greene KE, Henson PM (2003) By binding SIRPalpha or calreticulin/CD91, lung collectins act as dual function surveillance molecules to suppress or enhance inflammation. Cell 115(1):13–23

Garred P, Harboe M, Oettinger T, Koch C, Svejgaard A (1994) Dual role of mannan-binding protein in infections: another case of heterosis? Eur J Immunogenet 21:125–131

Garred P, Madsen HO, Balslev U, Hofmann B, Pedersen C, Gerstoft J, Svejgaard A (1997a) Susceptibility to HIV infection and progression of AIDS in relation to variant alleles of mannose-binding lectin. Lancet 349:236–240

Garred P, Richter C, Andersen AB, Madsen HO, Mtoni I, Svejgaard A, Shao J (1997b) Mannan-binding lectin in the sub-Saharan HIV and tuberculosis epidemics. Scand J Immunol 46:204–208

Garred P, Nielsen MA, Kurtzhals JA, Malhotra R, Madsen HO, Goka BQ, Akanmori BD, Sim RB, Hviid L (2003) Mannose-binding lectin is a disease modifier in clinical malaria and may function as opsonin for *Plasmodium falciparum*-infected erythrocytes. Infect Immun 71:5245–5253

Gaynor CD, Mccormack FX, Voelker DR, Mcgowan SE, Schlesinger LS (1995) Pulmonary surfactant protein A mediates enhanced phagocytosis of *Mycobacterium tuberculosis* by a direct interaction with human macrophages. J Immunol 155:5343–5351

Geertsma MF, Nibbering PH, Haagsman HP, Daha MR, van Furth R (1994) Binding of surfactant protein A to C1q receptors mediates phagocytosis of *Staphylococcus aureus* by monocytes. Am J Physiol 267(5 Pt 1):L578–L584

Geunes-Boyer S, Oliver TN, Janbon G, Lodge JK, Heitman J, Perfect JR, Wright JR (2009) Surfactant protein D increases phagocytosis of hypocapsular *Cryptococcus neoformans* by murine macrophages and enhances fungal survival. Infect Immun 77:2783–2794

Geunes-Boyer S, Beers MF, Perfect JR, Heitman J, Wright JR (2012) Surfactant protein D facilitates *Cryptococcus neoformans* infection. Infect Immun 80:2444–2453

Ghezzi MC, Raponi G, Angeletti S, Mancini C (1998) Serum-mediated enhancement of TNF-alpha release by human monocytes stimulated with the yeast form of *Candida albicans*. J Infect Dis 178:1743–1749

Ghildyal R, Hartley C, Varrasso A, Meanger J, Voelker DR, Anders EM, Mills J (1999) Surfactant protein A binds to the fusion glycoprotein of respiratory syncytial virus and neutralizes virion infectivity. J Infect Dis 180:2009–2013

Ghiran I, Barbashov SF, Klickstein LB, Tas SW, Jensenius JC, Nicholson-Weller A (2000) Complement receptor 1/CD35 is a receptor for mannan-binding lectin. J Exp Med 192:1797–1808

Giles SS, Zaas AK, Reidy MF, Perfect JR, Wright JR (2007) *Cryptococcus neoformans* is resistant to surfactant protein A mediated host defense mechanisms. PLoS ONE 2:e1370

Giraldo PC, Babula O, Goncalves AK, Linhares IM, Amaral RL, Ledger WJ, Witkin SS (2007) Mannose-binding lectin gene polymorphism, vulvovaginal candidiasis, and bacterial vaginosis. Obstet Gynecol 109:1123–1128

Gold JA, Hoshino Y, Tanaka N, Rom WN, Raju B, Condos R, Weiden MD (2004) Surfactant protein A modulates the inflammatory response in macrophages during tuberculosis. Infect Immun 72:645–650

Goyal S, Klassert TE, Slevogt H (2016) C-type lectin receptors in tuberculosis: what we know. Med Microbiol Immunol 205:513–535

Granell M, Urbano-Ispizua A, Suarez B, Rovira M, Fernandez-Aviles F, Martinez C, Ortega M, Uriburu C, Gaya A, Roncero JM, Navarro A, Carreras E, Mensa J, Vives J, Rozman C, Montserrat E, Lozano F (2006) Mannan-binding lectin pathway deficiencies and invasive fungal infections following allogeneic stem cell transplantation. Exp Hematol 34:1435–1441

Green PJ, Feizi T, Stoll MS, Thiel S, Prescott A, Mcconville MJ (1994) Recognition of the major cell surface glycoconjugates of *Leishmania parasites* by the human serum mannan-binding protein. Mol Biochem Parasitol 66:319–328

Gulati S, Sastry K, Jensenius JC, Rice PA, Ram S (2002) Regulation of the mannan-binding lectin pathway of complement on *Neisseria gonorrhoeae* by C1-inhibitor and alpha 2-macroglobulin. J Immunol 168:4078–4086

Hakozaki Y, Yoshiba M, Sekiyama K, Seike E, Iwamoto J, Mitani K, Mine M, Morizane T, Ohtani K, Suzuki Y, Wakamiya N (2002) Mannose-binding lectin and the prognosis of fulminant hepatic failure caused by HBV infection. Liver 22:29–34

Hansen S, Holmskov U (2002) Lung surfactant protein D (SP-D) and the molecular diverted descendants: conglutinin, CL-43 and CL-46. Immunobiology 205:498–517

Hansen S, Thiel S, Willis A, Holmskov U, Jensenius JC (2000) Purification and characterization of two mannan-binding lectins from mouse serum. J Immunol 164(5):2610–2618

Hansen S, Holm D, Moeller V, Vitved L, Bendixen C, Reid KB, Skjoedt K, Holmskov U (2002) CL-46, a novel collectin highly expressed in bovine thymus and liver. J Immunol 169(10):5726–5734

Hansen S, Selman L, Palaniyar N, Ziegler K, Brandt J, Kliem A, Jonasson M, Skjoedt MO, Nielsen O, Hartshorn K, Jorgensen TJ, Skjodt K, Holmskov U (2010) Collectin 11 (CL-11, CL-K1) is a MASP-1/3-associated plasma collectin with microbial-binding activity. J Immunol 185:6096–6104

Harrod KS, Trapnell BC, Otake K, Korfhagen TR, Whitsett JA (1999) SP-A enhances viral clearance and inhibits inflammation after pulmonary adenoviral infection. Am J Physiol 277:L580–L588

Hartley CA, Jackson DC, Anders EM (1992) Two distinct serum mannose-binding lectins function as beta inhibitors of influenza virus: identification of bovine serum beta inhibitor as conglutinin. J Virol 66:4358–4363

Hartshorn KL, Sastry K, Brown D, White MR, Okarma TB, Lee YM, Tauber AI (1993a) Conglutinin acts as an opsonin for influenza A viruses. J Immunol 151:6265–6273

Hartshorn KL, Sastry K, White MR, Anders EM, Super M, Ezekowitz RA, Tauber AI (1993b) Human mannose-binding protein functions as an opsonin for influenza A viruses. J Clin Invest 91:1414–1420

Hartshorn KL, Crouch EC, White MR, Eggleton P, Tauber AI, Chang D, Sastry K (1994) Evidence for a protective role of pulmonary surfactant protein D (SP-D) against influenza A viruses. J Clin Invest 94:311–319

Hartshorn KL, White MR, Voelker DR, Coburn J, Zaner K, Crouch EC (2000) Mechanism of binding of surfactant protein D to influenza A viruses: importance of binding to haemagglutinin to antiviral activity. Biochem J 351(Pt 2):449–458

Hartshorn KL, White MR, Tecle T, Holmskov U, Crouch EC (2006a) Innate defense against influenza A virus: activity of human neutrophil defensins and interactions of defensins with surfactant protein D. J Immunol 176(11):6962–6972

Hartshorn KL, Ligtenberg A, White MR, Van Eijk M, Hartshorn M, Pemberton L, Holmskov U, Crouch E (2006b) Salivary agglutinin and lung scavenger receptor cysteine-rich glycoprotein 340 have broad anti-influenza activities and interactions with surfactant protein D that vary according to donor source and sialylation. Biochem J 393:545–553

Haurum JS, Thiel S, Jones IM, Fischer PB, Laursen SB, Jensenius JC (1993) Complement activation upon binding of mannan-binding protein to HIV envelope glycoproteins. AIDS 7:1307–1313

Hawgood S, Brown C, Edmondson J, Stumbaugh A, Allen L, Goerke J, Clark H, Poulain F (2004) Pulmonary collectins modulate strain-specific influenza a virus infection and host responses. J Virol 78:8565–8572

Held K, Thiel S, Loos M, Petry F (2008) Increased susceptibility of complement factor B/C2 double knockout mice and mannan-binding lectin knockout mice to systemic infection with *Candida albicans*. Mol Immunol 45:3934–3941

Henning LN, Azad AK, Parsa KV, Crowther JE, Tridandapani S, Schlesinger LS (2008) Pulmonary surfactant protein A regulates TLR expression and activity in human macrophages. J Immunol 180(12):7847–7858

Henriksen ML, Brandt J, Andrieu JP, Nielsen C, Jensen PH, Holmskov U, Jorgensen TJ, Palarasah Y, Thielens NM, Hansen S (2013) Heteromeric complexes of native collectin kidney 1 and collectin liver 1 are found in the circulation with MASPs and activate the complement system. J Immunol 191:6117–6127

Herías MV, Hogenkamp A, van Asten AJ, Tersteeg MH, van Eijk M, Haagsman HP (2007) Expression sites of the collectin SP-D suggest its importance in first line host defence: power of combining in situ hybridisation, RT-PCR and immunohistochemistry. Mol Immunol 44(13):3324–3332 (Epub 2007 Apr 8)

Hickling TP, Bright H, Wing K, Gower D, Martin SL, Sim RB, Malhotra R (1999) A recombinant trimeric surfactant protein D carbohydrate recognition domain inhibits respiratory syncytial virus infection *in vitro* and *in vivo*. Eur J Immunol 29:3478–3484

Hickman-Davis JM, Lindsey JR, Zhu S, Matalon S (1998) Surfactant protein A mediates mycoplasmacidal activity of alveolar macrophages. Am J Physiol 274:L270–L277

Hillaire ML, Van Eijk M, Nieuwkoop NJ, Vogelzang-Van Trierum SE, Fouchier RA, Osterhaus AD, Haagsman HP, Rimmelzwaan GF (2012) The number and position of N-linked glycosylation sites in the hemagglutinin determine differential recognition of seasonal and 2009 pandemic H1N1 influenza virus by porcine surfactant protein D. Virus Res 169:301–305

Hogaboam CM, Takahashi K, Ezekowitz RA, Kunkel SL, Schuh JM (2004) Mannose-binding lectin deficiency alters the development of fungal asthma: effects on airway response, inflammation, and cytokine profile. J Leukoc Biol 75(5):805–814

Hohler T, Wunschel M, Gerken G, Schneider PM, Meyer Zum Buschenfelde KH, Rittner C (1998) No association between mannose-binding lectin alleles and susceptibility to chronic hepatitis B virus infection in German patients. Exp Clin Immunogenet 15:130–133

Holmberg V, Schuster F, Dietz E, Sagarriga Visconti JC, Anemana SD, Bienzle U, Mockenhaupt FP (2008) Mannose-binding lectin variant associated with severe malaria in young African children. Microbes Infect 10:342–348

Holmskov U, Malhotra R, Sim RB, Jensenius JC (1994) Collectins: collagenous C-type lectins of the innate immune defense system. Immunol Today 15(2):67–74

Holmskov U, Fischer PB, Rothmann A, Hojrup P (1996) Affinity and kinetic analysis of the bovine plasma C-type lectin collectin-43 (CL-43) interacting with mannan. FEBS Lett 393:314–316

Holmskov U, Lawson P, Teisner B, Tornoe I, Willis AC, Morgan C, Koch C, Reid KB (1997) Isolation and characterization of a new member of the scavenger receptor superfamily, glycoprotein-340 (gp-340), as a lung surfactant protein-D binding molecule. J Biol Chem 272(21):13743–13749

Holmskov U, Jensenius JC, Tornoe I, Lovendahl P (1998) The plasma levels of conglutinin are heritable in cattle and low levels predispose to infection. Immunology 93:431–436

Holmskov U, Mollenhauer J, Madsen J, Vitved L, Gronlund J, Tornoe I, Kliem A, Reid KB, Poustka A, Skjodt K (1999) Cloning of gp-340, a putative opsonin receptor for lung surfactant protein D. Proc Natl Acad Sci U S A 96(19):10794–10799

Hoppe HJ, Reid KB (1994) Collectins–soluble proteins containing collagenous regions and lectin domains–and their roles in innate immunity. Protein Sci 3(8):1143–1158 (Review)

Hussain S, Wright JR, Martin WJ 2 (2003) Surfactant protein A decreases nitric oxide production by macrophages in a tumor necrosis factor-alpha-dependent mechanism. Am J Respir Cell Mol Biol 28:520–527

Ihara S, Takahashi A, Hatsuse H, Sumitomo K, Doi K, Kawakami M (1991) Major component of Ra-reactive factor, a complement-activating bactericidal protein, in mouse serum. J Immunol 146:1874–1879

Ingram DG, Mitchell WR (1971) Conglutinin level in dairy cattle: changes associated with disease. Am J Vet Res 32:875–878

Iobst ST, Drickamer K (1994) Binding of sugar ligands to Ca^{2+}-dependent animal lectins. II. Generation of high-affinity galactose binding by site-directed mutagenesis. J Biol Chem 269(22):15512–15519

Ip WK, Lau YL (2004) Role of mannose-binding lectin in the innate defense against *Candida albicans*: enhancement of complement activation, but lack of opsonic function, in phagocytosis by human dendritic cells. J Infect Dis 190:632–640

Ip WK, Takahashi K, Moore KJ, Stuart LM, Ezekowitz RA (2008) Mannose-binding lectin enhances Toll-like receptors 2 and 6 signaling from the phagosome. J Exp Med 205:169–181

Ip WK, Takahashi K, Ezekowitz RA, Stuart LM (2009) Mannose-binding lectin and innate immunity. Immunol Rev 230:9–21

Jack DL, Dodds AW, Anwar N, Ison CA, Law A, Frosch M, Turner MW, Klein NJ (1998) Activation of complement by mannose-binding lectin on isogenic mutants of *Neisseria meningitidis* serogroup B. J Immunol 160:1346–1353

Jack DL, Jarvis GA, Booth CL, Turner MW, Klein NJ (2001) Mannose-binding lectin accelerates complement activation and increases serum killing of *Neisseria meningitidis* serogroup C. J Infect Dis 184:836–845

Jack DL, Lee ME, Turner MW, Klein NJ, Read RC (2005) Mannose-binding lectin enhances phagocytosis and killing of *Neisseria meningitidis* by human macrophages. J Leukoc Biol 77:328–336

Jacquet M, Cioci G, Fouet G, Bally I, Thielens NM, Gaboriaud C, Rossi V (2018) C1q and mannose-binding lectin interact with CR1 in the same region on CCP24-25 modules. Front Immunol 7(9):453

Jang S, Ohtani K, Fukuoh A, Yoshizaki T, Fukuda M, Motomura W, Mori K, Fukuzawa J, Kitamoto N, Yoshida I, Suzuki Y, Wakamiya N (2009) Scavenger receptor collectin placenta 1 (CL-P1) predominantly mediates zymosan phagocytosis by human vascular endothelial cells. J Biol Chem 284:3956–3965

Ji X, Olinger GG, Aris S, Chen Y, Gewurz H, Spear GT (2005) Mannose-binding lectin binds to Ebola and Marburg envelope glycoproteins, resulting in blocking of virus interaction with DC-SIGN and complement-mediated virus neutralization. J Gen Virol 86:2535–2542

Job ER, Deng YM, Tate MD, Bottazzi B, Crouch EC, Dean MM, Mantovani A, Brooks AG, Reading PC (2010) Pandemic H1N1 influenza A viruses are resistant to the antiviral activities of innate immune proteins of the collectin and pentraxin superfamilies. J Immunol 185:4284–4291

Kabha K, Schmegner J, Keisari Y, Parolis H, Schlepper-Schaeffer J, Ofek I (1997) SP-A enhances phagocytosis of Klebsiella by interaction with capsular polysaccharides and alveolar macrophages. Am J Physiol 272:L344–L352

Kalina M, Blau H, Riklis S, Kravtsov V (1995) Interaction of surfactant protein A with bacterial lipopolysaccharide may affect some biological functions. Am J Physiol 268:L144–L151

Kase T, Suzuki Y, Kawai T, Sakamoto T, Ohtani K, Eda S, Maeda A, Okuno Y, Kurimura T, Wakamiya N (1999) Human mannan-binding lectin inhibits the infection of influenza A virus without complement. Immunology 97:385–392

Kasper M, Sims G, Koslowski R, Kuss H, Thuemmler M, Fehrenbach H, Auten RL (2002) Increased surfactant protein D in rat airway goblet and Clara cells during ovalbumin-induced allergic airway inflammation. Clin Exp Allergy 32(8):1251–1258

Kaur S, Gupta VK, Shah A, Thiel S, Sarma PU, Madan T (2005) Plasma mannan-binding lectin levels and activity are increased in allergic patients. J Allergy Clin Immunol 116(6):1381–1383

Kaur S, Gupta VK, Shah A, Thiel S, Sarma PU, Madan T (2006) Elevated levels of mannan-binding lectin (MBL) and eosinophilia in patients of bronchial asthma with allergic rhinitis and allergic bronchopulmonary aspergillosis associate with a novel intronic polymorphism in MBL. Clin Exp Immunol 143(3):414–419

Kaur S, Gupta VK, Thiel S, Sarma PU, Madan T (2007) Protective role of mannan-binding lectin in a murine model of invasive pulmonary aspergillosis. Clin Exp Immunol 148:382–389

Kawasaki T, Etoh R, Yamashina I (1978) Isolation and characterization of a mannan-binding protein from rabbit liver. Biochem Biophys Res Commun 81(3):1018–1024

Kawasaki N, Kawasaki T, Yamashina I (1989) A serum lectin (mannan-binding protein) has complement-dependent bactericidal activity. J Biochem 106:483–489

Kerr MH, Paton JY (1999) Surfactant protein levels in severe respiratory syncytial virus infection. Am J Respir Crit Care Med 159:1115–1118

Keshi H, Sakamoto T, Kawai T, Ohtani K, Katoh T, Jang SJ, Motomura W, Yoshizaki T, Fukuda M, Koyama S, Fukuzawa J, Fukuoh A, Yoshida I, Suzuki Y, Wakamiya N (2006) Identification and characterization of a novel human collectin CL-K1. Microbiol Immunol 50(12):1001–1013

Kimberg M, Brown GD (2008) Dectin-1 and its role in antifungal immunity. Med Mycol 46:631–636

Kishore U, Wang JY, Hoppe HJ, Reid KB (1996) The alpha-helical neck region of human lung surfactant protein D is essential for the binding of the carbohydrate recognition domains to lipopolysaccharides and phospholipids. Biochem J 318(Pt 2):505–511

Kishore U, Madan T, Sarma PU, Singh M, Urban BC, Reid KB (2002) Protective roles of pulmonary surfactant proteins, SP-A and SP-D, against lung allergy and infection caused by *Aspergillus fumigatus*. Immunobiology 205(4–5):610–618

Kishore U, Greenhough TJ, Waters P, Shrive AK, Ghai R, Kamran MF, Bernal AL, Reid KB, Madan T, Chakraborty T (2006) Surfactant proteins SP-A and SP-D: structure, function and receptors. Mol Immunol 43(9):1293–1315

Kitz DJ, Stahl PD, Little JR (1992) The effect of a mannose binding protein on macrophage interactions with *Candida albicans*. Cell Mol Biol 38:407–412

Klabunde J, Berger J, Jensenius JC, Klinkert MQ, Zelck UE, Kremsner PG, Kun JF (2000) *Schistosoma mansoni*: adhesion of mannan-binding lectin to surface glycoproteins of cercariae and adult worms. Exp Parasitol 95:231–239

Klabunde J, Uhlemann AC, Tebo AE, Kimmel J, Schwarz RT, Kremsner PG, Kun JF (2002) Recognition of *Plasmodium falciparum* proteins by mannan-binding lectin, a component of the human innate immune system. Parasitol Res 88:113–117

Koneti A, Linke MJ, Brummer E, Stevens DA (2008) Evasion of innate immune responses: evidence for mannose binding lectin inhibition of tumor necrosis factor alpha production by macrophages in response to *Blastomyces dermatitidis*. Infect Immun 76:994–1002

Korir JC, Nyakoe NK, Awinda G, Waitumbi JN (2014) Complement activation by merozoite antigens of *Plasmodium falciparum*. PLoS ONE 9:e105093

Koziel H, Phelps DS, Fishman JA, Armstrong MY, Richards FF, Rose RM (1998) Surfactant protein-A reduces binding and phagocytosis of *Pneumocystis carinii* by human alveolar macrophages *in vitro*. Am J Respir Cell Mol Biol 18:834–843

Kozutsumi Y, Kawasaki T, Yamashina I (1980) Isolation and characterization of a mannan-binding protein from rabbit serum. Biochem Biophys Res Commun 95(2):658–664

Krarup A, Sorensen UB, Matsushita M, Jensenius JC, Thiel S (2005) Effect of capsulation of opportunistic pathogenic bacteria on binding of the pattern recognition molecules mannan-binding lectin, L-ficolin, and H-ficolin. Infect Immun 73:1052–1060

Kronborg G, Weis N, Madsen HO, Pedersen SS, Wejse C, Nielsen H, Skinhoj P, Garred P (2002) Variant mannose-binding lectin alleles are not associated with susceptibility to or outcome of invasive pneumococcal infection in randomly included patients. J Infect Dis 185:1517–1520

Kuan SF, Rust K, Crouch E (1992) Interactions of surfactant protein D with bacterial lipopolysaccharides. Surfactant protein D is an *Escherichia coli*-binding protein in bronchoalveolar lavage. J Clin Invest 90:97–106

Kuang Z, Hao Y, Hwang S, Zhang S, Kim E, Akinbi HT, Schurr MJ, Irvin RT, Hassett DJ, Lau GW (2011a) The *Pseudomonas aeruginosa* flagellum confers resistance to pulmonary surfactant protein-A by impacting the production of exoproteases through quorum-sensing. Mol Microbiol 79:1220–1235

Kuang Z, Hao Y, Walling BE, Jeffries JL, Ohman DE, Lau GW (2011b) *Pseudomonas aeruginosa* elastase provides an escape from phagocytosis by degrading the pulmonary surfactant protein-A. PLoS ONE 6:e27091

Kudo K, Sano H, Takahashi H, Kuronuma K, Yokota S, Fujii N, Shimada K, Yano I, Kumazawa Y, Voelker DR, Abe S, Kuroki Y (2004) Pulmonary collectins enhance phagocytosis of *Mycobacterium avium* through increased activity of mannose receptor. J Immunol 172:7592–7602

Kuhlman M, Joiner K, Ezekowitz RA (1989) The human mannose-binding protein functions as an opsonin. J Exp Med 169(5):1733–1745

Kurokawa K, Takahashi K, Lee BL (2016) The staphylococcal surface-glycopolymer wall teichoic acid (WTA) is crucial for complement activation and immunological defense against *Staphylococcus aureus* infection. Immunobiology 221:1091–1101

Kuronuma K, Sano H, Kato K, Kudo K, Hyakushima N, Yokota S, Takahashi H, Fujii N, Suzuki H, Kodama T, Abe S, Kuroki Y (2004) Pulmonary surfactant protein A augments the phagocytosis of *Streptococcus pneumoniae* by alveolar macrophages through a casein kinase 2-dependent increase of cell surface localization of scavenger receptor A. J Biol Chem 279:21421–21430

Lachmann PJ, Muller-Eberhard HJ (1968) The demonstration in human serum of "conglutinogen-activating factor" and its effect on the third component of complement. J Immunol 100:691–698

Laforce FM, Kelly WJ, Huber GL (1973) Inactivation of staphylococci by alveolar macrophages with preliminary observations on the importance of alveolar lining material. Am Rev Respir Dis 108:784–790

Laursen SB, Thiel S, Teisner B, Holmskov U, Wang Y, Sim RB, Jensenius JC (1994) Bovine conglutinin binds to an oligosaccharide determinant presented by iC3b, but not by C3, C3b or C3c. Immunology 81:648–654

Lawson PR, Reid KB (2000) The roles of surfactant proteins A and D in innate immunity. Immunol Rev 173:66–78

Lekkala M, Levine AM, Linke MJ, Crouch EC, Linders B, Brummer E, Stevens DA (2006) Effect of lung surfactant collectins on bronchoalveolar macrophage interaction with *Blastomyces dermatitidis*: inhibition of tumor necrosis factor alpha production by surfactant protein D. Infect Immun 74:4549–4556

Lemos MP, Mckinney J, Rhee KY (2011) Dispensability of surfactant proteins A and D in immune control of *Mycobacterium tuberculosis* infection following aerosol challenge of mice. Infect Immun 79:1077–1085

Leth-Larsen R, Zhong F, Chow VT, Holmskov U, Lu J (2007) The SARS coronavirus spike glycoprotein is selectively recognized by lung surfactant protein D and activates macrophages. Immunobiology 212:201–211

Levine AM, Gwozdz J, Stark J, Bruno M, Whitsett J, Korfhagen T (1999) Surfactant protein-A enhances respiratory syncytial virus clearance *in vivo*. J Clin Invest 103:1015–1021

Levine AM, Whitsett JA, Hartshorn KL, Crouch EC, Korfhagen TR (2001) Surfactant protein D enhances clearance of influenza A virus from the lung *in vivo*. J Immunol 167:5868–5873

Levine AM, Hartshorn K, Elliott J, Whitsett J, Korfhagen T (2002) Absence of SP-A modulates innate and adaptive defense responses to pulmonary influenza infection. Am J Physiol Lung Cell Mol Physiol 282:L563–L572

Levitz SM, Tabuni A, Treseler C (1993) Effect of mannose-binding protein on binding of *Cryptococcus neoformans* to human phagocytes. Infect Immun 61:4891–4893

Li D, Dong B, Tong Z, Wang Q, Liu W, Wang Y, Chen J, Xu L, Chen L, Duan Y (2012) MBL-mediated opsonophagocytosis of *Candida albicans* by human neutrophils is coupled with intracellular Dectin-1-triggered ROS production. PLoS ONE 7:e50589

Ligtenberg TJ, Bikker FJ, Groenink J, Tornoe I, Leth-Larsen R, Veerman EC, Nieuw Amerongen AV, Holmskov U (2001) Human salivary agglutinin binds to lung surfactant protein-D and is identical with scavenger receptor protein gp-340. Biochem J 359(Pt 1):243–248

Lillegard JB, Sim RB, Thorkildson P, Gates MA, Kozel TR (2006) Recognition of *Candida albicans* by mannan-binding lectin *in vitro* and *in vivo*. J Infect Dis 193:1589–1597

Lim BL, Holmskov U (1996) Expression of the carbohydrate recognition domain of bovine conglutinin and demonstration of its binding to iC3b and yeast mannan. Biochem Biophys Res Commun 218:260–266

Lim BL, Willis AC, Reid KB, Lu J, Laursen SB, Jensenius JC, Holmskov U (1994a) Primary structure of bovine collectin-43 (CL-43). Comparison with conglutinin and lung surfactant protein-D. J Biol Chem 269(16):11820–11824

Lim BL, Wang JY, Holmskov U, Hoppe HJ, Reid KB (1994b) Expression of the carbohydrate recognition domain of lung surfactant protein D and demonstration of its binding to lipopolysaccharides of Gram-negative bacteria. Biochem Biophys Res Commun 202:1674–1680

Limper AH, Crouch EC, O'riordan DM, Chang D, Vuk-Pavlovic Z, Standing JE, Kwon KY, Adlakha A (1995) Surfactant protein-D modulates interaction of *Pneumocystis carinii* with alveolar macrophages. J Lab Clin Med 126:416–422

Lin Z, deMello D, Phelps DS, Koltun WA, Page M, Floros J (2001) Both human SP-A1 and SP-A2 genes are expressed in small and large intestine. Pediatr Pathol Mol Med 20(5):367–386

Ling MT, Tu W, Han Y, Mao H, Chong WP, Guan J, Liu M, Lam KT, Law HK, Peiris JS, Takahashi K, Lau YL (2012) Mannose-binding lectin contributes to deleterious inflammatory response in pandemic H1N1 and avian H9N2 infection. J Infect Dis 205:44–53

Linke MJ, Harris CE, Korfhagen TR, Mccormack FX, Ashbaugh AD, Steele P, Whitsett JA, Walzer PD (2001) Immunosuppressed surfactant protein A-deficient mice have increased susceptibility to *Pneumocystis carinii* infection. J Infect Dis 183:943–952

Linke M, Ashbaugh A, Koch J, Tanaka R, Walzer P (2006) Efficient resolution of *Pneumocystis murina* infection in surfactant protein A-deficient mice following withdrawal of corticosteroid-induced immunosuppression. J Med Microbiol 55:143–147

Liu CF, Chen YL, Shieh CC, Yu CK, Reid KB, Wang JY (2005) Therapeutic effect of surfactant protein D in allergic inflammation of mite-sensitized mice. Clin Exp Allergy 35(4):515–521

Liu F, Liao Q, Liu Z (2006) Mannose-binding lectin and vulvovaginal candidiasis. Int J Gynaecol Obstet 92:43–47

Loveless RW, Feizi T, Childs RA, Mizuochi T, Stoll MS, Oldroyd RG, Lachmann PJ (1989) Bovine serum conglutinin is a lectin which binds non-reducing terminal N-acetylglucosamine, mannose and fucose residues. Biochem J 258:109–113

Lu J, Teh C, Kishore U, Reid KB (2002) Collectins and ficolins: sugar pattern recognition molecules of the mammalian innate immune system. Biochim Biophys Acta 1572(2–3):387–400

Lugo-Villarino G, Hudrisier D, Tanne A, Neyrolles O (2011) C-type lectins with a sweet spot for *Mycobacterium tuberculosis*. Eur J Microbiol Immunol (Bp) 1:25–40

Luty AJ, Kun JF, Kremsner PG (1998) Mannose-binding lectin plasma levels and gene polymorphisms in *Plasmodium falciparum* malaria. J Infect Dis 178:1221–1224

Ma YJ, Hein E, Munthe-Fog L, Skjoedt MO, Bayarri-Olmos R, Romani L, Garred P (2015) Soluble collectin-12 (CL-12) is a pattern recognition molecule initiating complement activation via the alternative pathway. J Immunol 195:3365–3373

Madan T, Eggleton P, Kishore U, Strong P, Aggrawal SS, Sarma PU, Reid KB (1997a) Binding of pulmonary surfactant proteins A and D to *Aspergillus fumigatus* conidia enhances phagocytosis and killing by human neutrophils and alveolar macrophages. Infect Immun 65:3171–3179

Madan T, Kishore U, Shah A, Eggleton P, Strong P, Wang JY, Aggrawal SS, Sarma PU, Reid KB (1997b) Lung surfactant proteins A and D can inhibit specific IgE binding to the allergens of *Aspergillus fumigatus* and block allergen-induced histamine release from human basophils. Clin Exp Immunol 110:241–249

Madan T, Kishore U, Singh M, Strong P, Clark H, Hussain EM, Reid KB, Sarma PU (2001a) Surfactant proteins A and D protect mice against pulmonary hypersensitivity induced by *Aspergillus fumigatus* antigens and allergens. J Clin Invest 107(4):467–475

Madan T, Kishore U, Singh M, Strong P, Hussain EM, Reid KB, Sarma PU (2001b) Protective role of lung surfactant protein D in a murine model of invasive pulmonary aspergillosis. Infect Immun 69:2728–2731

Madan T, Saxena S, Murthy KJ, Muralidhar K, Sarma PU (2002) Association of polymorphisms in the collagen region of human SP-A1 and SP-A2 genes with pulmonary tuberculosis in Indian population. Clin Chem Lab Med 40:1002–1008

Madan T, Reid KB, Singh M, Sarma PU, Kishore U (2005a) Susceptibility of mice genetically deficient in the surfactant protein (SP)-A or SP-D gene to pulmonary hypersensitivity induced by antigens and allergens of *Aspergillus fumigatus*. J Immunol 174(11):6943–6954

Madan T, Reid KB, Singh M, Sarma PU, Kishore U (2005b) Susceptibility of mice genetically deficient in the surfactant protein (SP)-A or SP-D gene to pulmonary hypersensitivity induced by antigens and allergens of *Aspergillus fumigatus*. J Immunol 174:6943–6954

Madan T, Reid KB, Clark H, Singh M, Nayak A, Sarma PU, Hawgood S, Kishore U (2010) Susceptibility of mice genetically deficient in SP-A or SP-D gene to invasive pulmonary aspergillosis. Mol Immunol 47(10):1923–1930

Madsen J, Kliem A, Tornoe I, Skjodt K, Koch C, Holmskov U (2000) Localization of lung surfactant protein D on mucosal surfaces in human tissues. J Immunol 164(11):5866–5870

Madsen J, Tornoe I, Nielsen O, Koch C, Steinhilber W, Holmskov U (2003) Expression and localization of lung surfactant protein A in human tissues. Am J Respir Cell Mol Biol 29(5):591–597

Malhotra R, Thiel S, Reid KB, Sim RB (1990) Human leukocyte C1q receptor binds other soluble proteins with collagen domains. J Exp Med 172(3):955–959. Sep 1

Malhotra R, Haurum J, Thiel S, Sim RB (1992) Interaction of C1q receptor with lung surfactant protein A. Eur J Immunol 22(6):1437–1445

Malhotra R, Willis AC, Jensenius JC, Jackson J, Sim RB (1993) Structure and homology of human C1q receptor (collectin receptor). Immunology 78(3):341–348

Malhotra R, Haurum JS, Thiel S, Sim RB (1994) Binding of human collectins (SP-A and MBP) to influenza virus. Biochem J 304(Pt 2):455–461

Malik S, Greenwood CM, Eguale T, Kifle A, Beyene J, Habte A, Tadesse A, Gebrexabher H, Britton S, Schurr E (2006) Variants of the SFTPA1 and SFTPA2 genes and susceptibility to tuberculosis in Ethiopia. Hum Genet 118:752–759

Mangano A, Rocco C, Marino SM, Mecikovsky D, Genre F, Aulicino P, Bologna R, Sen L (2008) Detrimental effects of mannose-binding lectin (MBL2) promoter genotype XA/XA on HIV-1 vertical transmission and AIDS progression. J Infect Dis 198:694–700

Matsushita M, Hijikata M, Ohta Y, Mishiro S (1998) Association of mannose-binding lectin gene haplotype LXPA and LYPB with interferon-resistant hepatitis C virus infection in Japanese patients. J Hepatol 29:695–700

Mcbride MO, Fischer PB, Sumiya M, Mcclure MO, Turner MW, Skinner CJ, Weber JN, Summerfield JA (1998) Mannose-binding protein in HIV-seropositive patients does not contribute to disease progression or bacterial infections. Int J STD AIDS 9:683–688

McCormack FX, Pattanajitvilai S, Stewart J, Possmayer F, Inchley K, Voelker DR (1997a) The Cys6 intermolecular disulfide bond and the collagen-like region of rat SP-A play critical roles in interactions with alveolar type II cells and surfactant lipids. J Biol Chem 272(44):27971–27979

McCormack FX, Festa AL, Andrews RP, Linke M, Walzer PD (1997b) The carbohydrate recognition domain of surfactant protein A mediates binding to the major surface glycoprotein of *Pneumocystis carinii*. Biochemistry 36:8092–8099

McCormack FX, Damodarasamy M, Elhalwagi BM (1999) Deletion mapping of N-terminal domains of surfactant protein A. The N-terminal segment is required for phospholipid aggregation and specific inhibition of surfactant secretion. J Biol Chem 274(5):3173–3181

McCormack FX, Gibbons R, Ward SR, Kuzmenko A, Wu H, Deepe GS Jr (2003) Macrophage-independent fungicidal action of the pulmonary collectins. J Biol Chem 278:36250–36256

McGreal EP, Ikewaki N, Akatsu H, Morgan BP, Gasque P (2002) Human C1qRp is identical with CD93 and the mNI-11 antigen but does not bind C1q. J Immunol 168(10):5222–5232

Mcneely TB, Coonrod JD (1994) Aggregation and opsonization of type A but not type B *Hemophilus influenzae* by surfactant protein A. Am J Respir Cell Mol Biol 11:114–122

Mehmood A, Kouser L, Kaur A, Holmskov U, Al-Ahdal MN, Sim RB, Kishore U, Tsolaki AG (2019) Complement dependent and independent interaction between Bovine Conglutinin and *Mycobacterium bovis* BCG: implications in bovine tuberculosis. Front Immunol 9:3159

Meschi J, Crouch EC, Skolnik P, Yahya K, Holmskov U, Leth-Larsen R, Tornoe I, Tecle T, White MR, Hartshorn KL (2005) Surfactant protein D binds to human immunodeficiency virus (HIV) envelope protein gp120 and inhibits HIV replication. J Gen Virol 86:3097–3107

Michelow IC, Lear C, Scully C, Prugar LI, Longley CB, Yantosca LM, Ji X, Karpel M, Brudner M, Takahashi K, Spear GT, Ezekowitz RA, Schmidt EV, Olinger GG (2011) High-dose mannose-binding lectin therapy for Ebola virus infection. J Infect Dis 203:175–179

Milanese M, Segat L, De Seta F, Pirulli D, Fabris A, Morgutti M, Crovella S (2008) MBL2 genetic screening in patients with recurrent vaginal infections. Am J Reprod Immunol 59:146–151

Miyamura K, Malhotra R, Hoppe HJ, Reid KB, Phizackerley PJ, Macpherson P, López Bernal A (1994) Surfactant proteins A (SP-A) and D (SP-D): levels in human amniotic fluid and localization in the fetal membranes. Biochim Biophys Acta 1210(3):303–307

Mogues T, Ota T, Tauber AI, Sastry KN (1996) Characterization of two mannose-binding protein cDNAs from rhesus monkey (*Macaca mulatta*): structure and evolutionary implications. Glycobiology 6(5):543–550

Moller-Kristensen M, Ip WK, Shi L, Gowda LD, Hamblin MR, Thiel S, Jensenius JC, Ezekowitz RA, Takahashi K (2006) Deficiency of mannose-binding lectin greatly increases susceptibility to postburn infection with *Pseudomonas aeruginosa*. J Immunol 176:1769–1775

Mullighan CG, Heatley S, Doherty K, Szabo F, Grigg A, Hughes TP, Schwarer AP, Szer J, Tait BD, Bik To L, Bardy PG (2002) Mannose-binding lectin gene polymorphisms are associated with major infection following allogeneic hemopoietic stem cell transplantation. Blood 99:3524–3529

Murakami S, Iwaki D, Mitsuzawa H, Sano H, Takahashi H, Voelker DR, Akino T, Kuroki Y (2002) Surfactant protein A inhibits peptidoglycan-induced tumor necrosis factor-alpha secretion in U937 cells and alveolar macrophages by direct interaction with toll-like receptor 2. J Biol Chem 277(9):6830–6837 (Epub 2001 Nov 27)

Murray E, Khamri W, Walker MM, Eggleton P, Moran AP, Ferris JA, Knapp S, Karim QN, Worku M, Strong P, Reid KB, Thursz MR (2002) Expression of surfactant protein D in the human gastric mucosa and during *Helicobacter pylori* infection. Infect Immun 70(3):1481–1487

Nadeem A, Chhabra SK, Masood A, Raj HG (2003) Increased oxidative stress and altered levels of antioxidants in asthma. J Allergy Clin Immunol 111(1):72–78

Nadesalingam J, Bernal AL, Dodds AW, Willis AC, Mahoney DJ, Day AJ, Reid KB, Palaniyar N (2003) Identification and characterization of a novel interaction between pulmonary surfactant protein D and decorin. J Biol Chem 278(28):25678–25687

Nadesalingam J, Reid KB, Palaniyar N (2005a) Collectin surfactant protein D binds antibodies and interlinks innate and adaptive immune systems. FEBS Lett 579(20):4449–4453

Nadesalingam J, Dodds AW, Reid KB, Palaniyar N (2005b) Mannose-binding lectin recognizes peptidoglycan via the N-acetyl glucosamine moiety, and inhibits ligand-induced pro-inflammatory effect and promotes chemokine production by macrophages. J Immunol 175:1785–1794

Nagy A, Kozma GT, Keszei M, Treszl A, Falus A, Szalai C (2003) The development of asthma in children infected with *Chlamydia pneumoniae* is dependent on the modifying effect of mannose-binding lectin. J Allergy Clin Immunol 112(4):729–734

Nauta AJ, Raaschou-Jensen N, Roos A, Daha MR, Madsen HO, Borrias-Essers MC, Ryder LP, Koch C, Garred P (2003) Mannose-binding lectin engagement with late apoptotic and necrotic cells. Eur J Immunol 33:2853–2863

Nayak A, Dodagatta-Marri E, Tsolaki AG, Kishore U (2012) An insight into the diverse roles of surfactant proteins, SP-A and SP-D in innate and adaptive immunity. Front Immunol 3:131

Neth O, Jack DL, Dodds AW, Holzel H, Klein NJ, Turner MW (2000) Mannose-binding lectin binds to a range of clinically relevant microorganisms and promotes complement deposition. Infect Immun 68:688–693

Neth O, Jack DL, Johnson M, Klein NJ, Turner MW (2002) Enhancement of complement activation and opsonophagocytosis by complexes of mannose-binding lectin with mannose-binding lectin-associated serine protease after binding to *Staphylococcus aureus*. J Immunol 169:4430–4436

Ng KK, Drickamer K, Weis WI (1996) Structural analysis of monosaccharide recognition by rat liver mannose-binding protein. J Biol Chem 271(2):663–674

Ng KK, Park-Snyder S, Weis WI (1998) Ca^{2+}-dependent structural changes in C-type mannose-binding proteins. Biochemistry 37(51):17965–17976

Nielsen SL, Andersen PL, Koch C, Jensenius JC, Thiel S (1995) The level of the serum opsonin, mannan-binding protein in HIV-1 antibody-positive patients. Clin Exp Immunol 100:219–222

Norsworthy PJ, Fossati-Jimack L, Cortes-Hernandez J, Taylor PR, Bygrave AE, Thompson RD, Nourshargh S, Walport MJ, Botto M (2004) Murine CD93 (C1qRp) contributes to the removal of apoptotic cells *in vivo* but is not required for C1q-mediated enhancement of phagocytosis. J Immunol. 172(6):3406–3414

Ogden CA, Decathelineau A, Hoffmann PR, Bratton D, Ghebrehiwet B, Fadok VA, Henson PM (2001) C1q and mannose binding lectin engagement of cell surface calreticulin and CD91 initiates macropinocytosis and uptake of apoptotic cells. J Exp Med 194:781–795

Ohtani K, Suzuki Y, Eda S, Kawai T, Kase T, Yamazaki H, Shimada T, Keshi H, Sakai Y, Fukuoh A, Sakamoto T, Wakamiya N (1999) Molecular cloning of a novel human collectin from liver (CL-L1). J Biol Chem 274(19):13681–13689

Ohtani K, Suzuki Y, Eda S, Kawai T, Kase T, Keshi H, Sakai Y, Fukuoh A, Sakamoto T, Itabe H, Suzutani T, Ogasawara M, Yoshida I, Wakamiya N (2001) The membrane-type collectin CL-P1 is a scavenger receptor on vascular endothelial cells. J Biol Chem 276(47):44222–44228

O'riordan DM, Standing JE, Kwon KY, Chang D, Crouch EC, Limper AH (1995) Surfactant protein D interacts with *Pneumocystis carinii* and mediates organism adherence to alveolar macrophages. J Clin Invest 95:2699–2710

Pandit H, Gopal S, Sonawani A, Yadav AK, Qaseem AS, Warke H, Patil A, Gajbhiye R, Kulkarni V, Al-Mozaini MA, Idicula-Thomas S, Kishore U, Madan T (2014) Surfactant protein D inhibits HIV-1 infection of target cells via interference with gp120-CD4 interaction and modulates pro-inflammatory cytokine production. PLoS ONE 9:e102395

Pandit H, Kale K, Yamamoto H, Thakur G, Rokade S, Chakraborty P, Vasudevan M, Kishore M, Madan T, Fichorova RN (2019) Surfactant Protein D reverses the gene signature of transepithelial HIV-1 passage and restricts the viral transfer across the vaginal barrier. Front Immunol 10:264

Pasula R, Wright JR, Kachel DL, Martin WJ 2 (1999) Surfactant protein A suppresses reactive nitrogen intermediates by alveolar macrophages in response to *Mycobacterium tuberculosis*. J Clin Invest 103:483–490

Pellis V, De Seta F, Crovella S, Bossi F, Bulla R, Guaschino S, Radillo O, Garred P, Tedesco F (2005) Mannose binding lectin and C3 act as recognition molecules for infectious agents in the vagina. Clin Exp Immunol 139:120–126

Perino J, Thielens NM, Crouch E, Spehner D, Crance JM, Favier AL (2013) Protective effect of surfactant protein D in pulmonary vaccinia virus infection: implication of A27 viral protein. Viruses 5:928–953

Phelps DS, Umstead TM, Rose RM, Fishman JA (1996) Surfactant protein-A levels increase during *Pneumocystis carinii* pneumonia in the rat. Eur Respir J 9:565–570

Piboonpocanun S, Chiba H, Mitsuzawa H, Martin W, Murphy RC, Harbeck RJ, Voelker DR (2005) Surfactant protein A binds *Mycoplasma pneumoniae* with high affinity and attenuates its growth by recognition of disaturated phosphatidylglycerols. J Biol Chem 280:9–17

Pikaar JC, Voorhout WF, Van Golde LM, Verhoef J, Van Strijp JA, Van Iwaarden JF (1995) Opsonic activities of surfactant proteins A and D in phagocytosis of gram-negative bacteria by alveolar macrophages. J Infect Dis 172:481–489

Polotsky VY, Fischer W, Ezekowitz RA, Joiner KA (1996) Interactions of human mannose-binding protein with lipoteichoic acids. Infect Immun 64:380–383

Polotsky VY, Belisle JT, Mikusova K, Ezekowitz RA, Joiner KA (1997) Interaction of human mannose-binding protein with *Mycobacterium avium*. J Infect Dis 175:1159–1168

Qu J, He L, Rong Z, Pan J, Chen X, Morrison DC, Li X (2001) Alteration of surfactant proteins A and D in bronchoalveolar lavage fluid of *Pneumocystis carinii* pneumonia. Chin Med J (Engl) 114:1143–1146

Ragas A, Roussel L, Puzo G, Riviere M (2007) The *Mycobacterium tuberculosis* cell-surface glycoprotein apa as a potential adhesin to colonize target cells via the innate immune system pulmonary C-type lectin surfactant protein A. J Biol Chem 282:5133–5142

Rappleye CA, Eissenberg LG, Goldman WE (2007) *Histoplasma capsulatum* alpha-(1,3)-glucan blocks innate immune recognition by the beta-glucan receptor. Proc Natl Acad Sci USA 104:1366–1370

Reading PC, Hartley CA, Ezekowitz RA, Anders EM (1995) A serum mannose-binding lectin mediates complement-dependent lysis of influenza virus-infected cells. Biochem Biophys Res Commun 217:1128–1136

Reading PC, Morey LS, Crouch EC, Anders EM (1997) Collectin-mediated antiviral host defense of the lung: evidence from influenza virus infection of mice. J Virol 71:8204–8212

Reading PC, Holmskov U, Anders EM (1998) Antiviral activity of bovine collectins against rotaviruses. J Gen Virol 79(Pt 9):2255–2263

Reichhardt MP, Loimaranta V, Thiel S, Finne J, Meri S, Jarva H (2012) The salivary scavenger and agglutinin binds MBL and regulates the lectin pathway of complement in solution and on surfaces. Front Immunol 16(3):205

Ross SC, Densen P (1984) Complement deficiency states and infection: epidemiology, pathogenesis and consequences of neisserial and other infections in an immune deficiency. Med (Baltimore) 63:243–273

Rosseau S, Guenther A, Seeger W, Lohmeyer J (1997) Phagocytosis of viable *Candida albicans* by alveolar macrophages: lack of opsonin function of surfactant protein A. J Infect Dis 175:421–428

Rosseau S, Hammerl P, Maus U, Gunther A, Seeger W, Grimminger F, Lohmeyer J (1999) Surfactant protein A down-regulates proinflammatory cytokine production evoked by *Candida albicans* in human alveolar macrophages and monocytes. J Immunol 163:4495–4502

Rothfuchs AG, Roffe E, Gibson A, Cheever AW, Ezekowitz RA, Takahashi K, Steindel M, Sher A, Bafica A (2012) Mannose-binding lectin regulates host resistance and pathology during experimental infection with *Trypanosoma cruzi*. PLoS ONE 7:e47835

Rothmann AB, Mortensen HD, Holmskov U, Højrup P (1997) Structural characterization of bovine collectin-43. Eur J Biochem 243(3):630–635

Roy S, Knox K, Segal S, Griffiths D, Moore CE, Welsh KI, Smarason A, Day NP, Mcpheat WL, Crook DW, Hill AV (2002) MBL genotype and risk of invasive pneumococcal disease: a case-control study. Lancet 359:1569–1573

Roy N, Ohtani K, Matsuda Y, Mori K, Hwang I, Suzuki Y, Inoue N, Wakamiya N (2016) Collectin CL-P1 utilizes C-reactive protein for complement activation. Biochim Biophys Acta 1860:1118–1128

Saifuddin M, Hart ML, Gewurz H, Zhang Y, Spear GT (2000) Interaction of mannose-binding lectin with primary isolates of human immunodeficiency virus type 1. J Gen Virol 81:949–955

Sano H, Sohma H, Muta T, Nomura S, Voelker DR, Kuroki Y (1999) Pulmonary surfactant protein A modulates the cellular response to smooth and rough lipopolysaccharides by interaction with CD14. J Immunol 163(1):387–395

Sano H, Chiba H, Iwaki D, Sohma H, Voelker DR, Kuroki Y (2000) Surfactant proteins A and D bind CD14 by different mechanisms. J Biol Chem 275:22442–22451

Santos IK, Costa CH, Krieger H, Feitosa MF, Zurakowski D, Fardin B, Gomes RB, Weiner DL, Harn DA, Ezekowitz RA, Epstein JE (2001) Mannan-binding lectin enhances susceptibility to visceral leishmaniasis. Infect Immun 69(8):5212–5215

Sasaki K, Tsutsumi A, Wakamiya N, Ohtani K, Suzuki Y, Watanabe Y, Nakayama N, Koike T (2000) Mannose-binding lectin polymorphisms in patients with hepatitis C virus infection. Scand J Gastroenterol 35:960–965

Sastry K, Zahedi K, Lelias JM, Whitehead AS, Ezekowitz RA (1991) Molecular characterization of the mouse mannose-binding proteins. The mannose-binding protein A but not C is an acute phase reactant. J Immunol 147(2):692–697

Sawada K, Ariki S, Kojima T, Saito A, Yamazoe M, Nishitani C, Shimizu T, Takahashi M, Mitsuzawa H, Yokota S, Sawada N, Fujii N, Takahashi H, Kuroki Y (2010) Pulmonary collectins

protect macrophages against pore-forming activity of *Legionella pneumophila* and suppress its intracellular growth. J Biol Chem 285:8434–8443
Saxena S, Madan T, Shah A, Muralidhar K, Sarma PU (2003) Association of polymorphisms in the collagen region of SP-A2 with increased levels of total IgE antibodies and eosinophilia in patients with allergic bronchopulmonary aspergillosis. J Allergy Clin Immunol 111:1001–1007
Schelenz S, Malhotra R, Sim RB, Holmskov U, Bancroft GJ (1995) Binding of host collectins to the pathogenic yeast *Cryptococcus neoformans*: human surfactant protein D acts as an agglutinin for acapsular yeast cells. Infect Immun 63:3360–3366
Selman L, Hansen S (2012) Structure and function of collectin liver 1 (CL-L1) and collectin 11 (CL-11, CL-K1). Immunobiology 217:851–863
Seppanen M, Lokki ML, Lappalainen M, Hiltunen-Back E, Rovio AT, Kares S, Hurme M, Aittoniemi J (2009) Mannose-binding lectin 2 gene polymorphism in recurrent herpes simplex virus 2 infection. Hum Immunol 70:218–221
Shi L, Takahashi K, Dundee J, Shahroor-Karni S, Thiel S, Jensenius JC, Gad F, Hamblin MR, Sastry KN, Ezekowitz RA (2004) Mannose-binding lectin-deficient mice are susceptible to infection with *Staphylococcus aureus*. J Exp Med 199:1379–1390
Shushimita S, van der Pol P, de Bruin RW, Ijzermans JN, van Kooten C, Dor FJ (2015) Mannan-binding lectin is involved in the protection against renal ischemia/reperfusion injury by dietary restriction. PLoS ONE. 10(9):e0137795
Singh M, Madan T, Waters P, Parida SK, Sarma PU, Kishore U (2003) Protective effects of a recombinant fragment of human surfactant protein D in a murine model of pulmonary hypersensitivity induced by dust mite allergens. Immunol Lett 86(3):299–307
Singh M, Madan T, Waters P, Sonar S, Singh SK, Kamran MF, Bernal AL, Sarma PU, Singh VK, Crouch EC, Kishore U (2009) Therapeutic effects of recombinant forms of full-length and truncated human surfactant protein D in a murine model of invasive pulmonary aspergillosis. Mol Immunol 46:2363–2369
Skjoedt MO, Roversi P, Hummelshøj T, Palarasah Y, Rosbjerg A, Johnson S, Lea SM, Garred P (2012) Crystal structure and functional characterization of the complement regulator mannose-binding lectin (MBL)/ficolin-associated protein-1 (MAP-1). J Biol Chem 287(39):32913–32921
Steffensen R, Thiel S, Varming K, Jersild C, Jensenius JC (2000) Detection of structural gene mutations and promoter polymorphisms in the mannan-binding lectin (MBL) gene by polymerase chain reaction with sequence-specific primers. J Immunol Methods 241:33–42
Stuart GR, Lynch NJ, Lu J, Geick A, Moffatt BE, Sim RB, Schwaeble WJ (1996) Localisation of the C1q binding site within C1q receptor/calreticulin. FEBS Lett 397(2–3):245–249
Sugar AM, Picard M (1988) Experimental blastomycosis pneumonia in mice by infection with conidia. J Med Vet Mycol 26:321–326
Swierzko AS, Bartlomiejczyk MA, Brzostek A, Lukasiewicz J, Michalski M, Dziadek J, Cedzynski M (2016) Mycobacterial antigen 85 complex (Ag85) as a target for ficolins and mannose-binding lectin. Int J Med Microbiol 306:212–221
Takahashi K, Ezekowitz RA (2005) The role of the mannose-binding lectin in innate immunity. Clin Infect Dis 41(Suppl 7):S440–S444
Takahashi K, Gordon J, Liu H, Sastry KN, Epstein JE, Motwani M, Laursen I, Thiel S, Jensenius JC, Carroll M, Ezekowitz RA (2002) Lack of mannose-binding lectin-A enhances survival in a mouse model of acute septic peritonitis. Microbes Infect 4(8):773–784
Takeda K, Miyahara N, Rha YH, Taube C, Yang ES, Joetham A, Kodama T, Balhorn AM, Dakhama A, Duez C, Evans AJ, Voelker DR, Gelfand EW (2003) Surfactant protein D regulates airway function and allergic inflammation through modulation of macrophage function. Am J Respir Crit Care Med 168(7):783–789 (Epub 2003 Jul 25)
Tan SM, Chung MC, Kon OL, Thiel S, Lee SH, Lu J (1996) Improvements on the purification of mannan-binding lectin and demonstration of its Ca$^{(2+)}$-independent association with a C1s-like serine protease. Biochem J 319(Pt 2):329–332
Tecle T, White MR, Crouch EC, Hartshorn KL (2007) Inhibition of influenza viral neuraminidase activity by collectins. Arch Virol 152:1731–1742

Teodorof C, Divakar S, Soontornniyomkij B, Achim CL, Kaul M, Singh KK (2014) Intracellular mannose binding lectin mediates subcellular trafficking of HIV-1 gp120 in neurons. Neurobiol Dis 69:54–64

Thawer S, Auret J, Schnoeller C, Chetty A, Smith K, Darby M, Roberts L, Mackay RM, Whitwell HJ, Timms JF, Madsen J, Selkirk ME, Brombacher F, Clark HW, Horsnell WG (2016) Surfactant protein-D is essential for immunity to Helminth infection. PLoS Pathog 12:e1005461

Thiel S, Vorup-Jensen T, Stover CM, Schwaeble W, Laursen SB, Poulsen K, Willis AC, Eggleton P, Hansen S, Holmskov U, Reid KB, Jensenius JC (1997) A second serine protease associated with mannan-binding lectin that activates complement. Nature 386(6624):506–510

Thiel S, Frederiksen PD, Jensenius JC (2006) Clinical manifestations of mannan-binding lectin deficiency. Mol Immunol 43:86–96

Thomas HC, Foster GR, Sumiya M, Mcintosh D, Jack DL, Turner MW, Summerfield JA (1996) Mutation of gene of mannose-binding protein associated with chronic hepatitis B viral infection. Lancet 348:1417–1419

Tino MJ, Wright JR (1999) Glycoprotein-340 binds surfactant protein-A (SP-A) and stimulates alveolar macrophage migration in an SP-A-independent manner. Am J Respir Cell Mol Biol 20:759–768

Tokunaga H, Ushirogawa H, Ohuchi M (2011) The pandemic (H1N1) 2009 influenza virus is resistant to mannose-binding lectin. Virol J 8:50

Troegeler A, Lugo-Villarino G, Hansen S, Rasolofo V, Henriksen ML, Mori K, Ohtani K, Duval C, Mercier I, Benard A, Nigou J, Hudrisier D, Wakamiya N, Neyrolles O (2015) Collectin CL-LK is a novel soluble pattern recognition receptor for *Mycobacterium tuberculosis*. PLoS ONE 10:e0132692

Uemura T, Sano H, Katoh T, Nishitani C, Mitsuzawa H, Shimizu T, Kuroki Y (2006) Surfactant protein A without the interruption of Gly-X-Y repeats loses a kink of oligomeric structure and exhibits impaired phospholipid liposome aggregation ability. Biochemistry 45(48):14543–14551

Uguz A, Berber Z, Coskun M, Halide Akbas S, Yegin O (2005) Mannose-binding lectin levels in children with asthma. Pediatr Allergy Immunol 16(3):231–235

Ushijima H, Schroder HC, Poznanovic S, Gasic MJ, Matthes E, Muller WE (1992) Inhibition of human immunodeficiency virus-1 infection by human conglutinin-like protein: *in vitro* studies. Jpn J Cancer Res 83:458–464

Vaid M, Kaur S, Madan T, Singh H, Gupta VK, Murthy KJR, Sarma PU (2006) Association of SP-D, MNL and I-NOS genetic variants with pulmonary tuberculosis. Indian J Hum Genet 12:105–110

Vaid M, Kaur S, Sambatakou H, Madan T, Denning DW, Sarma PU (2007) Distinct alleles of mannose-binding lectin (MBL) and surfactant proteins A (SP-A) in patients with chronic cavitary pulmonary aspergillosis and allergic bronchopulmonary aspergillosis. Clin Chem Lab Med 45:183–186

Valdimarsson H, Vikingsdottir T, Bang P, Saevarsdottir S, Gudjonsson JE, Oskarsson O, Christiansen M, Blou L, Laursen I, Koch C (2004) Human plasma-derived mannose-binding lectin: a phase I safety and pharmacokinetic study. Scand J Immunol 59:97–102

van Asbeck EC, Hoepelman AI, Scharringa J, Herpers BL, Verhoef J (2008) Mannose binding lectin plays a crucial role in innate immunity against yeast by enhanced complement activation and enhanced uptake of polymorphonuclear cells. BMC Microbiol 8:229

van De Wetering JK, Van Eijk M, Van Golde LM, Hartung T, Van Strijp JA, Batenburg JJ (2001) Characteristics of surfactant protein A and D binding to lipoteichoic acid and peptidoglycan, 2 major cell wall components of gram-positive bacteria. J Infect Dis 184:1143–1151

van de Wetering JK, van Golde LM, Batenburg JJ (2004a) Collectins: players of the innate immune system. Eur J Biochem 271(7):1229–1249

van De Wetering JK, Coenjaerts FE, Vaandrager AB, Van Golde LM, Batenburg JJ (2004b) Aggregation of *Cryptococcus neoformans* by surfactant protein D is inhibited by its capsular component glucuronoxylomannan. Infect Immun 72:145–153 van De Wetering JK, Van Remoortere A, Vaandrager AB, Batenburg JJ, Van Golde LM, Hokke CH, Van Hellemond JJ (2004c) Surfactant protein D binding to terminal alpha1-3-linked fucose residues and to *Schistosoma mansoni*. Am J Respir Cell Mol Biol 31:565–572

van Eijk M, White MR, Crouch EC, Batenburg JJ, Vaandrager AB, Van Golde LM, Haagsman HP, Hartshorn KL (2003) Porcine pulmonary collectins show distinct interactions with influenza A viruses: role of the N-linked oligosaccharides in the carbohydrate recognition domain. J Immunol 171:1431–1440

van Emmerik LC, Kuijper EJ, Fijen CA, Dankert J, Thiel S (1994) Binding of mannan-binding protein to various bacterial pathogens of meningitis. Clin Exp Immunol 97:411–416

van Iwaarden F, Welmers B, Verhoef J, Haagsman HP, van Golde LM (1990) Pulmonary surfactant protein A enhances the host-defense mechanism of rat alveolar macrophages. Am J Respir Cell Mol Biol 2(1):91–98

van Iwaarden JF, Van Strijp JA, Ebskamp MJ, Welmers AC, Verhoef J, Van Golde LM (1991) Surfactant protein A is opsonin in phagocytosis of herpes simplex virus type 1 by rat alveolar macrophages. Am J Physiol 261:L204–L209

van Iwaarden JF, Shimizu H, Van Golde PH, Voelker DR, Van Golde LM (1992a) Rat surfactant protein D enhances the production of oxygen radicals by rat alveolar macrophages. Biochem J 286(Pt 1):5–8

van Iwaarden JF, van Strijp JA, Visser H, Haagsman HP, Verhoef J, van Golde LM (1992b) Binding of surfactant protein A (SP-A) to herpes simplex virus type 1-infected cells is mediated by the carbohydrate moiety of SP-A. J Biol Chem 267(35):25039–25043

van Iwaarden JF, Pikaar JC, Storm J, Brouwer E, Verhoef J, Oosting RS, Van Golde LM, Van Strijp JA (1994) Binding of surfactant protein A to the lipid A moiety of bacterial lipopolysaccharides. Biochem J 303(Pt 2):407–411

van Rozendaal BA, Van Spriel AB, Van De Winkel JG, Haagsman HP (2000) Role of pulmonary surfactant protein D in innate defense against *Candida albicans*. J Infect Dis 182:917–922

Vandivier RW, Ogden CA, Fadok VA, Hoffmann PR, Brown KK, Botto M, Walport MJ, Fisher JH, Henson PM, Greene KE (2002) Role of surfactant proteins A, D, and C1q in the clearance of apoptotic cells *in vivo* and *in vitro*: calreticulin and CD91 as a common collectin receptor complex. J Immunol 169(7):3978–3986

Varga L, Szilágyi K, Lõrincz Z, Berrens L, Thiel S, Závodszky P, Daha MR, Thielens NM, Arlaud GJ, Nagy K, Späth P, Füst G (2003) Studies on the mechanisms of allergen-induced activation of the classical and lectin pathways of complement. Mol Immunol 39(14):839–846

Venkatraman Girija U, Furze CM, Gingras AR, Yoshizaki T, Ohtani K, Marshall JE, Wallis AK, Schwaeble WJ, El-Mezgueldi M, Mitchell DA, Moody PC, Wakamiya N, Wallis R (2015) Molecular basis of sugar recognition by collectin-K1 and the effects of mutations associated with 3MC syndrome. BMC Biol 17(13):27

Voorhout WF, Veenendaal T, Kuroki Y, Ogasawara Y, van Golde LM, Geuze HJ (1992) Immunocytochemical localization of surfactant protein D (SP-D) in type II cells, Clara cells, and alveolar macrophages of rat lung. J Histochem Cytochem 40(10):1589–1597

Voss T, Melchers K, Scheirle G, Schäfer KP (1991) Structural comparison of recombinant pulmonary surfactant protein SP-A derived from two human coding sequences: implications for the chain composition of natural human SP-A. Am J Respir Cell Mol Biol 4(1):88–94

Vuk-Pavlovic Z, Standing JE, Crouch EC, Limper AH (2001) Carbohydrate recognition domain of surfactant protein D mediates interactions with *Pneumocystis carinii* glycoprotein A. Am J Respir Cell Mol Biol 24:475–484

Vuk-Pavlovic Z, Mo EK, Icenhour CR, Standing JE, Fisher JH, Limper AH (2006) Surfactant protein D enhances Pneumocystis infection in immune-suppressed mice. Am J Physiol Lung Cell Mol Physiol 290:L442–L449

Walenkamp AM, Verheul AF, Scharringa J, Hoepelman IM (1999) Pulmonary surfactant protein A binds to *Cryptococcus neoformans* without promoting phagocytosis. Eur J Clin Invest 29:83–92

Wang JY, Kishore U, Reid KB (1995) A recombinant polypeptide, composed of the alpha-helical neck region and the carbohydrate recognition domain of conglutinin, self-associates to give a functionally intact homotrimer. FEBS Lett 376:6–10

Wang JY, Shieh CC, You PF, Lei HY, Reid KB (1998) Inhibitory effect of pulmonary surfactant proteins A and D on allergen-induced lymphocyte proliferation and histamine release in children with asthma. Am J Respir Crit Care Med 158(2):510–518

Wang M, Wang F, Yang J, Zhao D, Wang H, Shao F, Wang W, Sun R, Ling M, Zhai J, Song S (2013) Mannan-binding lectin inhibits *Candida albicans*-induced cellular responses in PMA-activated THP-1 cells through Toll-like receptor 2 and Toll-like receptor 4. PLoS ONE 8:e83517

Weikert LF, Edwards K, Chroneos ZC, Hager C, Hoffman L, Shepherd VL (1997) SP-A enhances uptake of bacillus Calmette-Guerin by macrophages through a specific SP-A receptor. Am J Physiol 272:L989–L995

Weikert LF, Lopez JP, Abdolrasulnia R, Chroneos ZC, Shepherd VL (2000) Surfactant protein A enhances mycobacterial killing by rat macrophages through a nitric oxide-dependent pathway. Am J Physiol Lung Cell Mol Physiol 279:L216–L223

Weis WI, Crichlow GV, Murthy HM, Hendrickson WA, Drickamer K (1991a) Physical characterization and crystallization of the carbohydrate-recognition domain of a mannose-binding protein from rat. J Biol Chem 266(31):20678–20686

Weis WI, Kahn R, Fourme R, Drickamer K, Hendrickson WA (1991b) Structure of the calcium-dependent lectin domain from a rat mannose-binding protein determined by MAD phasing. Science 254(5038):1608–1615

Weyer C, Sabat R, Wissel H, Kruger DH, Stevens PA, Prosch S (2000) Surfactant protein A binding to cytomegalovirus proteins enhances virus entry into rat lung cells. Am J Respir Cell Mol Biol 23:71–78

White RT, Damm D, Miller J, Spratt K, Schilling J, Hawgood S, Benson B, Cordell B (1985) Isolation and characterization of the human pulmonary surfactant apoprotein gene. Nature 317(6035):361–363

White CW, Greene KE, Allen CB, Shannon JM (2001) Elevated expression of surfactant proteins in newborn rats during adaptation to hyperoxia. Am J Respir Cell Mol Biol 25(1):51–59

White MR, Crouch E, Vesona J, Tacken PJ, Batenburg JJ, Leth-Larsen R, Holmskov U, Hartshorn KL (2005) Respiratory innate immune proteins differentially modulate the neutrophil respiratory burst response to influenza A virus. Am J Physiol Lung Cell Mol Physiol 289:L606–L616

Williams MD, Wright JR, March KL, Martin WJ 2 (1996) Human surfactant protein A enhances attachment of *Pneumocystis carinii* to rat alveolar macrophages. Am J Respir Cell Mol Biol 14:232–238

Wong SSW, Rani M, Dodagatta-Marri E, Ibrahim-Granet O, Kishore U, Bayry J, Latge JP, Sahu A, Madan T, Aimanianda V (2018) Fungal melanin stimulates surfactant protein D-mediated opsonization of and host immune response to *Aspergillus fumigatus* spores. J Biol Chem 293:4901–4912

Wu H, Kuzmenko A, Wan S, Schaffer L, Weiss A, Fisher JH, Kim KS, Mccormack FX (2003) Surfactant proteins A and D inhibit the growth of Gram-negative bacteria by increasing membrane permeability. J Clin Invest 111:1589–1602

Wu YP, Liu ZH, Wei R, Pan SD, Mao NY, Chen B, Han JJ, Zhang FS, Holmskov U, Xia ZL, De Groot PG, Reid KB, Xu WB, Sorensen GL (2009) Elevated plasma surfactant protein D (SP-D) levels and a direct correlation with anti-severe acute respiratory syndrome coronavirus-specific IgG antibody in SARS patients. Scand J Immunol 69:508–515

Xu J, Nakamura S, Islam MS, Guo Y, Ihara K, Tomioka R, Masuda M, Yoneyama H, Isogai E (2016) Mannose-binding lectin inhibits the motility of pathogenic salmonella by affecting the driving forces of motility and the chemotactic response. PLoS ONE 11:e0154165

Yang CH, Szeliga J, Jordan J, Faske S, Sever-Chroneos Z, Dorsett B, Christian RE, Settlage RE, Shabanowitz J, Hunt DF, Whitsett JA, Chroneos ZC (2005) Identification of the surfactant protein A receptor 210 as the unconventional myosin 18A. J Biol Chem 280(41):34447–34457

Yang HY, Li H, Wang YG, Xu CY, Zhao YL, Ma XG, Li XW, Chen H (2014) Correlation analysis between single nucleotide polymorphisms of pulmonary surfactant protein A gene and pulmonary tuberculosis in the Han population in China. Int J Infect Dis 26:31–36

Yong SJ, Vuk-Pavlovic Z, Standing JE, Crouch EC, Limper AH (2003) Surfactant protein D-mediated aggregation of *Pneumocystis carinii* impairs phagocytosis by alveolar macrophages. Infect Immun 71:1662–1671

Yuen MF, Lau CS, Lau YL, Wong WM (1999) h progression of liver disease in chronic hepatitis B infection. Hepatology 29:1248–1251

Zimmerman PE, Voelker DR, Mccormack FX, Paulsrud JR, Martin WJ 2 (1992) 120-kD surface glycoprotein of *Pneumocystis carinii* is a ligand for surfactant protein A. J Clin Invest 89:143–149

Chapter 5
Insect C-Type Lectins in Microbial Infections

Yibin Zhu, Xi Yu and Gong Cheng

Abstract C-type lectins (CTLs) are a family of carbohydrate-recognition domain (CRD)-containing proteins that bind to ligands in a calcium-dependent manner. CTLs act as important components of insect innate immune responses, such as pattern recognition, agglutination, encapsulation, melanization, phagocytosis and prophenoloxidase activation, as well as gut microbiome homeostasis maintenance, to defend against pathogens. Besides, some insect CTLs can facilitate pathogen infection and colonization. In this review, we describe the properties of insect CTLs and focus on explaining their role in viral, bacterial, parasitic and fungal infections.

Keywords C-type lectins · Infection · Insect

5.1 Introduction

Insects are the most diverse group of organisms on Earth, occupying an extremely important position in the ecosystem. Over long-term evolution in an adverse environment with various pathogens, insects have developed a unique innate immune system involving cellular immunity and humoral immunity (Sadd and Siva-Jothy 2006). Cellular immunity depends mainly on blood cells that engulf foreign antigens by phagocytosis. The humoral immune system is an open and complete defence system consisting of lectins, antimicrobial peptides (AMPs), antiviral factors, lysozyme and

---

Y Zhu and X Yu are equally contributed to this work.

Y. Zhu · X. Yu · G. Cheng (✉)
Tsinghua-Peking Joint Center for Life Sciences, Beijing Advanced Innovation Center for Structural Biology, School of Medicine, Tsinghua University, Beijing 100084, China
e-mail: gongcheng@mail.tsinghua.edu.cn

Y. Zhu · G. Cheng
Institute of Pathogenic Organisms, Shenzhen Center for Disease Control and Prevention, Shenzhen 518055, Guangdong, China

Y. Zhu · X. Yu
School of Life Science, Tsinghua University, Beijing 100084, China

S.-L. Hsieh (ed.), *Lectin in Host Defense Against Microbial Infections*,
Advances in Experimental Medicine and Biology 1204,
https://doi.org/10.1007/978-981-15-1580-4_5

protease inhibitors, and multifunctional blood cells (Hillyer 2015). Because of their inability to synthesize specific antibodies, invertebrates rely solely on non-specific immune responses to defend against pathogens. Upon pathogen invasion, insect immunity utilizes pattern-recognition proteins (PRPs) to help recognize pathogen-associated molecular patterns (PAMPs). There are many types of PRPs in insects, including peptidoglycan-recognition proteins (PGRPs), beta-1,3-glucan recognition proteins (bGRPs), scavenger receptors (SCRs), and C-type lectins (CTLs). Within the PRPs, CTLs play one of the most important roles and form the most diverse family (Zhang et al. 2015).

CTLs are a superfamily of carbohydrate-binding proteins and have a characteristic carbohydrate-recognition domain (CRD) (Dodd and Drickamer 2001) that binds to carbohydrates in a calcium-dependent manner (Drickamer 1993; Weis et al. 1998; Drickamer 1999). Some CRD-containing proteins are referred to as C-type lectin-like domain (CTLD) proteins, and with the continuous growth of the family, a few have been found that do not depend on calcium for binding activity (Zelensky and Gready 2005; Pees et al. 2016). Insect C-type lectins have various functional domains, including single CRDs, double CRDs (the immulectin group) and other functional domains (the CTL-X group). In the CTL-X group, functional domains containing complement control protein (CCP), immunoglobulin modules (Ig), an extracellular domain (CUB), epidermal growth factor-like domain (EGFL) and chitin binding domain (CBM) have been identified (Rao et al., 2015a, b; Waterhouse et al. 2007; Xia et al. 2015). Most insect CTLs have signal peptides, indicating that they are secreted proteins, while some CTLs contain transmembrane domains.

Recently, many works have reported the important position of CTLs in insect innate immune defence against microorganisms. Some insect CTLs play important roles as pattern-recognition receptors by binding invading pathogens and mediating immune responses, such as agglutination, encapsulation, melanization, phagocytosis and prophenoloxidase activation, to clear the pathogens. Certain other CTLs, however, have the opposite function; they facilitate pathogen infection. Thus, we will mainly focus on explaining the role of insect C-type lectins in viral, bacterial, parasitic and fungal infections.

5.2 Insect C-Type Lectin and Viral Infections

Multiple C-type lectins have been identified as susceptibility factors for viral infection in mosquitoes. Arthropod-borne viruses (arboviruses) are named for their unique method of transmission, via the bite of an arthropod vector, among which dengue virus (DENV), Zika virus (ZIKV), yellow fever virus (YFV), West Nile virus (WNV), Japanese encephalitis virus (JEV) and chikungunya virus (CHIKV) are transmitted by mosquito vectors and pose a major threat to human health (Zhu et al. 2017). These viruses enter the mosquito by an infectious blood meal and infect the vector in a sequential manner; first, the midgut epithelium is infected, and then the surrounding tissues become infected before the virus disseminates to the salivary gland (Guzman et al. 2016). During this process, viruses are able to utilize mosquito C-type lectins to facilitate infection. A previous study identified an *Aedes aegypti* C-type lectin, *mosGCTL-1*, which facilitates WNV infection in vitro and in vivo. The expression of *mosGCTL-1* is upregulated by WNV, and it interacts with WNV in a calcium-dependent manner. *mosGCTL-1* also plays an important role in WNV infection by interacting with a mosquito homologue of human CD45 (*mosPTP-1*), which enables viral attachment to cells and facilitates viral entry (Cheng et al. 2010). In addition, Liu and colleagues first identified 9 *Aedes aegypti mosGCTL* genes as key susceptibility factors for DENV-2 infection, among which *mosGCTL-3* was found to most significantly enhance viral infectivity by interacting with the DENV-2 envelope protein in vitro and in vivo. In contrast, antisera against multiple lectins efficiently reduced DENV-2 infection during a mosquito blood meal (Liu et al. 2014). Other studies have also shown that *mosGCTL-7* can bind to the JEV envelope protein via an N-glycan at N154 in a calcium-dependent manner. After binding, cell surface mosPTP-1 interacts with the mosGCTL-7-JEV complex and facilitates viral infection in vitro and in vivo (Liu et al. 2017).

In contrast to these viral infection facilitating roles in insects, C-type lectins can also induce protective responses during viral infections of insects. *Spodoptera exigua* C-type lectins (bracovirus-like lectins, Se-BLLs) are acquired by horizontal gene transfer and confer an adaptive advantage against baculovirus, a natural viral pathogen of *S. exigua*. Specifically, Se-BLL2 can reduce the pathogenicity of baculovirus, indicating that lectins can increase the resistance of *S. exigua* larvae to baculoviral infection (Gasmi et al. 2015). In addition, another lectin derived from *Spodoptera*, Se-BLL3, is able to reduce the mortality of *Spodoptera frugiperda* larvae during Junonia coenia densovirus (JcDV) infection, but no such effect is found in *S. exigua* larvae (Gasmi et al. 2018). Recently, two novel C-type lectins (MdCTLs) have been obtained from *Musca domestica* and show agglutination and antiviral properties in the laboratory. Interestingly, MdCTL-2 protein can inhibit the replication of influenza H1N1 virus, resembling the effect of ribavirin (Zhou et al. 2018).

Taken together, these studies show that insect C-type lectins play various roles in immune responses upon viral infection. Investigating the effect of C-type lectins during arbovirus infection may provide a better understanding of the replication of the virus in insects and shed light on new strategies for arbovirus transmission control. For instance, a transmission-blocking strategy may be feasible by blocking the C-type lectins that facilitate viral infection. Additionally, identifying novel insect C-type lectins may also lead to the discovery of new antiviral proteins.

5.3 Insect C-Type Lectins and Bacterial Infection

Insects utilize multiple immune systems to control bacterial overgrowth and maintain gut microbiome homeostasis. Specifically, C-type lectins are a large group of proteins in insects known for their ability to regulate bacterial survival and colonization. The role of insect C-type lectin in antibacterial immunity has been identified for *Manduca sexta* (Yu et al. 1999a, b). The *M. sexta* genome encodes 34 C-type lectins, among which is *M. sexta* immulectin-1 (*MsIML-1*), which was first cloned from the larval fat body. *MsIML-1* expression is induced by the injection of inactivated gram-positive and gram-negative bacteria or yeast, and it is highly expressed in insect immune tissues, such as fat body and haemolymph, after bacterial infection. Recombinant *MsIML-1* protein impairs bacterial growth by agglutinating bacteria and stimulating the activation of prophenoloxidase (PPO) (Yu et al. 1999a). Immulectin-2 from *M. sexta* (*MsIML-2*) was also isolated from the larval fat body and is induced by *Escherichia coli* and *Saccharomyces cerevisiae*. However, *MsIML-2* has not been measured in haemocytes after bacterial stimulation. *MsIML-2* can agglutinate *E. coli* in a calcium-dependent manner but has no effect on the agglutination of *S. cerevisiae* (Yu and Kanost 2000). *M. sexta* immulectin-3 (*MsIML-3*) and immulectin-4 (*MsIML-4*) cDNA has also been cloned from the larval fat body. By coating the proteins onto agarose beads, it has been shown that MsIML-4 can facilitate encapsulation and melanization by haemocytes in vitro, while MsIML-3 only enhances encapsulation in vitro (Yu et al. 2005, 2006). The silkworm *Bombyx mori* genome encodes 23 CTL genes (Rao et al. 2015b). A recent study revealed that C-type lectin-S3 from silkworm *B. mori* (*BmCTL-S3*) is expressed in various tissues, including the fat body, haemocytes, midgut, Malpighian tubule, ovary and embryo. The BmCTL-S3 protein facilitates the rapid clearance of *E. coli*, *S. aureus*, *Serratia marcescens* and *Bacillus subtilis* by binding to bacterial cell wall components (Zhan et al. 2016). Shahzad and colleagues also cloned a C-type lectin-S2 (*BmCTL-S2*) from the silkworm *B. mori* and performed functional analyses, which suggested that *BmCTL-S2* is a pattern-recognition receptor with antibacterial activity (Shahzad et al. 2017). In the mosquito field, Schnitger and colleagues identified two CTLs from *Anopheles gambiae*, which are required for the clearance of *E. coli* but not *S. aureus* (Schnitger et al. 2009). These results reveal that multifunctional CTLs in insects can promote the interaction between host and bacteria. A better understanding of the molecular mechanism of the antibacterial activity of CTLs may provide novel strategies for bacterial defence.

In addition, CTLs function as pattern-recognition receptors and exhibit antimicrobial activity in invertebrates other than insects. A recent study identified a novel soluble and bacteria-inducible PRR, Leulectin, containing a CTLD from the hepatopancreas of kuruma shrimp *Marsupenaeus japonicus*. This CTLD can agglutinate bacteria and promote haemocytic phagocytosis by recognizing LPS (Wang et al. 2017a, b). Another CTL, PmCLec, from the black tiger shrimp *Penaeus monodon*, was characterized. Purified recombinant PmCLec protein binds and agglutinates gram-positive bacteria, *Staphylococcus aureus* and *S. haemolyticus*, in the presence of calcium (Wongpanya et al. 2017). A mannose-binding CTL called FmLC3 was isolated from *Fenneropenaeus merguiensis* and exhibits agglutination activity against various microbial strains in a calcium-dependent manner (Runsaeng et al. 2017). Moreover, a CTL named *Mj*CC-CL in the kuruma shrimp (*Marsupenaeus japonicus*) contains both a CTLD and a coiled-coil domain (CCD). In an antibacterial response, the CTLD recognizes bacterial glycans leading to the interaction of the CCD with the surface receptor Domeless and subsequent activation of the JAK/STAT pathway and upregulation of AMP expression (Sun et al. 2017).

Most CTLs serve as immune factors suppressing bacterial growth in insects. However, some mosquito C-type lectins function as immune factors facilitating bacterial growth. Pang and colleagues revealed that blockade of mosGCTLs suppresses the growth of gut bacteria in *Aedes aegypti* and *Culex pipiens pallens*. It has been demonstrated that mosGCTLs function by directly coating bacterial cells and neutralizing antimicrobial peptide (AMP)-mediated elimination (Pang et al. 2016). In summary, insect C-type lectins mediate the regulation of insect microbiome homeostasis via various mechanisms and show pleiotropic effects in insect–microbial interactions.

5.4 Insect C-Type Lectins and Parasite Infections

Malaria parasites must go through a complex life cycle in an *Anopheles* mosquito to establish an infection before being transmitted to a new host. Osta and colleagues conducted a large-scale functional screen of mosquito genes and identified two C-type lectins (CTL4 and CTLMA2), which act as protective agonists for the development of *Plasmodium* ookinetes to oocysts and facilitate parasite susceptibility by inhibiting parasite melanization (Osta et al. 2004). Recently, Simoes and colleagues found that protection against parasite melanization by CTL4 and CTLMA2 in *A. gambiae* mosquitoes is dependent on infection intensity. Interestingly, CTL4 and CLTMA2 exert agonistic and antagonistic parasite regulation effects in different *Anopheles* species. (Simoes et al. 2017). However, silencing of CTL4 and CTLMA2 has no effect on *Plasmodium falciparum* development in humans (Cohuet et al. 2006). Additionally, Shin and colleagues identified a novel gene, *CLSP2*, encoding C-type lectin domains. Knockdown of *CLSP2* activates prophenoloxidase and thus inhibits parasite growth, suggesting that CLSP2 is an antagonist of Plasmodium parasite infection in mosquitoes (Shin et al. 2011).

Recently, another CTL was found in the cotton bollworm *Helicoverpa armigera* (*Ha*CTL3), which is involved in defence against parasites. HaCTL3 enhances haemocytic encapsulation and melanization by interacting with β-integrin. The expression of HaCTL3 is induced by 20-hydroxyecdysone (20E) treatment while silencing of the 20E receptor can abolish this response (Wang et al. 2017a, b).

Indeed, C-type lectins play various roles in host–parasite interactions. Investigating and understanding these dramatically different CTL mechanisms during the parasite defence process may lead to novel strategies to inhibit parasite transmission. Notably, Yoshida and colleagues generated transgenic mosquitoes expressing the C-type lectin from sea cucumber *Cucumaria echinata*, which impairs *Plasmodium berghei* oocyst formation. It is the first established *Anopheles* mosquito engineered to block parasite transmission (Yoshida et al. 2007).

5.5 Insect C-Type Lectins and Fungal Infection

Upon fungal infection, the innate immune system of insects must perform all defence functions because of the lack of adaptive immunity. Therefore, C-type lectins act as an important component in the defence against fungal infections in insects. Tian and colleagues identified a C-type lectin from *Helicoverpa armigera* (Halectin), which promotes haemocyte phagocytosis of pathogens and protects the insect from fungal infection (Tian et al. 2009). In addition, C-type lectin 14 (*CTL14*) from *H. armigera* is induced in the fifth larval stage by entomopathogen *Beauveria bassiana* infection specifically. Recombinant CTL14 protein can aggregate *B. bassiana* in vitro while silencing of CTL14 decreases the resistance to fungal infection in *H. armigera* (Cheng et al. 2018). Recently, to investigate the role of the JAK/STAT signalling pathway in the antifungal immune response of silkworm *Bombyx mori*, Geng and colleagues revealed that *B. mori* C-type lectin 5 (*BmCTL5*) is induced by *Beauveria bassiana* infection and showed that BmCTL5 might be a pattern-recognition receptor for the JAK/STAT signalling pathway in silkworms (Geng et al. 2016).

5.6 Conclusion

Insects utilize lectins to recognize pathogens, gain access to invaders through in vivo interactions, and eliminate exogenous invaders. Understanding the role of insect lectins in the insect immune system may provide insight into the interactions between pathogens and their hosts. Insect CTL genes regulate various innate immune responses, including pattern recognition, agglutination, encapsulation, melanization, phagocytosis and prophenoloxidase activation, as well as the maintenance of gut microbiome homeostasis. Some CTLs have specialized functions, such as stimulating cell proliferation (Table 5.1).

Mosquitoes are one of the major insect vectors of several human diseases in the world. The mosquito immune response regulates pathogen infection and transmission via multiple pathways. Specifically, different C-type lectins from *A. aegypti* have been found to facilitate the infection of mosquitoes by various arboviruses, including WNV, DENV and JEV (Cheng et al. 2010; Liu et al. 2014, 2017). Furthermore, other CTLs that regulate flavivirus infection in mosquitoes may be identified by using bioinformatics methods to analyse homologous functional domains of the CLTs in mosquitoes. The development of strategies to inhibit certain C-type lectins in the flavivirus life cycle may be a promising strategy for flavivirus transmission control. A better understanding of the role of C-type lectins in insect gut ecology, maintaining gut microbiome homeostasis or dysregulating the gut microbiota, may offer a novel target for vector-borne disease control in nature (Pang et al. 2016).

Unfortunately, research on insect lectins is still extremely limited. While a few studies have focused on certain lectins from a specific species of insect, the molecular mechanisms underlying the interactions between pathogens and most lectins are

Table 5.1 Function of C-type lectin in microbial infection in insect

Microbial		Species	Name	Functions	Calcium-dependent	References
Virus	WNV	*A. aegypti*	mosGCTL-1	Facilitates WNV infection by interacting with mosPTP-1 to enhance viral entry	+	Cheng et al. (2010)
	DENV	*A. aegypti*	mosGCTL-3	Interacts with DENV-2 envelope protein to facilitate virus infection	+	Liu et al. (2014)
	JEV	*A. aegypti*	mosGCTL-7	Binds to JEV envelope protein by a N-glycan at N154, promotes JEV infection	+	Liu et al. (2017)
	JcDv	*S. exigua*	Se-BLL2	Increases the resistance of *S. exigua* larvae to baculovirus infection	Unknown	Gasmi et al. (2015)
	JcDV	*S. frugiperda*	Se-BLL3	Reduces the mortality of *S. frugiperda* larvae	Unknown	Gasmi et al. (2018)
	HIN1 virus	*M. domestica*	MdCTL-2	Inhibition of HI Nl virus replication	Unknown	Jing et al. (2018)
Bacterial	Gram-negative bacteria	*M. sexta*	MsIML-1	Impaired bacteria growth by agglutinating bacteria and stimulating the activation of prophenolox	+	Yu et al. (1999a)
	E. coli	*M. sexta*	MsIML-2	Agglutinates *E. coli* in a calcium-dependent manner	+	Yu and Kanost (2000)

(continued)

Table 5.1 (continued)

Microbial		Species	Name	Functions	Calcium-dependent	References
	E. coli	*M. sexta*	MsIML-3	Agglutinates *E. coli* via a calcium-dependent manner and does not facilitate melanization	+	Yu et al. (2005)
	E. coli	*M. sexta*	MsIML-4	Binds to LPS in a calcium-dependent manner and enhances encapsulation and melanization	+	Yu et al. (2006)
	E. coli, S. aureus, S. marcescens, B. subtilis	*B. mori*	BmCTL-S3	Binds to bacterial cell wall to facilitate rapid clearance of bacteria	Unknown	Zhan et al. (2016)
	S. aureus, B. subtillis	*B. mori*	BmCTL-S2	Inhibits the growth of *B. subtilis* and *S. aureus*	+	Shahzad et al. (2017)
	E. coli	*A. gambiae*	mosGCTLs	Required for the clearance of *E. coli*	+	Schnitger et al. (2009)
	Symbiotic microbiome	*A. aegypti*	mosGCTLs	GCTLs directly coat the bacterial cells and inhibit antibacterial function of AMPs	+	Pang et al. (2016)
Parasite	*P. berghei*	*A. gambiae*	CTL4 and CTLMA2	Inhibit melanization to enhance parasite infection	+	Osta et al. (2004)
	P. gallinaceum	*A. gambiae*	CLSP2	Enhances parasite infection by impairing prophenoloxidase	+	Shin et al. (2011)

(continued)

Table 5.1 (continued)

Microbial		Species	Name	Functions	Calcium-dependent	References
	Parasite	*H. armigera*	HaCTL3	Regulated by the steroid hormone ecdysone to protect against parasites	Unknown	Wang et al. (2017a, b)
Fungi	*B. bassiana*	*H. armigera*	HaCTL14	Aggregates with *B. bassiana* to impair fungal infection	Unknown	Cheng et al. (2018)
	B. bassiana	*B. mori*	BmCTL	A pattern recognition receptor for JAK/STAT pathway in fungal infection	Unknown	Geng et al. (2016)

poorly understood. Current knowledge of CTLs suggests that, in addition to mediating innate immune responses, some insect CTLs have unique functions, such as antimicrobial activity and stimulating cell proliferation, while other CTLs have multiple functions. Current research on CTLs mainly focuses on the functional analysis of individual CTLs, while a general and extensive CTL functional investigation in insects should be considered. A better understanding of insect CTLs may promote the development of novel antibacterial or viral transmission-blocking strategies. Further investigation into the functions and mechanisms of insect CTLs in innate immunity is expected because it may contribute to the protection of beneficial insects as well as the biological control of pests and disease vectors.

Acknowledgements This work was funded by grants from the National Key Research and Development Plan of China (2018ZX09711003-004-003 and 2018YFA0507202), the Natural Science Foundation of China (81730063 and 31825001), the Shenzhen San-Ming Project for prevention and research on vector-borne diseases (SZSM201611064).

Competing Interests The authors declare that they have no competing financial interests.

References

Cheng G, Cox J, Wang P, Krishnan MN, Dai J, Qian F, Anderson JF, Fikrig E (2010) A C-type lectin collaborates with a CD45 phosphatase homolog to facilitate West Nile virus infection of mosquitoes. Cell 142(5):714–725. https://doi.org/10.1016/j.cell.2010.07.038

Cheng Y, Lin Z, Wang JM, Xing LS, Xiong GH, Zou Z (2018) CTL14, a recognition receptor induced in late stage larvae, modulates anti-fungal immunity in cotton bollworm *Helicoverpa armigera*. Dev Comp Immunol 84:142–152. https://doi.org/10.1016/j.dci.2018.02.010

Cohuet A, Osta MA, Morlais I, Awono-Ambene PH, Michel K, Simard F, Christophides GK, Fontenille D, Kafatos FC (2006) Anopheles and Plasmodium: from laboratory models to natural systems in the field. EMBO Rep 7(12):1285–1289. https://doi.org/10.1038/sj.embor.7400831

Dodd RB, Drickamer K (2001) Lectin-like proteins in model organisms: implications for evolution of carbohydrate-binding activity. Glycobiology 11(5):71R–79R

Drickamer K (1993) Evolution of Ca($^{2+}$)-dependent animal lectins. Prog Nucleic Acid Res Mol Biol 45:207–232

Drickamer K (1999) C-type lectin-like domains. Curr Opin Struct Biol 9(5):585–590

Gasmi L, Boulain H, Gauthier J, Hua-Van A, Musset K, Jakubowska AK, Aury JM, Volkoff AN, Huguet E, Herrero S et al (2015) Recurrent domestication by lepidoptera of genes from their parasites mediated by bracoviruses. PLoS Genet 11:e1005470

Gasmi L, Jakubowska AK, Ferre J, Ogliastro M, Herrero S (2018) Characterization of two groups of *Spodoptera exigua* Hubner (Lepidoptera: Noctuidae) C-type lectins and insights into their role in defense against the densovirus JcDV. Arch Insect Biochem Physiol 97(1). https://doi.org/10.1002/arch.21432

Geng T, Lv DD, Huang YX, Hou CX, Qin GX, Guo XJ (2016) JAK/STAT signaling pathway-mediated immune response in silkworm (*Bombyx mori*) challenged by *Beauveria bassiana*. Gene 595(1):69–76. https://doi.org/10.1016/j.gene.2016.09.043

Guzman MG, Gubler DJ, Izquierdo A, Martinez E, Halstead SB (2016) Dengue infection. Nat Rev Dis Primers 2(1)

Hillyer JF (2015) Integrated immune and cardiovascular function in pancrustacea: lessons from the insects. Integr Comp Biol 55(5):843–855. https://doi.org/10.1093/icb/icv021

Liu Y, Zhang F, Liu J, Xiao X, Zhang S, Qin C, Xiang Y, Wang P, Cheng G (2014) Transmission-blocking antibodies against mosquito C-type lectins for dengue prevention. PLoS Pathog 10(2):e1003931. https://doi.org/10.1371/journal.ppat.1003931

Liu K, Qian Y, Jung YS, Zhou B, Cao R, Shen T, Shao D, Wei J, Ma Z, Chen P, Zhu H, Qiu Y (2017) mosGCTL-7, a C-type lectin protein, mediates Japanese encephalitis virus infection in mosquitoes. J Virol 91(10). https://doi.org/10.1128/jvi.01348-16

Osta MA, Christophides GK, Kafatos FC (2004) Effects of mosquito genes on Plasmodium development. Science 303(5666):2030–2032. https://doi.org/10.1126/science.1091789

Pang X, Xiao X, Liu Y, Zhang R, Liu J, Liu Q, Wang P, Cheng G (2016) Mosquito C-type lectins maintain gut microbiome homeostasis. Nat Microbiol 1:16023. https://doi.org/10.1038/nmicrobiol.2016.23

Pees B, Yang W, Zarate-Potes A, Schulenburg H, Dierking K (2016) High innate immune specificity through diversified C-type lectin-like domain proteins in invertebrates. J Innate Immun 8(2):129–142. https://doi.org/10.1159/000441475

Rao XJ, Cao X, He Y, Hu Y, Zhang X, Chen YR, Blissard G, Kanost MR, Yu XQ, Jiang H (2015a) Structural features, evolutionary relationships, and transcriptional regulation of C-type lectin-domain proteins in *Manduca sexta*. Insect Biochem Mol Biol 62:75–85. https://doi.org/10.1016/j.ibmb.2014.12.006

Rao XJ, Shahzad T, Liu S, Wu P, He YT, Sun WJ, Fan XY, Yang YF, Shi Q, Yu XQ (2015b) Identification of C-type lectin-domain proteins (CTLDPs) in silkworm *Bombyx mori*. Dev Comp Immunol 53(2):328–338. https://doi.org/10.1016/j.dci.2015.07.005

Runsaeng P, Puengyam P, Utarabhand P (2017) A mannose-specific C-type lectin from *Fenneropenaeus merguiensis* exhibited antimicrobial activity to mediate shrimp innate immunity. Mol Immunol 92:87–98. https://doi.org/10.1016/j.molimm.2017.10.005

Sadd BM, Siva-Jothy MT (2006) Self-harm caused by an insect's innate immunity. Proc Biol Sci 273(1600):2571–2574. https://doi.org/10.1098/rspb.2006.3574

Schnitger AK, Yassine H, Kafatos FC, Osta MA (2009) Two C-type lectins cooperate to defend *Anopheles gambiae* against gram-negative bacteria. J Biol Chem 284(26):17616–17624. https://doi.org/10.1074/jbc.M808298200

Shahzad T, Zhan MY, Yang PJ, Yu XQ, Rao XJ (2017) Molecular cloning and analysis of a C-type lectin from silkworm *Bombyx mori*. Arch Insect Biochem Physiol 95(3). https://doi.org/10.1002/arch.21391

Shin SW, Zou Z, Raikhel AS (2011) A new factor in the Aedes aegypti immune response: CLSP2 modulates melanization. EMBO Rep 12(9):938–943. https://doi.org/10.1038/embor.2011.130

Simoes ML, Mlambo G, Tripathi A, Dong Y, Dimopoulos G (2017) Immune regulation of plasmodium is anopheles species specific and infection intensity dependent. MBio 8(5). https://doi.org/10.1128/mbio.01631-17

Sun JJ, Lan JF, Zhao XF, Vasta GR, Wang JX (2017) Binding of a C-type lectin's coiled-coil domain to the Domeless receptor directly activates the JAK/STAT pathway in the shrimp immune response to bacterial infection. PLoS Pathog 13(9):e1006626. https://doi.org/10.1371/journal.ppat.1006626

Tian YY, Liu Y, Zhao XF, Wang JX (2009) Characterization of a C-type lectin from the cotton bollworm, *Helicoverpa armigera*. Dev Comp Immunol 33(6):772–779. https://doi.org/10.1016/j.dci.2009.01.002

Wang P, Zhuo XR, Tang L, Liu XS, Wang YF, Wang GX, Yu XQ, Wang JL (2017a) C-type lectin interacting with beta-integrin enhances hemocytic encapsulation in the cotton bollworm, *Helicoverpa armigera*. Insect Biochem Mol Biol 86:29–40. https://doi.org/10.1016/j.ibmb.2017.05.005

Wang XW, Gao J, Xu YH, Xu JD, Fan ZX, Zhao XF, Wang JX (2017b) novel pattern recognition receptor protects shrimp by preventing bacterial colonization and promoting phagocytosis. J Immunol 198(8):3045–3057. https://doi.org/10.4049/jimmunol.1602002

Waterhouse RM, Kriventseva EV, Meister S, Xi Z, Alvarez KS, Bartholomay LC, Barillas-Mury C, Bian G, Blandin S, Christensen BM, Dong Y, Jiang H, Kanost MR, Koutsos AC, Levashina EA,

Li J, Ligoxygakis P, Maccallum RM, Mayhew GF, Mendes A, Michel K, Osta MA, Paskewitz S, Shin SW, Vlachou D, Wang L, Wei W, Zheng L, Zou Z, Severson DW, Raikhel AS, Kafatos FC, Dimopoulos G, Zdobnov EM, Christophides GK (2007) Evolutionary dynamics of immune-related genes and pathways in disease-vector mosquitoes. Science 316(5832):1738–1743. https://doi.org/10.1126/science.1139862

Weis WI, Taylor ME, Drickamer K (1998) The C-type lectin superfamily in the immune system. Immunol Rev 163:19–34

Wongpanya R, Sengprasert P, Amparyup P, Tassanakajon A (2017) A novel C-type lectin in the black tiger shrimp *Penaeus monodon* functions as a pattern recognition receptor by binding and causing bacterial agglutination. Fish Shellfish Immunol 60:103–113. https://doi.org/10.1016/j.fsi.2016.11.042

Xia X, Yu L, Xue M, Yu X, Vasseur L, Gurr GM, Baxter SW, Lin H, Lin J, You M (2015) Genome-wide characterization and expression profiling of immune genes in the diamondback moth, *Plutella xylostella* (L.). Sci Rep 5:9877. https://doi.org/10.1038/srep09877

Yoshida S, Shimada Y, Kondoh D, Kouzuma Y, Ghosh AK, Jacobs-Lorena M, Sinden RE (2007) Hemolytic C-type lectin CEL-III from sea cucumber expressed in transgenic mosquitoes impairs malaria parasite development. PLoS Pathog 3(12):e192. https://doi.org/10.1371/journal.ppat.0030192

Yu XQ, Kanost MR (2000) Immulectin-2, a lipopolysaccharide-specific lectin from an insect, *Manduca sexta*, is induced in response to gram-negative bacteria. J Biol Chem 275(48):37373–37381. https://doi.org/10.1074/jbc.M003021200

Yu XQ, Gan H, Kanost MR (1999a) Immulectin, an inducible C-type lectin from an insect, *Manduca sexta*, stimulates activation of plasma prophenol oxidase. Insect Biochem Mol Biol 29(7):585–597

Yu XQ, Prakash O, Kanost MR (1999b) Structure of a paralytic peptide from an insect, *Manduca sexta*. J Peptide Res Off J Am Peptide Soc 54(3):256–261

Yu XQ, Tracy ME, Ling E, Scholz FR, Trenczek T (2005) A novel C-type immulectin-3 from *Manduca sexta* is translocated from hemolymph into the cytoplasm of hemocytes. Insect Biochem Mol Biol 35(4):285–295. https://doi.org/10.1016/j.ibmb.2005.01.004

Yu XQ, Ling E, Tracy ME, Zhu Y (2006) Immulectin-4 from the tobacco hornworm *Manduca sexta* binds to lipopolysaccharide and lipoteichoic acid. Insect Mol Biol 15(2):119–128. https://doi.org/10.1111/j.1365-2583.2006.00618.x

Zelensky AN, Gready JE (2005) The C-type lectin-like domain superfamily. FEBS J 272(24):6179–6217. https://doi.org/10.1111/j.1742-4658.2005.05031.x

Zhan MY, Shahzad T, Yang PJ, Liu S, Yu XQ, Rao XJ (2016) A single-CRD C-type lectin is important for bacterial clearance in the silkworm. Dev Comp Immunol 65:330–339. https://doi.org/10.1016/j.dci.2016.08.004

Zhang X, He Y, Cao X, Gunaratna RT, Chen YR, Blissard G, Kanost MR, Jiang H (2015) Phylogenetic analysis and expression profiling of the pattern recognition receptors: Insights into molecular recognition of invading pathogens in *Manduca sexta*. Insect Biochem Mol Biol 62:38–50. https://doi.org/10.1016/j.ibmb.2015.02.001

Zhou J, Fang NN, Zheng Y, Liu KY, Mao B, Kong LN, Chen Y, Ai H (2018) Identification and characterization of two novel C-type lectins from the larvae of housefly, Musca domestica L. Arch Insect Biochem Physiol 98(3):e21467. https://doi.org/10.1002/arch.21467

Zhu Y, Zhang R, Zhang B, Zhao T, Wang P, Liang G, Cheng G (2017) Blood meal acquisition enhances arbovirus replication in mosquitoes through activation of the GABAergic system. Nat Commun 8(1):1262. https://doi.org/10.1038/s41467-017-01244-6

Chapter 6
Galectins in Host Defense Against Microbial Infections

Fang-Yen Li, Sheng-Fan Wang, Emerson S. Bernardes and Fu-Tong Liu

Abstract Galectins are differentially expressed in a variety of cell types, including immune cells, and characterized by the affinity for β-galactoside–containing glycans. There are fifteen galectin members in mammals. Galectins are primarily located intracellularly, but can be secreted outside the cells. They exhibit pivotal roles during microbial infection, such as pathogen recognition and innate and adaptive immunity, and this review aims to discuss the functions of endogenous galectins during infection by four main types of microbes (bacteria, fungi, viruses, and parasites). Extracellular galectins are known to exert a bacteriostatic effect on some bacteria via association with bacterial glycans, whereas cytosolic galectins are recognized to control antibacterial autophagy by binding to luminal host glycans of ruptured endo-lysosomes. With regard to fungal infections, most studies deal with galectin-3. Galectin-3 modulates fungal burdens, the adaptive immune responses, and mortality in fungi-infected mice, which has been shown to be associated with its ability to manipulate fungicidal functions in neutrophils and cytokine expression in dendritic cells. Some viral infections, such as human immunodeficiency virus (HIV) and influenza virus infections, can be regulated by galectin-1 and -3, and they affect various aspects of viral infections, including viral binding, replication, budding, transmission, and infection-associated inflammation. Functions of galectins during a number of different parasitic infections have been identified in studies using galectin-knockout mice. Different parasitic infections have consistently demonstrated a role of galectins in tuning T helper immune responses in infected hosts.

F.-Y. Li · F.-T. Liu (✉)
Institute of Biomedical Sciences, Academia Sinica, 128 Academia Road, Section 2, Taipei 11529, Taiwan
e-mail: ftliu@ibms.sinica.edu.tw

S.-F. Wang
Department of Medical Laboratory Science and Biotechnology, Kaohsiung Medical University, Kaohsiung, Taiwan
e-mail: wasf1234@kmu.edu.tw

E. S. Bernardes
Nuclear and Energy Research Institute - IPEN, São Paulo, Brazil
e-mail: ebernardes@ipen.br

S.-L. Hsieh (ed.), *Lectin in Host Defense Against Microbial Infections*,
Advances in Experimental Medicine and Biology 1204,
https://doi.org/10.1007/978-981-15-1580-4_6

Keywords Galectin · Bacterium · Virus · Fungus · Parasite · Immune response

6.1 Introduction

Galectins, β-galactoside-binding lectins, are found in a variety of cells and tissues, including immune cells and organs (Liu and Rabinovich 2005; Kasai and Hirabayashi 1996; Barondes et al. 1994). The basic structure of a galectin includes a carbohydrate recognition domain (CRD; about 130 amino acids) that interacts with β-galactosides. Presently, 15 galectins have been identified in mammals (Vasta 2012). They are nucleocytoplasmic proteins synthesized without a classical signal sequence, but can be secreted via yet unidentified pathway(s). Several studies have focused on the extracellular functions of these proteins. Most of them utilized recombinant galectins added to single cell type systems and concluded that galectins modulate the cellular functions by binding to certain cell surface glycoconjugates. In vivo evidence is lacking to firmly establish this mode of action. Administration of recombinant galectins into experimental animals has also been utilized to corroborate certain extracellular functions; however, galectins do not possess specific receptors and may induce various cellular responses in different cell types by binding to an array of cell-surface glycoconjugates. Therefore, presumably, the observed functions of exogenously delivered galectins may differ from those of endogenous galectins in vivo, especially intracellular ones.

A fair number of studies have been reported that utilize the cells with given galectins knocked down by siRNA or knocked out, such as those isolated from genetically engineered mice. These studies addressed the functions of endogenous galectins and many were correlated with the phenotypes associated with knocking out of selected galectins, specifically in mice. These approaches often concluded that these proteins function intracellularly; however, eventually, this mode of action requires in vivo validation as well.

Through these various approaches, numerous galectins have been linked to host–pathogen interactions. These include direct binding to the microbes, which affects their survival or functions, modulating the innate or adaptive responses of various immune cells against microbes either via extra- or intracellular mode of action. In this review, we elaborate on the previously reported studies addressing the functions of endogenous galectins.

6.2 Bacterial Infections

Numerous different bacterial species are known to modulate galectin expression in the mammalian cells. Several studies have revealed a direct antimicrobial effect of galectins on the bacteria. Galectin-3 binds to *Helicobacter pylori* via recognition of LPS O-antigen (Fowler et al. 2006). The binding results in the shape change of *H.*

pylori from spiral to coccoid form, and subsequently, leads to bacterial cell death (Park et al. 2016). Both galectin-4 and galectin-8 recognize blood group B antigen (BGB) on *Escherichia coli* via their C-terminal domains, and both are effective in elimination of BGB-expressing *E. coli* (Stowell et al. 2010). Similarly, recombinant galectin-3 suppresses the growth of *Streptococcus pneumoniae* in vitro (Farnworth et al. 2008). Compared with wild-type mice, galectin-3 knockout mice develop more severe pneumococcal pneumonia. Administration of recombinant galectin-3 into galectin-3 knockout mice reduces *S. pneumoniae* cell count in blood and alleviates the degree of infection-induced sepsis, confirming a protective role of extracellular galectin-3 against *S. pneumoniae* infection in vivo.

In the aforementioned studies, recombinant galectins were utilized, and in some of them, employed at micromolar concentrations. It is uncertain whether such high levels of galectins are present under physiological or pathological circumstances. Additional studies would be necessary to establish the mode of interaction between galectins and pathogens.

6.2.1 Intracellular Galectins Regulate the Host Autophagic Response Against Bacterial Invasion

Intracellular Pathogens

Intriguingly, galectins are presently well-recognized as markers for vacuole lysis induced by intracellular bacteria. After internalization into the host cells, some invasive bacteria, such as *Salmonella* Typhimurium and *Listeria monocytogenes*, may disrupt the endocytic vesicles surrounding them, which renders the luminal side of these vesicles, including glycans initially displayed on the cell surface, exposed to the cytosol. Intracellular galectins, including galectin-3, galectin-8, and galectin-9, were found to be accumulated in the vicinity of internalized bacteria, and this accumulation phenomenon requires the presence of host mature glycans (Paz et al. 2010; Thurston et al. 2012). This suggests that the destruction of endocytic vesicles by intracellular bacteria allows the exposure of luminal host glycans to cytosolic galectins for recognition, thus leading to their accumulation at the sites of intracellular bacteria in host cells during an infection.

As reported by Thurston et al., the recruitment of galectin-8 to intracellular bacteria in the host cells activates antibacterial autophagy (Thurston et al. 2012). In this study, it was demonstrated that galectin-8 can interact with NDP52, an autophagy adaptor. Upon detection of host glycans displayed on damaged phagosomes that initially contain *Salmonella* Typhimurium, galectin-8 promotes autophagic clearance of bacteria-containing vesicles by recruiting NDP52 and its downstream autophagy machinery. Notably, the recruitment of NDP52 to intracellular bacteria is also governed by another bacteria-coating signal, ubiquitin. NDP52, which can bind to both galectin-8 and ubiquitin, was demonstrated to be recruited toward intravacuolar bacteria in two phases. Galectin-8-mediated NDP52 recruitment is early, but transient, whereas ubiquitin-mediated NDP52 recruitment occurs later, but is persistent. Therefore, either of the two "eat-me signals," galectin-8 and ubiquitin, on invading bacteria was revealed to suffice to target intracellular bacteria for autophagic elimination (Thurston et al. 2016).

A number of other studies have demonstrated that after sensing the intravacuolar rupture, galectin-8 binds to several other autophagy-related proteins for autophagic activation. In Group A *Streptococcus* (GAS)-infected cells, galectin-8 was found to interact with parkin, an E3 ubiquitin ligase, and mediate its recruitment to intracellular GAS (Cheng et al. 2017). Depletion of galectin-8 or parkin in the host cells suppresses the coating of ubiquitin on GAS and enhances intracellular GAS replication. This suggests that galectin-8 can also trigger antibacterial autophagy through parkin. mTOR, a serine/threonine kinase, is well-known as an essential autophagy regulator (Kim and Guan 2015). Upon lysosomal damage, mTOR is inhibited for autophagy activation, and this inhibition was found to be controlled by galectins. Galectin-8 and galectin-9 suppress the mTOR activity by interacting with the Ragulator-Rag SLC38A9 system and AMPK, respectively, which contributes to the autophagic elimination of lysosome-damaging intracellular bacteria, such as *Mycobacterium tuberculosis* (Jia et al. 2018).

Due to its implication in antibacterial autophagy, galectin-8 was also suggested to play a pivotal role in noncanonical inflammasome activation during bacterial invasion (Meunier et al. 2014). Cytoplasmic LPS was reported to trigger caspase-11-mediated noncanonical inflammasome activation (Hagar et al. 2013). Since autophagy may limit the release of LPS into the cytosol by sequestering the invading bacteria that have escaped from endocytic vesicles, galectin-8-mediated autophagy stimulation presumably results in a reduction in the intracellular LPS amount and thus its downstream caspase-11 activation.

In addition to galectin-8 and galectin-9, galectin-3 is also implicated in autophagy regulation during intracellular bacterial infection. Upon endomembrane injury, galectin-3 recognizes exposed β-galactoside glycoconjugates on damaged membranes and in turn associates with TRIM16 (Chauhan et al. 2016). This galectin-3-TRIM16 complex acts as a platform to recruit other autophagy regulators, including Beclin1, ULK1, and ATG16L1, for autophagy initiation, which is known to be critical for protecting the cells from *M. tuberculosis* invasion. Furthermore, IFN-inducible guanylate binding proteins (GBPs), which are crucial in host defense against intracellular bacteria, also interact with galectin-3 (Feeley et al. 2017). In a similar manner to its effect on TRIM16, galectin-3 directs GBP1 and GBP2 to the intracellular vacuoles damaged by *Yersinia pseudotuberculosis* or *Legionella pneumophila* and facilitates the targeting of these bacteria by ubiquitin and p62, an autophagy receptor.

In contrast to these reports, Cheng et al. (2017) reported a detrimental role of galectin-3 in the elimination of invading bacteria: galectin-3 decreases the recruitment of galectin-8 and ubiquitin to intracellular GAS, thus resulting in increased GAS replication in the host cells. This antagonistic function of galectin-3 on galectin-8 was also demonstrated in our recent study. We observed that galectin-3 recruited to damaged *L. monocytogenes*-containing phagosomes dampens antibacterial autophagy in a host *N*-glycan-dependent manner (Weng et al. 2018). Therefore, although both galectin-3 and galectin-8 are accumulated on damaged endocytic vesicles, they may exert contrasting effects on antibacterial autophagy under certain circumstances.

Interestingly, galectin-3 and galectin-8 exhibit a distinct binding affinity for sialic acid-containing glycans. Galectin-3 has only one CRD, and its binding to β-galactoside-containing glycoconjugates is blocked by terminal alpha2,6-sialylation (Zhuo and Bellis 2011). In contrast, galectin-8 comprises two CRDs, and its N-terminal CRD exhibits a strong affinity for sialylated galactosides (Carlsson et al. 2007). Thus, an alteration in the extent of terminal sialylation on host glycans may result in a differential recruitment of galectin-3 and galectin-8 to the intracellular bacteria. This notion has been proven in our study revealing that removal of terminal sialic acids on host glycans by sialidase leads to increased and decreased accumulation of galectin-3 and galectin-8 on *L. monocytogenes*-damaged phagosomes, respectively, which presumably contributes to attenuated antibacterial autophagy, and thus, augmented replication of intracellular bacteria (Weng et al. 2018).

Lysosomal Damage Caused by Bacterial Toxins

When secreted extracellularly, several bacterial pore-forming toxins are known to trigger lysosomal membrane permeabilization (LMP), including listeriolysin O from *L. monocytogenes*, perfringolysin O from *Clostridium perfringens*, and pneumolysin from *S. pneumoniae* (Malet et al. 2017). *H. pylori* vacuolating cytotoxin A (VacA) is another secreted bacterial toxin with a pore-forming activity and was suspected to disrupt the endocytic vesicles (Palframan et al. 2012). In our recent study, we suggested a role of *H. pylori* in promoting lysosomal damage in the gastric epithelial cells as we observed that intracellular galectin-8 aggregation is augmented after infection, and that these aggregates are mainly colocalized with the lysosomes (Li et al. 2019). Furthermore, we found that VacA-deficient *H. pylori* provokes less galectin-8 aggregation than its wild-type counterpart, thus supporting the possibility that VacA induces lysosomal injury in *H. pylori*-infected cells.

6.2.2 Discussion

There are several layers of defense to initiate antibacterial autophagy against intracellular bacteria, including diacylglycerol localization to the bacteria-containing phagosomes, ubiquitination of cytosol-invading bacteria, and the recognition of intravacuolar damage by galectins. Galectin-3 depletion increased ubiquitination of intracellular bacteria, such as GAS (Cheng et al. 2017). This suggests that galectins may regulate the decoration of cytosolic bacteria by ubiquitin; however, whether an inverse regulation could occur is unknown. Similarly, whether galectins control the accumulation of diacylglycerol on invading bacteria or vice versa is undetermined. Thus, it may be worth clarifying whether diacylglycerol, ubiquitin, and galectins affect each other's recruitment to cytosolic bacteria during an infection.

Intracellular galectins regulate the antibacterial autophagy by modulating the recruitment of autophagy proteins to the sites of intracellular bacteria. In addition to this mode of action, other possible mechanisms were observed by which galectins manipulate autophagy activation. For instance, NADPH oxidases stimulate antibacterial autophagy by generating reactive oxygen species (ROS) (Huang et al. 2009). Galectin-3 is known to suppress ROS-dependent killing of *Candida albicans* by neutrophils (Wu et al. 2017) and dampen the phorbol myristate acetate-induced ROS generation in neutrophils after *Toxoplasma gondii* infection (Alves et al. 2010). Hence, it is possible that the functional effect of galectins on antibacterial autophagy involves the regulation of cellular ROS production.

There are 15 galectin family members in mammals, but the luminal host glycans on damaged endocytic vesicles are selectively recognized by galectin-1, galectin-3, galectin-8, and galectin-9. The reason for this difference remains unclear. In addition, whether these galectins can compete with each other for binding to intravacuolar host glycans is worth addressing. In our study of the effects of galectin-3 and galectin-8 in antibacterial autophagy in macrophages infected with *Listeria monocytogenes*, we found these two galectins do not compete with each other in binding to glycans displayed on phagosomes that are damaged by escaping *Listeria*. Additional investigations should provide more insights. The recruitment of galectin-3, galectin-8, and galectin-9 to bacteria-containing vesicles is known to regulate antibacterial autophagy, whereas the functional consequences of galectin-1 recruitment to the injured endocytic vesicles are poorly understood. Since various galectin inhibitors (such as TD139) were designed to target the glycan-binding pocket of galectins and may serve as promising drugs for treatment of some human diseases, including idiopathic pulmonary fibrosis, presumably, we should be aware whether these drugs hamper the sensing of vacuolar damage by galectins and in turn affect the autophagic clearance of invading bacteria.

6.3 Fungal Infections

6.3.1 Candida albicans

C. albicans is a commensal fungus that colonizes the skin and mucosal surfaces of the gut. Galectin-3 is known to exhibit several roles during *C. albicans* infection. In *C. albicans*-exposed mouse neutrophils, galectin-3 executes a function in attenuating ROS production by downregulating Syk activation (Wu et al. 2017). Thus, compared to wild-type mouse neutrophils, galectin-3 knockout counterparts exhibited increased ROS-dependent killing of *C. albicans*. In accordance with this detrimental effect of galectin-3 on fungal clearance, galectin-3 knockout mice with systemic candidiasis depicted lesser fungal load, renal pathology, and mortality.

On the other hand, galectin-3 can modulate cytokine production from *C. albicans*-exposed dendritic cells and in turn affect the T helper immune response in vivo. Our recent study revealed that infected galectin-3 knockout dendritic cells produce more Th17-axis cytokines, IL-6 and IL-23, than their infected wild-type counterparts, and that adoptive transfer of these infected knockout cells into wild-type mice induces stronger Th17 responses (Fermin Lee et al. 2013). Such higher Th17 responses may confer protection against *C. albicans* infection due to the observation that fungal load in mice receiving infected galectin-3 knockout dendritic cells is lower than that in mice receiving infected wild-type dendritic cells.

6.3.2 Histoplasma capsulatum

Histoplasma capsulatum is a dimorphic fungal pathogen that causes histoplasmosis. Similar to that observed in *C. albicans*-infected dendritic cells, galectin-3 depletion enhanced the IL-23 production in *H. capsulatum*-infected dendritic cells (Wu et al. 2013). In a systemic histoplasmosis mouse model, galectin-3 knockout mice presented lesser fungal load and higher Th17 responses than wild-type mice. Administration of an IL-17A-neutralizing antibody increased the extent of fungal load in galectin-3 knockout mice to a level comparable to that in wild-type mice. Hence, it was inferred that galectin-3 suppresses the ability of the host to clear *H. capsulatum* via negative regulation of IL-17A response.

6.3.3 Paracoccidioides brasiliensis

Paracoccidioides brasiliensis, a dimorphic fungus, is the causal agent of paracoccidioidomycosis. Different from those found in *C. albicans*- and *H. capsulatum*-infected mice, galectin-3 deficiency leads to increased fungal load and less survival in *P. brasiliensis*-infected mice (Ruas et al. 2009). This greater susceptibility to *P. brasiliensis* infection in galectin-3 knockout mice was suggested to be associated with an impaired delayed-type hypersensitivity (DTH) reaction and enhanced Th2 immune responses.

6.3.4 Discussion

Although galectin-3 affects the T helper cell responses by tuning cytokine production in dendritic cells during fungal infection, the detailed mechanism of this cytokine regulation remains unclear. The Notch signaling pathway plays a crucial role in T helper cell differentiation (Amsen et al. 2015), and its activation can be controlled by galectin-3. During *Leishmania major* infection, the absence of galectin-3 triggers Notch signaling activation and cytokine expression in dendritic cells and affects T helper polarization (Fermino et al. 2016). Thus, it may be worth examining whether galectin-3 modulates cytokine production from fungus-infected dendritic cells via the Notch signaling pathway in future studies.

In addition to bacterial infection, fungal infection can mediate the disruption of endocytic vesicles. *C. albicans* is a polymorphic fungus that can switch between yeast and hyphal forms (Berman 2006). Following phagocytosis, *C. albicans* yeast cells transit to the hyphal form within phagosomes, and the intraphagosomal hyphal growth has been found to stretch the phagosomal membrane and eventually cause the rupture of phagosomes (Westman et al. 2018). Since such endomembrane damage is presumably recognized by galectins, it will be interesting to address whether cytosolic galectins are recruited toward intracellular *C. albicans* hyphae via binding to glycans on the remnant of ruptured phagosomes around the fungus, thus affecting fungal growth via autophagy regulation.

6.4 Viral Infections

Information regarding the role of galectins in viral infection, replication, and dissemination is limited. The reported studies mainly focused on the effects of galectins on the extracellular aspects, which are mediated via their CRD interaction with glycans on the envelope or capsid proteins of the virions or host cell surfaces. Several effects were demonstrated by the use of recombinant proteins added in vitro to viruses or to mammalian cells in a single cell type system. The pitfalls of this approach are mentioned in the Introduction section. We need to consider the real effects in the in vivo environment. Functions of endogenous galectins have gradually gained attention and the findings suggested that they can exert regulatory effects by functioning intracellularly, where they interact with other intracellular proteins via protein–protein interaction rather than protein–glycan interaction. Various reports indicated that galectin-1 and galectin-3 present either positive or negative regulation on virus infections, whereas galectin-9 normally displays negative regulatory effects, suggesting antagonizing effects or therapeutic applications of these galectins in the viral infections. In this section, we focus on studies in which the role of given galectins was investigated by using systems where galectins were knocked down or knocked out, thus aiming to address the functions of endogenous galectins. Several studies have reported the modulation of galectin expression by viruses; however, these are not discussed in this review.

6.4.1 DNA Virus Infection

6.4.1.1 Minute Virus

The minute virus of mice (MVM), comprising a single-stranded DNA genome, is the prototype virus of the *Protoparvovirus* genus within the *Parvoviridae* family. Infection of MVM can cause multiple organ damage in mice during fetal development or shortly after birth (Baker 1998; Lopez-Bueno et al. 2008). MVM is one of the most common viral pathogens in laboratory mice, occurring both in conventionally housed mice and those maintained in a strict barrier system. Parvoviruses remain a major threat to laboratory mouse colonies (Lopez-Bueno et al. 2008) and are monitored by extensive health surveillance programs. A study found that galectin-3 and Mgat5 (a glycosyltransferase associated the synthesis of oligosaccharides recognized by this galectin) are both necessary for efficient cell entry and infection by the MVM prototype strain (MVMp). Since MVM can be a contaminant of cell lines and transplantable tumors (Nicklas et al. 1993), the study also demonstrated that human cancer cells with higher expression of galectin-3 exhibited higher susceptibility to infection by MVMp than those with lower levels of galectin-3. The results indicate that galectin-3 and Mgat5 are linked to MVMp infection and that galectin-3 is a factor of MVMp oncotropism (Garcin et al. 2015).

6.4.1.2 Adenovirus

Adenoviruses are non-enveloped double-stranded DNA viruses belonging to the genus *Mastadenovirus*. These are majorly responsible for respiratory illness; however, they can also infect numerous human tissues and cause several other illnesses, including hepatitis, conjunctivitis, myocarditis, and gastroenteritis (Ghebremedhin 2014). Galectins are known to participate in cellular defense against bacterial infections via binding to glycans on damaged intracellular vacuoles. These vacuoles initially contain phagocytosed bacteria and are damaged as the bacteria escape from the vacuoles. This process can be followed by autophagic activation, as is in the case of galectin-8, leading to the formation of autophagosomes and bacterial destruction (Thurston et al. 2012); however, galectin-3 can suppress autophagic activation in these situations (Weng et al. 2018). It has recently been reported that adenoviruses were also able to damage the endosomal membrane in order to penetrate into the cytoplasm by its capsid PPxY motif via interacting with intracellular galectin-8. Similarly to bacterial infection, this further causes sequestration of the autophagic receptors NDP52 and p62 (Montespan et al. 2017).

6.4.2 RNA Virus Infection

6.4.2.1 Human Immunodeficiency Virus

Human immunodeficiency virus (HIV) belongs to the family *Reoviridae* and genus *Lentivirus*. HIV is classified into HIV-1 and HIV-2 based on the genetic characteristics and differences in the viral antigens. The major propagation route of HIV-1 is via cell-to-cell contact without the involvement of viruses freely present in the extracellular space during the transmission process (Bracq et al. 2018; Poli 2013). In recent years, a number of studies demonstrated the role of galectins in HIV-1 infection. Galectin-1 is abundantly expressed in the thymus and lymph nodes and secreted by various immune cells, such as activated CD8 T lymphocytes. Human recombinant galectin-1 facilitates HIV-1 infection of macrophages and CD4 T cells, via accelerating HIV-1 binding to the susceptible cells in a β-galactoside-dependent manner (Ouellet et al. 2005); however, as aforementioned, this approach might not address the function of endogenous galectin-1 in vivo. In addition, the exosomes derived from dendritic cells (DC) were found to contain galectin-3 and promote HIV-1 infection. The authors demonstrated that β-lactose (an inhibitor of galectin-3) blocked DC exosome-mediated HIV-1 infection of T-cells (Kulkarni and Prasad 2017), suggesting that galectin-3 in exosomes contributes to the effect of exosomes in a CRD-dependent manner. Nevertheless, exosomes are extracellular vesicles derived from multivesicular endosomes and contain cytosolic proteins, lipids, and nucleotides, in their lumen. Galectin-3 present in the lumen of DC exosomes may not be able to interact with cell surface glycoconjugates via its CRD. In contrast, exosome-associated galectin-3 may exert intracellular functions in the recipient cells, after the exosomes are taken up by these cells.

A study on HIV-infected macrophages (as the viral reservoir) revealed that galectin-3 expression in HIV-1-infected THP-1-derived macrophages induces cell death in a caspase-independent pathway. This suggests the role of galectin-3 in eradication of HIV-1 infected macrophages (Xue et al. 2017). This was also demonstrated with the use of a recombinant protein added exogenously to the cells. In contrast, our recent study revealed that endogenous galectin-3 promotes HIV-1 budding via stabilization of the apoptosis-linked gene 2 (ALG-2)-interacting protein, Alix (formerly known as AIP1), a protein that harbors binding sites for both ESCRT-I and ESCRT-III as well as $p6^{Gag}$ (Wang et al. 2014). Our unpublished data also suggest that endogenous galectin-3 is involved in regulation of virological synapse formation as well as mediating cell-to-cell transmission of HIV-1. The contribution of galectin-3 in HIV-1 infection is further strengthened by the finding that HIV-1 infection of CD4 T lymphocytes induced galectin-3 expression; this occurs through the HIV-1 transcription factor that binds to the galectin-3 promoter (Fogel et al. 1999).

6.4.2.2 Enterovirus

Enteroviruses are positive-sense single-stranded RNA viruses linked with various diseases in mammals, including humans. While there are many serotypes of this virus, enterovirus type 71 (EV71) is more widely studied and known to be one of the major causative agents of hand-foot-and-mouth disease. These may also cause aseptic meningitis, brainstem encephalitis (BE), and acute flaccid paralysis that is indistinguishable from poliomyelitis (Wang et al. 2003), and may be associated with a high mortality rate (26%) (Wang et al. 1999). Infection of EV71 begins with the attachment of the virion to the surface of the target cells. In a model cell system, galectin-3 was found to positively regulate EV71 replication, as the viral load was significantly reduced when galectin-3 was knocked out in the host cells (Huang et al. 2016).

6.4.2.3 Influenza Virus

Influenza viruses belong to the family *Orthomyxoviridae*. They comprise an envelope and negative-sense single-strand RNA segments (Arbeitskreis Blut UBBK 2009). There are four major serotypes of influenza virus, including A, B, C, and D. Influenza A and B virus infections often cause severe illness in humans. Influenza A virus (IAV) has been reported to cause pandemics via reassortment of eight RNA segmented genomes, which are also defined as the antigenic shift. All subtypes of IAVs have been found in the avian population, whereas influenza A, B, and C are known to infect humans (Spackman 2008). Considering the role of galectins in influenza virus infection, a study indicated that treatment of galectin-1 induced influenza A H1N1-infected BEAS-2B (the human bronchial epithelial cells) to be arrested at the G0/G1 phase in vitro (Fang et al. 2014), suggesting that galectin-1 regulates influenza virus-infected cell via apoptosis. Another study found that galectin-1 was upregulated in the lungs of mice inoculated with influenza A/WSN/33 H1N1, indicating a positive correlation between galectin-1 levels and viral loads during the acute phase of viral infection in vivo (Yang et al. 2011). This study also found that cells treated with galectin-1 presented lower viral yields when exposed to influenza virus in vitro, suggesting that galectin-1 is able to ameliorate influenza virus pathogenesis (Yang et al. 2011). Another study reported that aloe-emodin treatment both helped the cell resist IAV infection and upregulated galectin-3 expression. The authors proposed that galectin-3 is responsible for the therapeutic effect of this agent, because they also showed that recombinant galectin-3 inhibited IAV replication in infected cells (Li et al. 2014). However, additional experiments are required to provide evidence that the inhibitory effects of this agent is indeed through its upregulating of galectin-3, and through the induced galectin-3 being secreted and exerting its effect extracellularly.

Highly pathogenic avian influenza A H5N1 virus is known to cause pneumonia and acute respiratory distress syndrome in humans, inducing an excessive inflammatory response contributing to develop severe disease and high mortality rates (Chen et al. 2018). Recently, we demonstrated that endogenous galectin-3 enhanced the effects of H5N1 infection by promoting host inflammatory responses in the mouse lungs, which was due to upregulation of IL-1β production and activation of NLRP3 inflammasome (Chen et al. 2018). Severe inflammation can cause cellular and tissue damages and contribute to morbidity and mortality associated with influenza virus infection. Therefore, endogenous galectin-3 might contribute to the high mortality rate caused by H5N1.

6.4.3 Discussion

Viral infection is one of the most serious microbial infections since most viruses hijack host cell components to assist their life cycle including replication, budding and transmission, subsequently leading to damage or kill the cells, which cause illness. Another reason for viral infections being serious is that treatment for most of the viral infection can only help alleviate the general symptoms while the host immune system is activated to fight the virus. Therefore, the immune system is a key component for the host to against virus infection. In addition, some host factors have been known to be generated during virus infections and these factors may show inhibitory or enhancing capabilities via their intracellular or extracellular functions.

For instance, endogenous galectins as cellular factors are known to regulate the immune responses as well as viral infections. Despite a number of studies done on galectins effects on viral infection, most of them emphasize on extracellular functions of galectins on viral infection using treatment with recombinant galectins. Nevertheless, the intracellular function of galectins has gradually drawn attention in the virological field. The effects of endogenous galectins on regulating viral infection or dissemination might act indirectly interacting with cellular components which are important factors that are hijacked by virus or directly on virus replication related components. As described above, in a number of cases, endogenous galectins have been shown to help the virus, for example galectin-3 on MVMp, HIV-1, and EV71 infections. In addition, galectin-3 was reported to trigger NLRP3 inflammasome to induce severe inflammation in mice lung after receiving H5N1 avian influenza infection (Zhuo and Bellis 2011). Thus, galectin-3 may regulate immune response and subsequently affect virus infection.

Overall, galectins may have big roles when it comes to viral infection, due to the potential to help or counteract viral infection. Currently, there is still a great deal to be studied regarding the roles of galectin in virus binding, replication, budding and transmission. This will ultimately require studies in the in vivo settings, including the use of animal models. Continued studies in this direction promise to result in pharmaceutical agents that are useful in treatment of viral infections by targeting galectins.

6.5 Parasitic Infections

It has been long known that both unicellular and multicellular parasites, display on their surface and/or shed into the host circulation, glycans structures similar to those recognized by host galectins. In accordance with this, galectins interact with parasite glycoconjugates and play a pivotal role in pathogen recognition, host cell invasion, and parasite escape. More importantly, studies conducted with galectin-deficient mice using different models of parasitic infections have revealed a role of endogenous galectins in modifying the innate and adaptive immune responses against parasites.

The parasitic infections are mainly caused by protozoans (unicellular eukaryotic organisms) and helminths (multicellular worms). The major classes of protozoan parasites (*Plasmodium*, *Entamoeba histolytica*, *Leishmania*, *Trypanosoma*, and *Toxoplasma*) and the three classes of helminths (nematodes, trematodes, and cestodes) are still a leading cause of morbidity and mortality in the developing countries. Therefore, research on the role of galectins in establishing a host–parasite interaction can provide novel insights into the molecular pathology of these parasites and also generate novel therapeutic targets to control the parasitic infections. Even though parasites possess galectin-like molecules capable of recognizing the host glycome, parasite glycan-binding proteins are not included in this chapter.

6.5.1 Helminths: Multicellular Parasites

6.5.1.1 Schistosoma

Schistosomiasis is a major parasitic disease caused by parasitic trematodes. Humans are affected by three *Schistosoma* species (*Schistosoma japonicum*, *Schistosoma mansoni*, and *Schistosoma haematobium*) and the disease is mainly characterized by chronic hepatic damage due to the formation of granulomas around tissue-trapped eggs. It has been previously reported that schistosome glycoconjugates containing terminal GalNAcβ1–4GlcNAc (LacdiNAc (LDN)) glycans or their fucosylated derivatives containing GalNAcβ1–4(Fucα1–3)GlcNAc (LDNF) are abundant on the surfaces of worms and eggs (van Remoortere et al. 2000). These glycans are highly antigenic and can be also found in the circulation of the infected host. Indeed, it was demonstrated that inert beads coated with terminal LacdiNAc or Galβ1-4GlcNAc (LacNAc, LN) can trigger granulomatous lesions similar to the ones caused by *Schistosoma* eggs during chronic schistosomiasis (Van de Vijver et al. 2006). Galectin-3 but not galectin-1 was identified as a major LDN-binding protein in immune cells, *Schistosoma* eggs and in liver granulomas of *S. mansoni*-infected hamsters (van den Berg et al. 2004). Soon after these discoveries, numerous experimental studies using *S. mansoni*-infected galectin-3-knockout mice compared with infected wild-type mice revealed that, in the absence of galectin-3, the granulomas were smaller in diameter and displayed remarkable changes in the monocyte/macrophage, eosinophil, and B lymphocyte subpopulations (Oliveira et al. 2007). Similar results independently reported by other laboratories revealed that *S. mansoni*-infected galectin-3-knockout mice mounted a biased cellular and humoral Th1 response and developed a reduced number of liver granulomas both in the acute and chronic phases of experimental infection. This indicates that galectin-3 can modulate the immune/inflammatory responses during *S. mansoni* infection (Breuilh et al. 2007). In addition, during the course of *S. mansoni* infection, galectin-3-knockout mice displayed an increase in plasma cell numbers in the bone marrow, spleen, and mesenteric lymph nodes and atypical peritoneal IgM +/IgA+ B1a and B1b lymphocytes compared to galectin-3-knockout mice (Oliveira et al. 2016).

Interestingly, besides a direct role for galectin-3 in the inflammatory/immune response mounted against the parasite, the above-mentioned studies also indicated a role for galectin-3 in B cell differentiation/activation, which was further confirmed in conditions not related to microbial infection (de Oliveira et al. 2018). Despite some controversies found in the literature regarding the role of galectin-3 in the Schistosomiasis (Bickle and Helmby 2007), undoubtedly, the host-like glycans expressed/secreted by this parasite may be recognized by endogenous galectin-3. Nevertheless, to date it has been difficult to quantify and verify the extent to which the interaction of extracellular galectin-3 with these glycans accounts for all the phenotypes reported in *S. mansoni*-infected galectin-3-knockout mice. Indeed, similar immune response signature has also been reported in other infection models (Fermino et al. 2016; da Silva et al. 2017; Fermino et al. 2013; Bernardes et al. 2006), in which the recognition of glycans by extracellular galectin-3 is irrelevant. This indicates that these phenotypes are directly linked to the intracellular functions of galectin-3 rather than an extracellular role.

6.5.1.2 *Leishmania* sp.

Leishmaniasis is an infection caused by an intracellular parasite of the genus *Leishmania* and is transmitted by the bite of a sandfly (genera *Phlebotomus* and *Lutzomyia*), which is endemic to tropical and subtropical regions. Clinical manifestation of Leishmaniasis can be classified into four types: cutaneous, diffuse cutaneous, monocutaneous, and visceral leishmaniasis (Kevric et al. 2015). The disease is driven by a classic T-helper (Th) cell type response, resulting in expanded Th1, Th2, and T regulatory (Treg) cell populations. While Th1 $CD4^+$ response leads to protozoal killing, the predominance of Th2 $CD4^+$ response results in persistence and spread of infection. Treg cells, in contrast, modulate the immune response against the parasite and limit the damage to the host (Kevric et al. 2015).

The roles of galectins in innate and adaptive immune responses and host–pathogen interactions during Leishmania infection have been investigated over the past 15 years. Pelletier and Sato (Pelletier and Sato 2002) first demonstrated that galectin-3 may play a pivotal role in the infection by *Leishmania major*. These researchers reported that *L. major* surface lipophosphoglycans interact with extracellular galectin-3 produced by the immune cells at the site of infection. After binding, galectin-3 becomes susceptible to the cleavage by zinc metalloproteases, resulting in truncated galectin-3 lacking the N-terminal domain implicated in lectin oligomerization and not being able to exert immunomodulatory functions. The same group also demonstrated that galectin-3 secreted at the infection site stimulates the migration of neutrophils, which helps restrict the parasite spread during acute *L. major* infection (Bhaumik et al. 2013). These results were corroborated in the work by Fermino et al. (Fermino et al. 2013), who revealed that *L. major*-infected galectin-3-knockout mice displayed increased susceptibility to *L. major* infection and higher parasite load when compared with wild-type mice. In addition, Fermino et al. (2016) also demonstrated that endogenous galectin-3 may modulate the Th1/Th2 immune response, and consequently provide immunity against *L. major*, by interacting with the Notch signaling components. These authors found that the absence of galectin-3 in dendritic cells leads to increased Jagged-1/Notch pathway signaling and proinflammatory cytokines, which could be reverted by the addition of exogenous galectin-3 (Fermino et al. 2016). Corroborating these in vitro findings, the draining lymph nodes and footpad lesion cells from *L. major*-infected galectin-3-knockout mice also presented enhanced levels of Notch signaling components, which contribute to the augmented and mixed Th1/Th2 responses observed in these mice (Fermino et al. 2016).

6.5.1.3 *Toxoplasma gondii*

Toxoplasmosis is caused by a protozoan *T. gondii*, an obligate intracellular parasite capable of infecting mammals and birds (Calero-Bernal and Gennari 2019). Experimental infection of mice with *T. gondii* provides a model to study mechanisms of T helper 1 cell-mediated response to the intracellular infections. In general, resistance to *T. gondii* infection has been associated with the establishment of a polarized Th1 response, characterized by the high production of IFN-γ directed by IL-12 from dendritic cells. More recently, it has been reported that IFN-γ produced by cells involved in innate immunity (NK cells and neutrophils) is a crucial determinant of parasite resistance (Sturge et al. 2013).

Studies using galectin-3-knockout mice provided evidence that galectin-3 is crucial in the experimental infection by *T. gondii* (Bernardes et al. 2006). *T. gondii*-infected galectin-3-knockout mice displayed a heightened Th1 response driven by increased amounts of IL-12 produced by dendritic cells, suggesting that endogenous galectin-3 acts as a negative modulator of IL-12 production by dendritic cells. As a consequence, this regulation influences the assembly of the adaptive immune response against this parasite (Bernardes et al. 2006). Further, it was demonstrated that activation of Notch signaling in bone marrow dendritic cells (BMDCs) upon stimulation with Jagged1 was more pronounced in galectin-3-knockout BMDCs, leading to an increased cytokine production (Fermino et al. 2016). Interestingly, the surface plasmon resonance assay revealed that galectin-3 binds to Jagged1 and reduces the turnover of this protein in endothelial cells (Dos Santos et al. 2017). Thus, it has been proposed that cell surface galectin-3 binds to Jagged1 in BMDCs, thus controlling the activation threshold of this cell type, although an intracellular role of galectin-3 cannot be excluded.

The expression of galectin-1 and galectin-3 was also upregulated in both genetically susceptible (C57BL/6) and resistant (Balb/c) mice after infection with *T. gondii*, whereas galectin-9 expression was increased only in the susceptible mice (Chen et al. 2017). Despite this differential expression, the blockade of potential galectin-binding-partners on the cell surface using lactose (galectins' inhibitor) was not enough to change the ocular pathology of resistant mice (Chen et al. 2017). The results speak against an exclusive cell surface role of galectin-3 and point out that intracellular galectin-3 might account for several functions attributed to this protein.

Currently, it is also known that galectin-3 binds to the surface glycosylphosphatidylinositols (GPIs) of *T. gondii*. GPI–galectin-3 interaction, collectively with Toll-like receptors (TLR2 and TLR4), is important for *T. gondii* recognition by macrophages (Debierre-Grockiego et al. 2010) and also stimulates the production of matrix metalloproteinase-9 by macrophages, which in turn degrades the extracellular galectin-3 (Debierre-Grockiego et al. 2010). Similar to what has been demonstrated in other experimental infection models, endogenous galectin-3 has been also associated with enhanced recruitment and activation of neutrophils during *T. gondii* infection (Alves et al. 2010, 2013). These data reinforce a role for endogenous galectins in orchestrating innate immunity, which does not necessarily depend on whether galectin binds to glycans on the parasite surface.

6.5.1.4 *Plasmodium* sp.

Malaria is a disease caused by a protozoan belonging to the *Plasmodium* genus and is transmitted by the bite of Anopheles female mosquitoes infected with the parasite. The severe form of malaria is characterized by cerebral malaria, severe anemia, hypoglycemia, pulmonary edema, acute kidney injury, and jaundice (White 2018).

Galectins, in particular galectin-3 and galectin-9, have been found to display varying roles in the experimental malaria models; however, the precise mechanism behind these functions remains unclear. Lack of galectin-3 in susceptible mice (C57Bl/6) resulted in partial but significant protection from experimental cerebral malaria by *Plasmodium berghei*, associated with a minor increase in parasitemia in these mice (Oakley et al. 2009). Different results were reported by Toscano et al. (Toscano et al. 2012), where the absence of galectin-3 in C57BL/6 mice did not change the systemic parasitemia of mice infected with *P. berghei*. In fact, the roles of galectin-3 in experimental malaria seem to be species-specific, since the lack of this lectin decreased parasitemia during *P. yoelii* infection, but did not change the course of infection caused by *P. berghei* and *P. chabaudi* (Toscano et al. 2012).

A potential role of galectin-9 during experimental malaria has also been proposed. Galectin-9 expression was significantly upregulated throughout the infection, in the lung, liver, and spleen, after *P. berghei* infection. In the meantime, there was some upregulation of galectin-3 and galectin-8 at the beginning of the infection, but no change was observed in galectin-1 expression (Xiao et al. 2016; Liu et al. 2016). In an independent study, it was demonstrated that patients with acute *P. falciparum* malaria comprised high concentrations of serum galectin-9, which was associated with disease severity (Dembele et al. 2016).

Despite the apparent increase in the expression of galectin-9 throughout the experimental infection by *P. berghei*, specific mechanisms of action in the pathophysiology of malaria have not been proposed or demonstrated in these studies. In another study, α-lactose (a galectin inhibitor) was injected into the *P. berghei*-infected mice and this treatment resulted in increased parasitemia and mortality (Liu et al. 2016). Based on these results, the authors concluded that galectin-9 (and its binding partners) plays a major protective role in the immunopathology caused by *P. berghei*. However, notably, α-lactose is not specific for galectin-9 and can bind to other galectins, and might have effects on cellular responses not associated with galectins. Essentially, most of the functions attributed to galectin-9 are associated with the interaction with its extracellular binding partners, Tim-3. However, galectin-9 as well all the other members of the galectin family, are crucial role in controlling immune response and inflammatory tissue damage by acting intracellularly, which has been overlooked in some of these studies.

6.5.1.5 *Trypanosoma cruzi*

The hemoflagellate protozoan *Trypanosoma cruzi* is the agent of Chagas disease, a parasitic vector-borne disease transmitted by arthropods from Triatominae subfamily. Chagas disease is the major cause of infectious myocarditis and affects numerous people in Central and Latin America (Cunha-Neto and Chevillard 2014; Duran-Rehbein et al. 2014). Trypomastigote and amastigote forms of *T. cruzi* coexist in the infected mammals. While a small amount of trypomastigote forms circulate through the bloodstream, enabling its spread and transmission, the amastigote forms remain in the cytoplasm of a wide variety of cell types, where they replicate and cause tissue damage (Tarleton 2015).

The initial studies aimed at identifying *T. cruzi* glycans were initiated more than four decades ago (Alves and Colli 1975) and it took a few years until Araújo Jorge and de Souza (de Araujo Jorge and de Souza 1984) first demonstrated that neuraminidase treatment increased the uptake of epi- and trypomastigote forms by macrophages and this effect could be blocked by galactose. Notwithstanding, only 16 years later Moody et al. (Moody et al. 2000) demonstrated unequivocally for the first time that exogenous galectin-3 could bind to *T. cruzi* surface glycans and promote the binding of this parasite to matrix glycoproteins. Later, the same authors extended their findings and reported that endogenous galectin-3 expression is required for *T. cruzi* adhesion to human cells, allowing the parasite entry (Kleshchenko et al. 2004). In addition, endogenous galectin-3 accumulates around the amastigotes, thus contributing to the process of invasion and intracellular trafficking, by accumulating around the parasites shortly after they escape the parasitophore vacuoles inside the cell (Machado et al. 2014; Reignault et al. 2014). Nonetheless, despite the fact that studies with *T. cruzi*-infected galectin-3-knockout mice corroborated the role of galectin-3 in fine-tuning the innate and adaptive immune response (da Silva et al. 2017; Pineda et al. 2015), they did not support the role of galectin-3 in *T. cruzi* adhesion and invasion. Indeed, *T. cruzi* replicated even more in the galectin-3-knockout than in wild-type mice. These results may indicate at least two points of view: (1) galectins are redundant and these findings point to a possible role of other galectin family members in *T. cruzi* adhesion and invasion; (2) in in vivo studies using Lgals3$^{-/-}$ mice, the intracellular functions of galectin-3 in all cell-types combined overcome the extracellular role in some particular cell type, which makes it difficult to confirm certain expected phenotypes in vivo. In contrast, Acosta-Rodriguez et al. (Acosta-Rodriguez et al. 2004) blocked the expression of galectin-3 by using antisense nucleotides and demonstrated that B cell differentiation toward a plasma cell phenotype was favored. Furthermore, the role of galectin-3 in favoring a memory cell phenotype was also confirmed by the use of galectin-3-knockout mice (Oliveira et al. 2009).

Besides galectin-3, several other members of the galectin's family have been found to bind to surface proteins of *T. cruzi* parasite, including galectin-1, galectin-4, galectin-7, galectin-8, and galectin-14 (Pineda et al. 2015). It has been reported that galectin-1 is upregulated in the cardiac tissue from patients with Chagas disease (Giordanengo et al. 2001), B cells from *T. cruzi*-infected mice (Zuniga et al. 2001), *T. cruzi*-infected macrophages (both from in vitro and in vivo infection) (Zuniga et al. 2001), and cardiac cells after in vitro infection (Benatar et al. 2015). Although in vitro studies demonstrated that exogenously added galectin-1 can inhibit replication of the parasite inside macrophages and cardiac cell line (Zuniga et al. 2001; Benatar et al. 2015), *T-cruzi*-infected galectin-1-knockout mice presented lower mortality and lower cardiac parasite load (Poncini et al. 2015). In contrast, the in vitro studies highlighted that galectin-1 was capable of inducing apoptosis in T cells (Perillo et al. 1995; Toscano et al. 2007). Corroborating these in vitro data, *T-cruzi*-infected galectin-1 knockout mice presented an increased number of CD8(+) T cells and higher frequency of IFN-γ-producing CD4(+) T cells in the muscle tissues and draining lymph nodes (Poncini et al. 2015), indicating that at least in vivo the main role of endogenous galectin-1 is inducing death of the activated T cells. Although different galectin-family members can interact in vitro with glycans on the surface of *T. cruzi* and control parasite invasion and replication, the data obtained using galectin-deficient mice have revealed that galectins may control the parasitic infections mainly by regulating the innate and adaptive immune responses.

6.6 Discussion

Most galectins are bivalent or multivalent and can mediate cell-cell and cell-extracellular matrix interaction, and therefore, it was hypothesized that these galectins could also mediate parasite–host cell and parasite–host extracellular matrix adhesion. Indeed, the initial studies on the role of galectins in the host–parasite relationship emphasized on this specific ability of few members of the galectin family. However, despite numerous studies reporting that different galectins could bind to a wide range of parasitic organisms and favor their binding and cell invasion, studies using galectin-deficient mice revealed that parasite binding and invasion were favored in the absence of specific galectins instead. After parasite invasion and considering that galectins can be found in the cytoplasm, it was also predictable that the cytosolic galectins could still interact with parasite glycans, affecting parasite escape from the parasitophorous vacuole and multiplication. Although it cannot be ruled out that cell surface and/or cytosolic galectins may interact with parasite glycans, these mechanisms do not seem essential for the parasite to establish successful infection.

In contrast, galectin-deficient mice revealed a role for galectins in modulating the innate and adaptive immune responses. As a result, it has been reported that galectins control the immune cell migration, activation, and differentiation, which directly or indirectly will determine the type, intensity, and duration of the inflammatory stage and parasite-specific immune response. In vitro studies have revealed that galectins can form structures called lattice by binding several glycan receptors on the cell surface. By doing so, galectins can influence the turnover rate of cell surface receptors, either by increasing or decreasing the activation threshold of immune cells. Indeed, it has been reported that immune cells from galectin-deficient mice display an increased or decreased expression of different types of receptors, which are associated with the development of a deficient or heightened immune response against a parasitic infection. In a similar manner, but acting intracellularly, it has also been demonstrated that galectins control the expression of cell surface receptors and modulate the activation of immune cells. Therefore, as far as host/parasite relationship is concerned, it seems that interaction of galectins with self-glycans/receptors, expressed either on the cell surface or intracellularly, play a pivotal role in determining the outcome of the parasitic infection, overshadowing the role of galectin-parasite glycan interaction.

6.7 Conclusions

The functions of various endogenous galectins in bacterial, viral, fungal, and parasitic infections are summarized in Fig. 6.1. Although in this review we focus on the functions of endogenous galectins, we also include some studies dealing with the effects of exogenously added galectins. In the latter, we added few comments on the possible pitfalls with the approaches. Regardless, a great deal of knowledge has been accumulated with regard to the galectin function in host–microbe interactions; however, it is evident a great deal remains to be investigated.

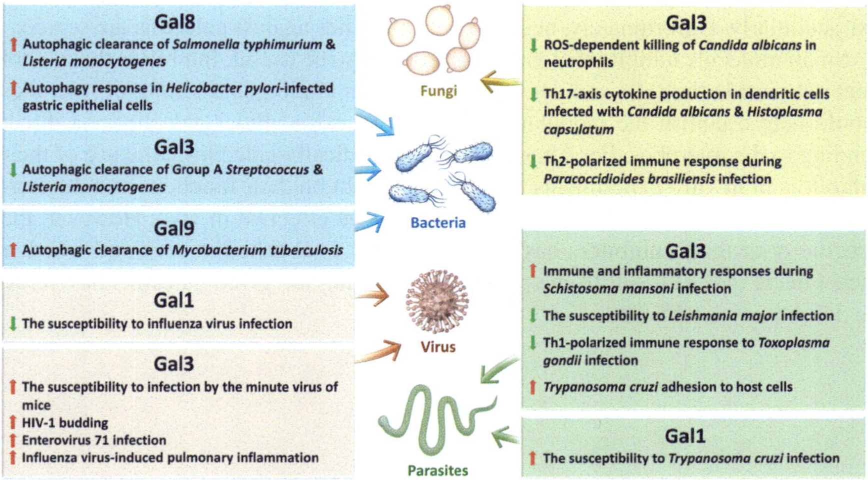

Fig. 6.1 Functions of endogenous galectins

Galectins can be found in the nucleus, cytoplasm, and are secreted via an uncharacterized mechanism. They can play a role in the host–microbe relationship by acting in three different ways:

(1) Extracellular/cell surface-bound galectin can interact with the glycans on the microbes and affect the survival of the microbes or promote microbe adhesion on and invasion of host cells
(2) Extracellular galectin can interact with the host glycans and control the response of the host cells infected by the microbes
(3) Intracellular galectin can interact with the intracellular host proteins and control the response of cells to microbial infection, in a fashion that may not be dependent on glycans.

In addition, recent reports have shown that intracellular/cytoplasmic galectin can interact with the microbial glycans. Moreover, an intriguing host–microbe interaction that is gaining significant attention is the galectin-mediated modulation of host cell responses by binding to glycans in the microbe-containing phagosomes. These glycans initially reside in the lumen but subsequently are exposed to the cytosol, when the phagosomes were damaged, as the microbes escape into the cytosol.

In relation to these mode of interactions, studies reporting the functions with the use of exogenously added galectins would inevitably conclude that galectins function extracellularly. In contrast, studies of endogenous galectins often led to the conclusion that galectins function intracellularly. Additional studies are clearly needed to firmly establish the mode actions.

Specific inhibitors targeting the individual galectins are needed to elucidate the mechanism of galectin-mediated host–microbe interaction. Neutralizing antibodies against various galectins would be useful to establish the mode of interaction between galectins and pathogens and revealing whether galectins function intracellularly or extracellularly. Unfortunately, neutralizing antibodies against galectins are scarce.

Small molecule inhibitors of galectins would also be useful. Inhibitors that do not penetrate inside the cells and yet suppress galectin binding and killing of microbes would help establish the modulatory function of extracellular galectins via direct binding to the microbes. For some galectins, specifically galectin-3, the use of their inhibitors in in vitro experiments has also shed light on their functions and, furthermore, some of these inhibitors were reported to be effective in vivo. However, the specificity of these inhibitors has not been firmly established in vivo and whether their effects indeed result from specifically targeting the given galectins has yet to be definitively established.

References

Acosta-Rodriguez EV, Montes CL, Motran CC, Zuniga EI, Liu FT, Rabinovich GA, Gruppi A (2004) Galectin-3 mediates IL-4-induced survival and differentiation of B cells: functional cross-talk and implications during *Trypanosoma cruzi* infection. J Immunol 172(1):493–502

Alves MJ, Colli W (1975) Glycoproteins from *Trypanosoma cruzi*: partial purification by gel chromatography. FEBS Lett 52(2):188–190

Alves CM, Silva DA, Azzolini AE, Marzocchi-Machado CM, Carvalho JV, Pajuaba AC, Lucisano-Valim YM, Chammas R, Liu FT, Roque-Barreira MC, Mineo JR (2010) Galectin-3 plays a modulatory role in the life span and activation of murine neutrophils during early *Toxoplasma gondii* infection. Immunobiology 215(6):475–485. https://doi.org/10.1016/j.imbio.2009.08.001

Alves CM, Silva DA, Azzolini AE, Marzocchi-Machado CM, Lucisano-Valim YM, Roque-Barreira MC, Mineo JR (2013) Galectin-3 is essential for reactive oxygen species production by peritoneal neutrophils from mice infected with a virulent strain of *Toxoplasma gondii*. Parasitology 140(2):210–219. https://doi.org/10.1017/S0031182012001473

Amsen D, Helbig C, Backer RA (2015) Notch in T cell differentiation: all things considered. Trends Immunol 36(12):802–814. https://doi.org/10.1016/j.it.2015.10.007

Arbeitskreis Blut UBBK (2009) Influenza virus. Transfus Med Hemotherapy 36(1):32–39. https://doi.org/10.1159/000197314

Baker DG (1998) Natural pathogens of laboratory mice, rats, and rabbits and their effects on research. Clin Microbiol Rev 11(2):231–266

Barondes SH, Cooper DN, Gitt MA, Leffler H (1994) Galectins. Structure and function of a large family of animal lectins. J Biol Chem 269 (33):20807–20810

Benatar AF, Garcia GA, Bua J, Cerliani JP, Postan M, Tasso LM, Scaglione J, Stupirski JC, Toscano MA, Rabinovich GA, Gomez KA (2015) Galectin-1 prevents infection and damage induced by *Trypanosoma cruzi* on cardiac cells. PLoS Negl Trop Dis 9(10):e0004148. https://doi.org/10.1371/journal.pntd.0004148

Berman J (2006) Morphogenesis and cell cycle progression in *Candida albicans*. Curr Opin Microbiol 9(6):595–601. https://doi.org/10.1016/j.mib.2006.10.007

Bernardes ES, Silva NM, Ruas LP, Mineo JR, Loyola AM, Hsu DK, Liu FT, Chammas R, Roque-Barreira MC (2006) *Toxoplasma gondii* infection reveals a novel regulatory role for galectin-3 in the interface of innate and adaptive immunity. Am J Pathol 168(6):1910–1920. https://doi.org/10.2353/ajpath.2006.050636

Bhaumik P, St-Pierre G, Milot V, St-Pierre C, Sato S (2013) Galectin-3 facilitates neutrophil recruitment as an innate immune response to a parasitic protozoa cutaneous infection. J Immunol 190(2):630–640. https://doi.org/10.4049/jimmunol.1103197

Bickle Q, Helmby H (2007) Lack of galectin-3 involvement in murine intestinal nematode and schistosome infection. Parasite Immunol 29(2):93–100. https://doi.org/10.1111/j.1365-3024.2006.00923.x

Bracq L, Xie M, Benichou S, Bouchet J (2018) Mechanisms for cell-to-cell transmission of HIV-1. Front Immunol 9:260. https://doi.org/10.3389/fimmu.2018.00260

Breuilh L, Vanhoutte F, Fontaine J, van Stijn CM, Tillie-Leblond I, Capron M, Faveeuw C, Jouault T, van Die I, Gosset P, Trottein F (2007) Galectin-3 modulates immune and inflammatory responses during helminthic infection: impact of galectin-3 deficiency on the functions of dendritic cells. Infect Immun 75(11):5148–5157. https://doi.org/10.1128/IAI.02006-06

Calero-Bernal R, Gennari SM (2019) Clinical toxoplasmosis in dogs and cats: an update. Front Vet Sci 6:54. https://doi.org/10.3389/fvets.2019.00054

Carlsson S, Oberg CT, Carlsson MC, Sundin A, Nilsson UJ, Smith D, Cummings RD, Almkvist J, Karlsson A, Leffler H (2007) Affinity of galectin-8 and its carbohydrate recognition domains for ligands in solution and at the cell surface. Glycobiology 17(6):663–676. https://doi.org/10.1093/glycob/cwm026

Chauhan S, Kumar S, Jain A, Ponpuak M, Mudd MH, Kimura T, Choi SW, Peters R, Mandell M, Bruun JA, Johansen T, Deretic V (2016) TRIMs and galectins globally cooperate and TRIM16 and galectin-3 co-direct autophagy in endomembrane damage homeostasis. Dev Cell 39(1):13–27. https://doi.org/10.1016/j.devcel.2016.08.003

Chen SJ, Zhang YX, Huang SG, Lu FL (2017) Galectins expressed differently in genetically susceptible C57BL/6 and resistant BALB/c mice during acute ocular *Toxoplasma gondii* infection. Parasitology 144(8):1064–1072. https://doi.org/10.1017/S0031182017000270

Chen YJ, Wang SF, Weng IC, Hong MH, Lo TH, Jan JT, Hsu LC, Chen HY, Liu FT (2018) Galectin-3 enhances Avian H5N1 influenza A virus-induced pulmonary inflammation by promoting NLRP3 inflammasome activation. Am J Pathol 188(4):1031–1042. https://doi.org/10.1016/j.ajpath.2017.12.014

Cheng YL, Wu YW, Kuo CF, Lu SL, Liu FT, Anderson R, Lin CF, Liu YL, Wang WY, Chen YD, Zheng PX, Wu JJ, Lin YS (2017) Galectin-3 inhibits galectin-8/parkin-mediated ubiquitination of group A streptococcus. MBio 8(4). https://doi.org/10.1128/mbio.00899-17

Cunha-Neto E, Chevillard C (2014) Chagas disease cardiomyopathy: immunopathology and genetics. Mediators Inflamm 2014:683230. https://doi.org/10.1155/2014/683230

da Silva AA, Teixeira TL, Teixeira SC, Machado FC, Dos Santos MA, Tomiosso TC, Tavares PCB, Brigido R, Martins FA, Silva NSL, Rodrigues CC, Roque-Barreira MC, Mortara RA, Lopes DS, Avila VMR, da Silva CV (2017) Galectin-3: a friend but not a foe during *Trypanosoma cruzi* experimental infection. Front Cell Infect Microbiol 7:463. https://doi.org/10.3389/fcimb.2017.00463

de Araujo Jorge TC, de Souza W (1984) Effect of carbohydrates, periodate and enzymes in the process of endocytosis of *Trypanosoma cruzi* by macrophages. Acta Trop 41(1):17–28

de Oliveira FL, Dos Santos SN, Ricon L, da Costa TP, Pereira JX, Brand C, Fermino ML, Chammas R, Bernardes ES, El-Cheikh MC (2018) Lack of galectin-3 modifies differentially Notch ligands in bone marrow and spleen stromal cells interfering with B cell differentiation. Sci Rep 8(1):3495. https://doi.org/10.1038/s41598-018-21409-7

Debierre-Grockiego F, Niehus S, Coddeville B, Elass E, Poirier F, Weingart R, Schmidt RR, Mazurier J, Guerardel Y, Schwarz RT (2010) Binding of *Toxoplasma gondii* glycosylphosphatidylinositols to galectin-3 is required for their recognition by macrophages. J Biol Chem 285(43):32744–32750. https://doi.org/10.1074/jbc.M110.137588

Dembele BP, Chagan-Yasutan H, Niki T, Ashino Y, Tangpukdee N, Shinichi E, Krudsood S, Kano S, Hattori T (2016) Plasma levels of Galectin-9 reflect disease severity in malaria infection. Malar J 15(1):403. https://doi.org/10.1186/s12936-016-1471-7

Dos Santos SN, Sheldon H, Pereira JX, Paluch C, Bridges EM, El-Cheikh MC, Harris AL, Bernardes ES (2017) Galectin-3 acts as an angiogenic switch to induce tumor angiogenesis via Jagged-1/Notch activation. Oncotarget 8(30):49484–49501. https://doi.org/10.18632/oncotarget.17718

Duran-Rehbein GA, Vargas-Zambrano JC, Cuellar A, Puerta CJ, Gonzalez JM (2014) Mammalian cellular culture models of *Trypanosoma cruzi* infection: a review of the published literature. Parasite 21:38. https://doi.org/10.1051/parasite/2014040

Fang S, Zhang K, Wang T, Wang X, Lu X, Peng B, Wu W, Zhang R, Chen S, Zhang R, Xue H, Yu M, Cheng J (2014) Primary study on the lesions and specific proteins in BEAS-2B cells induced with the 2009 A (H1N1) influenza virus. Appl Microbiol Biotechnol 98(23):9691–9701. https://doi.org/10.1007/s00253-014-5852-y

Farnworth SL, Henderson NC, Mackinnon AC, Atkinson KM, Wilkinson T, Dhaliwal K, Hayashi K, Simpson AJ, Rossi AG, Haslett C, Sethi T (2008) Galectin-3 reduces the severity of pneumococcal pneumonia by augmenting neutrophil function. Am J Pathol 172(2):395–405. https://doi.org/10.2353/ajpath.2008.070870

Feeley EM, Pilla-Moffett DM, Zwack EE, Piro AS, Finethy R, Kolb JP, Martinez J, Brodsky IE, Coers J (2017) Galectin-3 directs antimicrobial guanylate binding proteins to vacuoles furnished with bacterial secretion systems. Proc Natl Acad Sci U S A 114(9):E1698–E1706. https://doi.org/10.1073/pnas.1615771114

Fermin Lee A, Chen HY, Wan L, Wu SY, Yu JS, Huang AC, Miaw SC, Hsu DK, Wu-Hsieh BA, Liu FT (2013) Galectin-3 modulates Th17 responses by regulating dendritic cell cytokines. Am J Pathol 183(4):1209–1222. https://doi.org/10.1016/j.ajpath.2013.06.017

Fermino ML, Dias FC, Lopes CD, Souza MA, Cruz AK, Liu FT, Chammas R, Roque-Barreira MC, Rabinovich GA, Bernardes ES (2013) Galectin-3 negatively regulates the frequency and function of CD4(+) CD25(+) Foxp3(+) regulatory T cells and influences the course of Leishmania major infection. Eur J Immunol 43(7):1806–1817. https://doi.org/10.1002/eji.201343381

Fermino ML, Dylon LS, Cecilio NT, Santos SN, Toscano MA, Dias-Baruffi M, Roque-Barreira MC, Rabinovich GA, Bernardes ES (2016) Lack of galectin-3 increases Jagged1/Notch activation in bone marrow-derived dendritic cells and promotes dysregulation of T helper cell polarization. Mol Immunol 76:22–34. https://doi.org/10.1016/j.molimm.2016.06.005

Fogel S, Guittaut M, Legrand A, Monsigny M, Hebert E (1999) The tat protein of HIV-1 induces galectin-3 expression. Glycobiology 9(4):383–387

Fowler M, Thomas RJ, Atherton J, Roberts IS, High NJ (2006) Galectin-3 binds to *Helicobacter pylori* O-antigen: it is upregulated and rapidly secreted by gastric epithelial cells in response to H. pylori adhesion. Cell Microbiol 8(1):44–54. https://doi.org/10.1111/j.1462-5822.2005.00599.x

Garcin PO, Nabi IR, Pante N (2015) Galectin-3 plays a role in minute virus of mice infection. Virology 481:63–72. https://doi.org/10.1016/j.virol.2015.02.019

Ghebremedhin B (2014) Human adenovirus: viral pathogen with increasing importance. Eur J Microbiol Immunol 4(1):26–33. https://doi.org/10.1556/EuJMI.4.2014.1.2

Giordanengo L, Gea S, Barbieri G, Rabinovich GA (2001) Anti-galectin-1 autoantibodies in human Trypanosoma cruzi infection: differential expression of this beta-galactoside-binding protein in cardiac Chagas' disease. Clin Exp Immunol 124(2):266–273

Hagar JA, Powell DA, Aachoui Y, Ernst RK, Miao EA (2013) Cytoplasmic LPS activates caspase-11: implications in TLR4-independent endotoxic shock. Science 341(6151):1250–1253. https://doi.org/10.1126/science.1240988

Huang J, Canadien V, Lam GY, Steinberg BE, Dinauer MC, Magalhaes MA, Glogauer M, Grinstein S, Brumell JH (2009) Activation of antibacterial autophagy by NADPH oxidases. Proc Natl Acad Sci U S A 106(15):6226–6231. https://doi.org/10.1073/pnas.0811045106

Huang WC, Chen HL, Chen HY, Peng KP, Lee Y, Huang LM, Chang LY, Liu FT (2016) Galectin-3 and its genetic variation rs4644 modulate enterovirus 71 infection. PLoS ONE 11(12):e0168627. https://doi.org/10.1371/journal.pone.0168627

Jia J, Abudu YP, Claude-Taupin A, Gu Y, Kumar S, Choi SW, Peters R, Mudd MH, Allers L, Salemi M, Phinney B, Johansen T, Deretic V (2018) Galectins control mTOR in response to endomembrane damage. Mol Cell 70(1):120–135 e128. https://doi.org/10.1016/j.molcel.2018.03.009

Kasai K, Hirabayashi J (1996) Galectins: a family of animal lectins that decipher glycocodes. J Biochem 119(1):1–8

Kevric I, Cappel MA, Keeling JH (2015) New world and old world leishmania infections: a practical review. Dermatol Clin 33(3):579–593. https://doi.org/10.1016/j.det.2015.03.018

Kim YC, Guan KL (2015) mTOR: a pharmacologic target for autophagy regulation. J Clin Invest 125(1):25–32. https://doi.org/10.1172/JCI73939

Kleshchenko YY, Moody TN, Furtak VA, Ochieng J, Lima MF, Villalta F (2004) Human galectin-3 promotes *Trypanosoma cruzi* adhesion to human coronary artery smooth muscle cells. Infect Immun 72(11):6717–6721. https://doi.org/10.1128/IAI.72.11.6717-6721.2004

Kulkarni R, Prasad A (2017) Exosomes derived from HIV-1 infected DCs mediate viral trans-infection via fibronectin and Galectin-3. Sci Rep 7(1):14787. https://doi.org/10.1038/s41598-017-14817-8

Li S-W, Yang T-C, Lai C-C, Huang S-H, Liao J-M, Wan L, Lin Y-J, Lin C-W (2014) Antiviral activity of aloe-emodin against influenza A virus via galectin-3 up-regulation. Eur J Pharmacol 738:125–132. https://doi.org/10.1016/j.ejphar.2014.05.028

Li FY, Weng IC, Lin CH, Kao MC, Wu MS, Chen HY, Liu FT (2019) *Helicobacter pylori* induces intracellular galectin-8 aggregation around damaged lysosomes within gastric epithelial cells in a host O-glycan-dependent manner. Glycobiology 29(2):151–162. https://doi.org/10.1093/glycob/cwy095

Liu FT, Rabinovich GA (2005) Galectins as modulators of tumour progression. Nat Rev Cancer 5(1):29–41 (nrc1527 [pii]) https://doi.org/10.1038/nrc1527

Liu J, Xiao S, Huang S, Pei F, Lu F (2016a) Upregulated Tim-3/galectin-9 expressions in acute lung injury in a murine malarial model. Parasitol Res 115(2):587–595. https://doi.org/10.1007/s00436-015-4775-6

Liu J, Huang S, Su XZ, Song J, Lu F (2016b) Blockage of galectin-receptor interactions by alpha-lactose exacerbates *Plasmodium berghei*-induced pulmonary immunopathology. Sci Rep 6:32024. https://doi.org/10.1038/srep32024

Lopez-Bueno A, Segovia JC, Bueren JA, O'Sullivan MG, Wang F, Tattersall P, Almendral JM (2008) Evolution to pathogenicity of the parvovirus minute virus of mice in immunodeficient mice involves genetic heterogeneity at the capsid domain that determines tropism. J Virol 82(3):1195–1203. https://doi.org/10.1128/JVI.01692-07

Machado FC, Cruz L, da Silva AA, Cruz MC, Mortara RA, Roque-Barreira MC, da Silva CV (2014) Recruitment of galectin-3 during cell invasion and intracellular trafficking of *Trypanosoma cruzi* extracellular amastigotes. Glycobiology 24(2):179–184. https://doi.org/10.1093/glycob/cwt097

Malet JK, Cossart P, Ribet D (2017) Alteration of epithelial cell lysosomal integrity induced by bacterial cholesterol-dependent cytolysins. Cell Microbiol 19(4). https://doi.org/10.1111/cmi.12682

Meunier E, Dick MS, Dreier RF, Schurmann N, Kenzelmann Broz D, Warming S, Roose-Girma M, Bumann D, Kayagaki N, Takeda K, Yamamoto M, Broz P (2014) Caspase-11 activation requires lysis of pathogen-containing vacuoles by IFN-induced GTPases. Nature 509(7500):366–370. https://doi.org/10.1038/nature13157

Montespan C, Marvin SA, Austin S, Burrage AM, Roger B, Rayne F, Faure M, Campell EM, Schneider C, Reimer R, Grünewald K, Wiethoff CM, Wodrich H (2017) Multi-layered control of Galectin-8 mediated autophagy during adenovirus cell entry through a conserved PPxY motif in the viral capsid. PLoS Pathog 13(2):e1006217. https://doi.org/10.1371/journal.ppat.1006217

Moody TN, Ochieng J, Villalta F (2000) Novel mechanism that *Trypanosoma cruzi* uses to adhere to the extracellular matrix mediated by human galectin-3. FEBS Lett 470(3):305–308

Nicklas W, Kraft V, Meyer B (1993) Contamination of transplantable tumors, cell lines, and monoclonal antibodies with rodent viruses. Lab Anim Sci 43(4):296–300

Oakley MS, Majam V, Mahajan B, Gerald N, Anantharaman V, Ward JM, Faucette LJ, McCutchan TF, Zheng H, Terabe M, Berzofsky JA, Aravind L, Kumar S (2009) Pathogenic roles of CD14, galectin-3, and OX40 during experimental cerebral malaria in mice. PLoS ONE 4(8):e6793. https://doi.org/10.1371/journal.pone.0006793

Oliveira FL, Frazao P, Chammas R, Hsu DK, Liu FT, Borojevic R, Takiya CM, El-Cheikh MC (2007) Kinetics of mobilization and differentiation of lymphohematopoietic cells during experimental murine schistosomiasis in galectin-3 -/- mice. J Leukoc Biol 82(2):300–310. https://doi.org/10.1189/jlb.1206747

Oliveira FL, Chammas R, Ricon L, Fermino ML, Bernardes ES, Hsu DK, Liu FT, Borojevic R, El-Cheikh MC (2009) Galectin-3 regulates peritoneal B1-cell differentiation into plasma cells. Glycobiology 19(11):1248–1258. https://doi.org/10.1093/glycob/cwp120

Oliveira FL, Bernardes ES, Brand C, dos Santos SN, Cabanel MP, Arcanjo KD, Brito JM, Borojevic R, Chammas R, El-Cheikh MC (2016) Lack of galectin-3 up-regulates IgA expression by peritoneal B1 lymphocytes during B cell differentiation. Cell Tissue Res 363(2):411–426. https://doi.org/10.1007/s00441-015-2203-y

Ouellet M, Mercier S, Pelletier I, Bounou S, Roy J, Hirabayashi J, Sato S, Tremblay MJ (2005) Galectin-1 acts as a soluble host factor that promotes HIV-1 infectivity through stabilization of virus attachment to host cells. J Immunol 174(7):4120. https://doi.org/10.4049/jimmunol.174.7.4120

Palframan SL, Kwok T, Gabriel K (2012) Vacuolating cytotoxin A (VacA), a key toxin for *Helicobacter pylori* pathogenesis. Front Cell Infect Microbiol 2:92. https://doi.org/10.3389/fcimb.2012.00092

Park AM, Hagiwara S, Hsu DK, Liu FT, Yoshie O (2016) Galectin-3 plays an important role in innate immunity to gastric infection by *Helicobacter pylori*. Infect Immun 84(4):1184–1193. https://doi.org/10.1128/IAI.01299-15

Paz I, Sachse M, Dupont N, Mounier J, Cederfur C, Enninga J, Leffler H, Poirier F, Prevost MC, Lafont F, Sansonetti P (2010) Galectin-3, a marker for vacuole lysis by invasive pathogens. Cell Microbiol 12(4):530–544. https://doi.org/10.1111/j.1462-5822.2009.01415.x

Pelletier I, Sato S (2002) Specific recognition and cleavage of galectin-3 by Leishmania major through species-specific polygalactose epitope. J Biol Chem 277(20):17663–17670. https://doi.org/10.1074/jbc.M201562200

Perillo NL, Pace KE, Seilhamer JJ, Baum LG (1995) Apoptosis of T cells mediated by galectin-1. Nature 378(6558):736–739. https://doi.org/10.1038/378736a0

Pineda MA, Cuervo H, Fresno M, Soto M, Bonay P (2015a) Lack of Galectin-3 prevents cardiac fibrosis and effective immune responses in a murine model of *Trypanosoma cruzi* infection. J Infect Dis 212(7):1160–1171. https://doi.org/10.1093/infdis/jiv185

Pineda MA, Corvo L, Soto M, Fresno M, Bonay P (2015b) Interactions of human galectins with *Trypanosoma cruzi*: binding profile correlate with genetic clustering of lineages. Glycobiology 25(2):197–210. https://doi.org/10.1093/glycob/cwu103

Poli G (2013) Cell-to-cell vs. cell-free HIV-1 transmission from macrophages to CD4+ T lymphocytes: lessons from the virology textbook. AIDS 27(14):2307–2308. https://doi.org/10.1097/qad.0b013e328363619a

Poncini CV, Ilarregui JM, Batalla EI, Engels S, Cerliani JP, Cucher MA, van Kooyk Y, Gonzalez-Cappa SM, Rabinovich GA (2015) *Trypanosoma cruzi* infection imparts a regulatory program in dendritic cells and T cells via Galectin-1-dependent mechanisms. J Immunol 195(7):3311–3324. https://doi.org/10.4049/jimmunol.1403019

Reignault LC, Barrias ES, Soares Medeiros LC, de Souza W, de Carvalho TM (2014) Structures containing galectin-3 are recruited to the parasitophorous vacuole containing *Trypanosoma cruzi* in mouse peritoneal macrophages. Parasitol Res 113(6):2323–2333. https://doi.org/10.1007/s00436-014-3887-8

Ruas LP, Bernardes ES, Fermino ML, de Oliveira LL, Hsu DK, Liu FT, Chammas R, Roque-Barreira MC (2009) Lack of galectin-3 drives response to *Paracoccidioides brasiliensis* toward a Th2-biased immunity. PLoS One 4(2):e4519. https://doi.org/10.1371/journal.pone.0004519

Spackman E (2008) A brief introduction to the avian influenza virus. Methods Mol Biol 436:1–6. https://doi.org/10.1007/978-1-59745-279-3_1

Stowell SR, Arthur CM, Dias-Baruffi M, Rodrigues LC, Gourdine JP, Heimburg-Molinaro J, Ju T, Molinaro RJ, Rivera-Marrero C, Xia B, Smith DF, Cummings RD (2010) Innate immune lectins kill bacteria expressing blood group antigen. Nat Med 16(3):295–301. https://doi.org/10.1038/nm.2103

Sturge CR, Benson A, Raetz M, Wilhelm CL, Mirpuri J, Vitetta ES, Yarovinsky F (2013) TLR-independent neutrophil-derived IFN-gamma is important for host resistance to intracellular pathogens. Proc Natl Acad Sci U S A 110(26):10711–10716. https://doi.org/10.1073/pnas.1307868110

Tarleton RL (2015) CD8 + T cells in *Trypanosoma cruzi* infection. Semin Immunopathol 37(3):233–238. https://doi.org/10.1007/s00281-015-0481-9

Thurston TL, Wandel MP, von Muhlinen N, Foeglein A, Randow F (2012) Galectin 8 targets damaged vesicles for autophagy to defend cells against bacterial invasion. Nature 482(7385):414–418. https://doi.org/10.1038/nature10744

Thurston TL, Boyle KB, Allen M, Ravenhill BJ, Karpiyevich M, Bloor S, Kaul A, Noad J, Foeglein A, Matthews SA, Komander D, Bycroft M, Randow F (2016) Recruitment of TBK1 to cytosol-invading Salmonella induces WIPI2-dependent antibacterial autophagy. EMBO J 35(16):1779–1792. https://doi.org/10.15252/embj.201694491

Toscano MA, Bianco GA, Ilarregui JM, Croci DO, Correale J, Hernandez JD, Zwirner NW, Poirier F, Riley EM, Baum LG, Rabinovich GA (2007) Differential glycosylation of TH1, TH2 and TH-17 effector cells selectively regulates susceptibility to cell death. Nat Immunol 8(8):825–834. https://doi.org/10.1038/ni1482

Toscano MA, Tongren JE, de Souza JB, Liu FT, Riley EM, Rabinovich GA (2012) Endogenous galectin-3 controls experimental malaria in a species-specific manner. Parasite Immunol 34(7):383–387. https://doi.org/10.1111/j.1365-3024.2012.01366.x

Van de Vijver KK, Deelder AM, Jacobs W, Van Marck EA, Hokke CH (2006) LacdiNAc- and LacNAc-containing glycans induce granulomas in an in vivo model for schistosome egg-induced hepatic granuloma formation. Glycobiology 16(3):237–243. https://doi.org/10.1093/glycob/cwj058

van den Berg TK, Honing H, Franke N, van Remoortere A, Schiphorst WE, Liu FT, Deelder AM, Cummings RD, Hokke CH, van Die I (2004) LacdiNAc-glycans constitute a parasite pattern for galectin-3-mediated immune recognition. J Immunol 173(3):1902–1907

van Remoortere A, Hokke CH, van Dam GJ, van Die I, Deelder AM, van den Eijnden DH (2000) Various stages of schistosoma express Lewis(x), LacdiNAc, GalNAcbeta1-4 (Fucalpha1-3)GlcNAc and GalNAcbeta1-4(Fucalpha1-2Fucalpha1-3)GlcNAc carbohydrate epitopes: detection with monoclonal antibodies that are characterized by enzymatically synthesized neoglycoproteins. Glycobiology 10(6):601–609

Vasta GR (2012) Galectins as pattern recognition receptors: structure, function, and evolution. Adv Exp Med Biol 946:21–36. https://doi.org/10.1007/978-1-4614-0106-3_2

Wang SM, Liu CC, Tseng HW, Wang JR, Huang CC, Chen YJ, Yang YJ, Lin SJ, Yeh TF (1999) Clinical spectrum of enterovirus 71 infection in children in southern Taiwan, with an emphasis on neurological complications. Clin Infect Dis 29(1):184–190. https://doi.org/10.1086/520149

Wang SM, Lei HY, Huang KJ, Wu JM, Wang JR, Yu CK, Su IJ, Liu CC (2003) Pathogenesis of enterovirus 71 brainstem encephalitis in pediatric patients: roles of cytokines and cellular immune activation in patients with pulmonary edema. J Infect Dis 188(4):564–570. https://doi.org/10.1086/376998

Wang SF, Tsao CH, Lin YT, Hsu DK, Chiang ML, Lo CH, Chien FC, Chen P, Arthur Chen YM, Chen HY, Liu FT (2014) Galectin-3 promotes HIV-1 budding via association with Alix and Gag p6. Glycobiology 24(11):1022–1035. https://doi.org/10.1093/glycob/cwu064

Weng IC, Chen HL, Lo TH, Lin WH, Chen HY, Hsu DK, Liu FT (2018) Cytosolic galectin-3 and -8 regulate antibacterial autophagy through differential recognition of host glycans on damaged phagosomes. Glycobiology 28(6):392–405. https://doi.org/10.1093/glycob/cwy017

Westman J, Moran G, Mogavero S, Hube B, Grinstein S (2018) *Candida albicans* hyphal expansion causes phagosomal membrane damage and luminal alkalinization. MBio 9(5). https://doi.org/10.1128/mbio.01226-18

White NJ (2018) Anaemia and malaria. Malar J 17(1):371. https://doi.org/10.1186/s12936-018-2509-9

Wu SY, Yu JS, Liu FT, Miaw SC, Wu-Hsieh BA (2013) Galectin-3 negatively regulates dendritic cell production of IL-23/IL-17-axis cytokines in infection by *Histoplasma capsulatum*. J Immunol 190(7):3427–3437. https://doi.org/10.4049/jimmunol.1202122

Wu SY, Huang JH, Chen WY, Chan YC, Lin CH, Chen YC, Liu FT, Wu-Hsieh BA (2017) Cell intrinsic galectin-3 attenuates neutrophil ROS-dependent killing of Candida by modulating CR26 downstream syk activation. Front Immunol 8:48. https://doi.org/10.3389/fimmu.2017.00048

Xiao S, Liu J, Huang S, Lu F (2016) Increased Gal-9 and Tim-3 expressions during liver damage in a murine malarial model. Parasitol Res 115(2):663–672. https://doi.org/10.1007/s00436-015-4784-5

Xue J, Fu C, Cong Z, Peng L, Peng Z, Chen T, Wang W, Jiang H, Wei Q, Qin C (2017) Galectin-3 promotes caspase-independent cell death of HIV-1-infected macrophages. FEBS J 284(1):97–113. https://doi.org/10.1111/febs.13955

Yang ML, Chen YH, Wang SW, Huang YJ, Leu CH, Yeh NC, Chu CY, Lin CC, Shieh GS, Chen YL, Wang JR, Wang CH, Wu CL, Shiau AL (2011) Galectin-1 binds to influenza virus and ameliorates influenza virus pathogenesis. J Virol 85(19):10010–10020. https://doi.org/10.1128/JVI.00301-11

Zhuo Y, Bellis SL (2011) Emerging role of alpha2,6-sialic acid as a negative regulator of galectin binding and function. J Biol Chem 286(8):5935–5941. https://doi.org/10.1074/jbc.R110.191429

Zuniga E, Rabinovich GA, Iglesias MM, Gruppi A (2001a) Regulated expression of galectin-1 during B-cell activation and implications for T-cell apoptosis. J Leukoc Biol 70(1):73–79

Zuniga E, Gruppi A, Hirabayashi J, Kasai KI, Rabinovich GA (2001b) Regulated expression and effect of galectin-1 on *Trypanosoma cruzi*-infected macrophages: modulation of microbicidal activity and survival. Infect Immun 69(11):6804–6812. https://doi.org/10.1128/IAI.69.11.6804-6812.2001

Chapter 7
Galectins in Host–Pathogen Interactions: Structural, Functional and Evolutionary Aspects

Gerardo R. Vasta

Abstract Galectins are a family of ß-galactoside-binding lectins characterized by a unique sequence motif in the carbohydrate recognition domain, and evolutionary and structural conservation from fungi to invertebrates and vertebrates, including mammals. Their biological roles, initially understood as limited to recognition of endogenous ("self") carbohydrate ligands in embryogenesis and early development, dramatically expanded in later years by the discovery of their roles in tissue repair, cancer, adipogenesis, and regulation of immune homeostasis. In recent years, however, evidence has also accumulated to support the notion that galectins can bind ("non-self") glycans on the surface of potentially pathogenic microbes, and function as recognition and effector factors in innate immunity. Thus, this evidence has established a new paradigm by which galectins can function not only as pattern recognition receptors but also as effector factors, by binding to the microbial surface and inhibiting adhesion and/or entry into the host cell, directly killing the potential pathogen by disrupting its surface structures, or by promoting phagocytosis, encapsulation, autophagy, and pathogen clearance from circulation. Strikingly, some viruses, bacteria, and protistan parasites take advantage of the aforementioned recognition roles of the vector/host galectins, for successful attachment and invasion. These recent findings suggest that galectin-mediated innate immune recognition and effector mechanisms, which throughout evolution have remained effective for preventing or fighting viral, bacterial, and parasitic infection, have been "subverted" by certain pathogens by unique evolutionary adaptations of their surface glycome to gain host entry, and the acquisition of effective mechanisms to evade the host's immune responses.

Keywords Galectin · Infection · Recognition · Effector · Subversion of galectin function

G. R. Vasta (✉)
Department of Microbiology and Immunology, University of Maryland School of Medicine, UMB, Baltimore, USA
e-mail: gvasta@som.umaryland.edu

Institute of Marine and Environmental Technology, Columbus Center, University of Maryland, 701 East Pratt Street, Baltimore, MD 21202, USA

S.-L. Hsieh (ed.), *Lectin in Host Defense Against Microbial Infections*, Advances in Experimental Medicine and Biology 1204,
https://doi.org/10.1007/978-981-15-1580-4_7

7.1 Introduction

Based on unique sequence motifs in the carbohydrate recognition domain (CRD) and their structural fold, lectins from protistans, fungi, invertebrates, and vertebrates have been classified into families that include galectins (formerly S-type lectins), C-type, F-type, X-type, R-type, P-type, and others (reviewed in Vasta and Ahmed 2008). Rigorous structural and phylogenetic analyses of individual lectin families among extant animal species have yielded critical information about their evolutionary history and biological role(s). For example, C- and F-type lectins and galectins are of ubiquitous taxonomic distribution, and highly diversified from the functional standpoint (Zelensky and Gready 2005; Vasta et al. 2012). However, while C- and F-type lectins are structurally heterogeneous lectin families (Vasta et al. 2017), galectins are generally considered as structurally and evolutionary conserved (Cooper 2002; Cummings et al. 2017).

Galectins constitute a family of ß-galactoside-binding proteins characterized by a canonical sequence motif in their CRDs, that are widely distributed in eukaryotic taxa, including fungi, sponges, and both invertebrates and vertebrates (Cooper 2002). Most galectins are non-glycosylated soluble proteins, although exceptions with transmembrane domains have been reported (Lipkowitz et al. 2004). Galectins are synthesized in the cytosol, and can be translocated into the nucleus where they can form part of the spliceosome (Cho and Cummings 1995a, b; Tsay et al. 1999) (Fig. 7.1a). Moreover, although galectins lack a typical secretion signal peptide, they can be secreted into the extracellular space by direct translocation across the plasma membrane (Cleves et al. 1996).

Although galectins have been evolutionarily conserved from a structural standpoint (Houzelstein et al. 2004), the galectin repertoire in any given mammalian species is diversified, and constituted by multiple galectin types, subtypes, and isoforms (Vasta and Ahmed 2008; Cooper 2002). The distinct domain organization of mammalian galectins has led to their classification into three major types: "proto", "chimera", and "tandem-repeat" (TR) (Hirabayashi and Kasai 1993) (Fig. 7.1b). The peptide subunits of the proto-type galectins contain a single CRD and can form non-covalently linked homodimers, with a dimerization equilibrium Kd of 7 μM (Cho and Cummings 1995a, b). The chimera galectins also house a single CRD, but display an N-terminal domain rich in proline and glycine that enables the formation of oligomers. Finally, TR galectins display two CRDs that are bridged by a functional linker peptide that can range from 5 to over 50 amino acids in length. Within each galectin type, up to 15 distinct galectin subtypes have been identified, numbered following the order of their discovery. Among these, galectin-1, -2, -5, -7, -10, -11, -13, -14, and -15 are proto-type, galectin-3 is a chimera-type, and galectin-4, -6, -8, -9, and -12 are TR-type (Hirabayashi and Kasai 1993). In invertebrate species, galectins may have unique structural features, including multiple CRDs *per* polypeptide, such as the 4-CRD galectins described in clams, oysters, and snails (Tasumi and Vasta 2007; Feng et al. 2013, 2015a, b; Kurz et al.

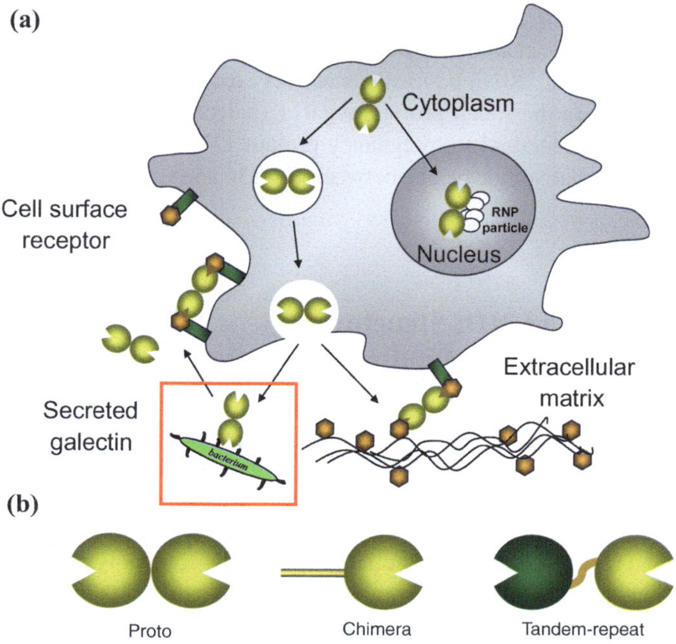

Fig. 7.1 Expression, secretion, and functions of galectins, and galectin types. **a** Galectins are synthesized in the cytosol, and can be translocated into the nucleus and interact with ribonucleoprotein (RNP) particles. Galectins can be transported and secreted to the extracellular space, where they interact with glycans in the extracellular matrix, bridge cell surface receptors, or recognize microbial glycans (red box). **b** Galectins are classified into three main types: proto, chimera, and tandem-repeat types. Proto-type galectins contain one carbohydrate-recognition domain (CRD) per subunit and are usually homodimers of non-covalently linked subunits (galectin-1, -2, -5, -7, -10, -11, -13, -14, and -15). In contrast, chimera-type galectins are monomeric with a carboxy-terminal CRD, joined to an amino-terminal peptide that contains a collagen-like sequence rich in proline and glycine, and can oligomerize as trimers (galectin3). In the tandem-repeat galectins, two CRDs are covalently joined by a linker peptide (galectin-4, -6, -8, -9, and -12)

2013; Vasta et al. 2015), or the shrimp galectin MjGal that resembles a chimera-type galectin, although the CRD is located at the N-terminal end (Shi et al. 2014).

In addition to the three major galectin types, other galectin-related proteins that do not follow the canonical structure of the typical family members have been described. Among these, two notable exceptions have been reported to display transmembrane domains (Lipkowitz et al. 2004; Gorski et al. 2002). In addition, soluble galectin-like proteins, such as the eye lens galectin-related inter-fiber protein (GRIFIN), the galectin-related protein (GRP) (previously HSPC159; hematopoietic stem cell precursor), and the Charcot-Leyden crystals (galectin-10), have been identified, and their structural and functional aspects are characterized to various levels of detail (Su et al.

2018; Zhou et al. 2008; Ogden et al. 1998). Mammalian GRIFIN lacks carbohydrate-binding activity (Ogden et al. 1998), but the fish (Ahmed and Vasta 2008) and chicken (García Caballero et al. 2016) equivalents actively recognize ß-galactosides. Other unusual galectins are the sheep protein ovgal11 (galectin-11; Preston et al. 2015), and the proto-type galectin-13 ("pregnancy protein 13") which in solution forms a dimer stabilized by a disulfide bridge between Cys136 and Cys138 (Than et al. 1999, 2004).

7.2 The Galectin CRD: Structure and Carbohydrate-Binding Properties

The structure of galectin-1 (Gal1) in complex with *N*-acetyl-lactosamine (LacNAc) enabled the identification of the specific amino acid residues of the CRD that interact directly or indirectly, through water molecules and hydroxyl groups on the disaccharide, as well as the nature of the interactions (hydrogen bonds, electrostatic, and van der Waals) (Liao et al. 1994). The 135 amino acids-long polypeptide subunit of Gal1 folds into a β-sandwich that comprises two anti-parallel β-sheets of five and six strands each (S1–S6 and F1–F5). In the Gal1 dimer, the subunits are related by a two-fold rotational axis perpendicular to the plane of the β-sheets. The single carbohydrate recognition cleft is formed by three continuous concave strands (S4–S6) in which H^{44}, B^{46}, R^{48}, H^{52}, B^{61}, W^{68}, E^{71}, and R^{73} establish direct interactions with LacNAc, and determine the carbohydrate specificity of galectin-1 (Liao et al. 1994). The non-reducing terminal galactose ring is maintained in place by a hydrophobic interaction with W^{68}, while additional water-mediated interactions between H^{52}, D^{54}, and R^{73} and the nitrogen of the *N*-acetyl group determine the higher affinity for LacNAc over lactose. The dissociation constants of bovine Gal1 for lactose, LacNAc, and thiodigalactoside (TDG) measured by microcalorimetry are in the range of 10^{-5} M, with two binding sites per Gal1 dimer (Schwarz et al. 1998). The crystal structures of additional galectin types (human galectin-2, -3, and -7) in complex with mono- or disaccharides, or biantennary oligosaccharides were later resolved (Seetharaman et al. 1998; Caldararu et al. 2019; Si et al. 2016; Ramaswamy et al. 2015). More recently, the structures of the individual N- and C-terminal CRDs of TR galectins, such as galectins-4, -8, and -9, were resolved by either crystallization or NMR spectroscopy (reviewed in Di Lella et al. 2011).

In addition to the primary binding cleft in the galectin CRD described above, additional carbohydrate-binding areas (extended binding site) can enhance affinity for larger or more complex glycans. In the galectin-3 (Gal3) CRD, for example, the carbohydrate-binding site is shaped as a cleft open at both ends, exposing the GlcNAc of the LacNAc to the solvent (Seetharaman et al. 1998). This extended binding site in Gal3 results in increased affinity for polylactosamines and ABH blood group oligosaccharides [Fucα1, 2; GalNAcα1,3(Fucα1,2); and Galα1,3(Fucα1,2)] (Seetharaman et al. 1998). For TR galectins, the two CRDs are structurally similar but show either different affinities for the same ligand such as galectin-4 (Gal4), or different fold and specificities altogether, such as galectin-8 (Krejcirikova et al. 2011; Ideo et al. 2003, 2011). For galectins from invertebrates, such as the *Caenorhabditis elegans* 16-kDa galectin and the oyster (*Crassostrea virginica*) galectins CvGal1 and CvGal2, their binding specificity for blood group oligosaccharides is determined by a shorter loop 4 in the primary binding cleft in the galectin CRD (Ahmed et al. 2002; Feng et al. 2013; Vasta et al. 2015).

The dimerization of proto-type galectins such as Gal1 is critical for their function in mediating interactions between cells, or cells and glycans in the extracellular matrix (ECM) (Gabius 1997). The binding of galectins to multivalent glycans on the cell surface may promote lattice formation, as supported by the structure of Gal1 complexed with a biantennary glycan, in which the ligand is cross-linked between two Gal1 dimers (Bourne et al. 1994). Similar interactions via the N terminus domain leading to the formation of oligomers (trimers and pentamers) have been proposed for Gal3 (Morris et al. 2004; Fortuna-Costa et al. 2014). For the vertebrate TR galectins and the multi-CRD galectins from invertebrates, the carbohydrate specificity of the CRDs present in the polypeptide is similar but not identical (Carlsson et al. 2007; Nagae et al. 2009; Vasta et al. 2015; Houzelstein et al. 2004; Krejcirikova et al. 2011). This is supported by the capacity of TR galectins to cross-link cells with different synthetic glycoconjugates (Tomizawa et al. 2005; Ideo et al. 2011). The multivalency of galectins attained by the presence of multiple distinct CRDs in a single subunit polypeptide as described above for the TR galectins, or by oligomerization of the galectin subunits, such as in the proto or chimera types, enables cross-linking of two or more cells, and adhesion of cells to glycosylated surfaces, as well as the formation of lattices at the cell surface that are critical for signaling or receptor endocytosis (Nabi et al. 2015; Kutzner et al. 2019; Garner and Baum 2008; Rabinovich et al. 2007a, b) (Fig. 7.2a).

The compact β-sandwich structure of the galectin CRD in the presence of bound ligand determines the resistance of Gal1 to protease activity (Liao et al. 1994), while the resistance to oxidative inactivation in the extracellular environment can be rationalized by changes in the dimerization equilibrium which are determined by the oxidation state of cysteine sulfhydryl groups (Stowell et al. 2009). Six key cysteine residues, some of which are located on the surface of the molecule on the face opposite to the CRD, are potentially susceptible to oxidation (DiLella et al. 2010; Liao et al. 1994; Lobsanov et al. 1993). Upon secretion to the oxidative extracellular environment, some of these cysteines establish intramolecular disulfide bonds, causing conformational changes that hinder dimerization (Lopez-Lucendo et al. 2004). Therefore, the oxidation state of cysteine sulfhydryl groups, the presence of carbohydrate ligand, and the dimerization equilibrium play critical roles in a dynamic interplay in which specific binding to glycan ligands enhances dimerization of Gal1, and reduces its sensitivity to oxidative inactivation (Stowell et al. 2009).

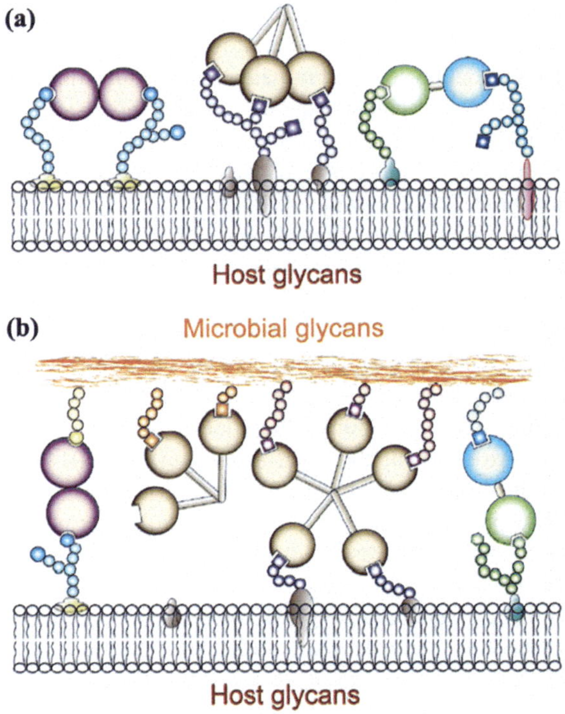

Fig. 7.2 Recognition of "self" and "non-self" glycans by galectins. **a** In the extracellular space, galectins form multivalent oligomers that cross-link cell surface glycoproteins and glycolipids, form microdomains and lattices, and activate signaling pathways. **b** Proto, chimera, and tandem-repeat galectins can function as pattern recognition receptors (PRRs) and establish trans-interactions with the host cell surface and microbial glycans

7.3 Functional Aspects: Recognition of Endogenous ("Self") Glycans

While proto and chimera type galectin subunits possess a single CRD, they can form oligomeric structures that can interact with and cross-link multivalent ligands, either soluble glycoproteins or glycolipids, and complex glycans on the cell surface or ECM with increased avidity (Dam and Brewer 2008; Cho and Cummings 1995a, b). TR galectins possess two CRDs in a single polypeptide, and can function similarly. The density and presentation of the cell surface glycans modulates affinity of the CRD–ligand interaction via negative cooperativity and thus, multivalent galectins can cross-link them into lattices that induce their clustering into lipid raft microdomains (Dam and Brewer 2008; Cho and Cummings 1995a, b; Nabi et al. 2015; Kutzner et al. 2019). Therefore, galectin–ligand interactions can modulate cell function by inducing reorganization or association of cell surface components, regulating turnover of endocytic receptors, and activating or attenuating signaling pathways (Dam and Brewer 2008; Garner and Baum 2008; Rabinovich et al. 2007a, b). Moreover, because the various galectin types and subtypes exhibit differences in carbohydrate specificity and affinity, and bind a broad range of glycans that display the requisite topologies, the galectin repertoire is endowed of substantial diversity in recognition properties. This, together with the galectins' unique tissue-specific expression, distribution, and local concentrations, supports extensive functional diversification (Cooper 2002; Vasta and Ahmed 2008; Vasta 2009). Accordingly, the biological function of a particular galectin may vary among cells, tissues, and fluids depending on their concentration, the redox properties of the intra- or extracellular environments, and the availability and multivalent presentation of carbohydrate ligands at the cell surface or ECM (Vasta and Ahmed 2008; Vasta 2009).

The observation that galectins in chicken muscle were developmentally regulated suggested that their biological roles were related to embryogenesis and early development. The finding that these galectins preferentially recognized polylactosamines present on the myoblast surface and the ECM led to hypothesize that galectins promote myoblast fusion (reviewed in Cummings et al. 2017). Subsequent studies on murine Gal1 and Gal3 revealed key roles in the development of notochord, skeletal muscle, and central nervous system (Colnot et al. 1997, 2001; Georgiadis et al. 2007). More recently, the increasing availability of null mice for selected galectins enabled analyses of developmental phenotypes, and the elucidation of the unique functions of the multiple galectin types and subtypes in the complex mammalian galectin repertoires. In the past few years, genetically tractable models such as *Drosophila, C. elegans,* and zebrafish (*Danio rerio*) have become useful systems for addressing the biological roles of galectins (Pace et al. 2002; Ahmed et al. 2002; Vasta et al. 2004; Nemoto-Sasaki et al. 2008; Feng et al. 2015a, b; Nita-Lazar et al. 2016). For example, anti-sense knockdown approaches in zebrafish revealed key roles of galectins in early differentiation and development of the myotome (Ahmed et al. 2009a, b), and retinal repair and regeneration (Craig et al. 2010; Eastlake et al. 2017).

The multiple roles of galectins in cancer have been addressed with increasing interest over the past two decades (Reviewed in Méndez-Huergo et al. 2017). Melanoma, prostate, and ovarian cancer may overexpress galectin-1, -3, -7, -8 and -9, and in some cases their expression profiles can be associated with malignancy stage or metastatic potential (Blidner et al. 2015; Hill et al. 2010). Expression of Gal1 in the vascular endothelium promotes tumor angiogenesis, by a mechanism that involves binding of Gal1 to complex *N*-glycans on VEGF receptor 2 (VEGFR2) and activation of VEGF-like signaling (Croci et al. 2014). In the early stages of prostate adenocarcinoma, Gal3 expression can be silenced by promoter methylation (Ahmed et al. 2009a, b; Ahmed and Al Sadek 2015), but in later stages, together with its preferred ligand on the cell surface—the Thomsen-Friedenreich disaccharide (Galβ1,3GalNAc)—Gal3 has key roles in tumor angiogenesis, tumor-endothelial cell adhesion, metastasis, and evasion of immune surveillance by killing of activated T cells (Guha et al. 2013). Since the 1990s, the roles of galectins as regulators of both innate and adaptive immune homeostasis have firmly established and characterized in detail (Di Lella et al. 2011). Galectins are ubiquitously expressed and distributed in mammalian tissues, including most cells of the innate (dendritic cells, macrophages, mast cells, natural killer cells, gamma/delta T cells, and B-1 cells) and adaptive (activated B and T cells) immune system, and as in other cell types (Stowell et al. 2008; Rabinovich et al. 2007a, b). Immune challenge by viruses, bacteria, and eukaryotic parasites, however, can significantly alter their expression and secretion (Rabinovich et al. 2012;

Stowell et al. 2008). Endogenous glycans recognized by galectins on the surface of immune cells and other cell types include β-integrins, CD45, GM1, CD44, Tim3, MUC1, podoplanin, CD166, ABH-type oligosaccharides CD43, CD45, CD7, CD71, CD44, TIM3, CTLA4, MUC1, MUC16, and MerTK (Rabinovich and Toscano 2009; Guzman-Aranguez et al. 2009; Hirabayashi et al. 2002; Wu et al. 2002; Krzeminski et al. 2011; Zhu et al. 2005). Galectins can function as pro- or anti-inflammatory factors in innate immune responses. For example, Gal1 can block or attenuate signaling that promote leukocyte infiltration, migration, and recruitment (Stowell et al. 2008). Gal3 expression in epithelia, macrophages, and dendritic cells is upregulated during inflammation, and can promote macrophage recruitment and anti-microbial activity (Liu et al. 2012; Toledo et al. 2014). Galectin-9 (Gal9) functions as a chemoattractant for eosinophils and further promotes their activation, oxidative activity, and degranulation (Hirashima et al. 2002).

The functions of galectins as modulators of lymphocyte development and adaptive immune responses have been the focus of intense research in the past few years (Rabinovich et al. 2012; Liu et al. 2012). In the bone marrow and thymic compartments lymphocyte precursors interact with stromal cells via Gal1, which is critical to their development, selection, and migration to the periphery (Rossi et al. 2006). Additionally, Gal1 can either drive apoptosis or enhance proliferation of T-cells, depending on the microenvironment in which the exposure takes place, as well as the T cell developmental stage and activation status. For Gal3, however, pro- or anti-apoptotic effects on T-cells are determined by whether the exposure is extracellular or intracellular, respectively (Hsu and Liu 2008). Galectins can also modulate cytokine synthesis and secretion by T cells, and determine the Th1/Th2 balance of the immune (Rabinovich et al. 2012; Liu et al. 2012; Hsu and Liu 2008). More recently, it was shown that Gal1 can induce tolerogenic phenotypes in dendritic cells leading to expansion of regulatory T cells that can promote feto-maternal tolerance and suppress autoimmune neuroinflammation (Blois et al. 2007, 2019). Thus, it has become firmly established that galectins can modulate immune homeostasis with either beneficial or detrimental effects on pathological conditions that result from depressed or exacerbated immune function, such as cancer, inflammation, allergy, and autoimmune disorders (Rabinovich et al. 2012; Liu et al. 2012; Hsu and Liu 2008).

In recent years, associations of type 2 diabetes, obesity, and inflammation with adipogenic roles of galectin-12 (Gal12) and Gal3 have been identified and characterized in detail (Yang et al. 2011a, b, c; Hsu et al. 2018; Pejnovic et al. 2013; Pang et al. 2013). Gal12 was shown to function as a negative regulator of lipolysis and insulin sensitivity (Yang et al. 2011a, b, c; Hsu et al. 2018). In contrast, Gal3 null mice exhibit increased adiposity, systemic inflammation, dysregulated glucose metabolism, diabetes-associated kidney damage, and diet-induced atherogenesis (Pejnovic et al. 2013; Pang et al. 2013).

7.4 Role(s) of Galectins in Infection: Recognition of Exogenous ("Non-self") Glycans and Effector Functions

The recent finding that galectins can bind exogenous ("non-self") glycans on the surface of viruses, bacteria, protistan parasites, and fungi has led to a new paradigm about their potential roles in innate immunity as pattern recognition receptors (PRRs) (reviewed in Vasta 2009) (Fig. 7.2b). For example, Gal1 can bind to complex-type N-linked oligosaccharide on the HIV-1 gp120 envelope glycoprotein and to *Trichomonas vaginalis* lipophosphoglycan, whereas Gal3 recognizes both terminal and internal *N*-acetyllactosamine units in lipopolysaccharides from meningococcus (*Neisseria meningitidis*), gonococcus (*N. gonorrhoeae*), *Haemophilus influenzae*, and *Helicobacter pylori*, polysaccharide type XIV from pneumococcus (*Streptococcus pneumoniae*), LacdiNAc from *Schistosoma mansoni*, and oligomannans from *Candida albicans*. The lipophosphoglycan from *Leishmania major* is recognized by both Gal3 and Gal9, while Gal4 recognizes *Escherichia coli* strains that display blood group B oligosaccharides. The substantial diversity of the galectin repertoire(s), including the presence of isoforms, and unique specificity of each galectin subtype toward glycan ligands, suggests a broad recognition capacity for non-self carbohydrate moieties. In addition to the TR galectins that have at least two CRDs (up to four CRDs in invertebrate galectins; Vasta et al. 2015), the proto- and chimera-type galectins can form oligomers with two or more CRDs per molecule. Thus, all three galectin types are endowed with multivalent binding properties that enable not only the formation of lattices at the cell surface but also the capacity to cross-link cells. These properties enable galectins to participate in direct recognition of pathogens and parasites, as well as effector factors in downstream processes that lead to modulation of innate and adaptive immune responses. As will be discussed below, binding of host galectins to surface glycans either on the surface of pathogens and parasites or to their receptors on the host cell surface can lead to various outcomes beneficial to the host, including hindering or blocking their attachment to or entry into the host cells, direct killing of the pathogen, opsonization followed by phagocytosis and intracellular killing, encapsulation, or granuloma formation. In most cases, these galectin-mediated innate immune defense mechanisms take place simultaneously or subsequently upon challenge of any given pathogen. Furthermore, as multiple lectin families and other innate immune receptors such as TLRs are present in any single species, cooperative/synergic defense functions of galectins with other receptors have been described (Esteban et al. 2011; Jouault et al. 2006).

It should be noted that as invertebrates lack the adaptive immune response typical of vertebrates characterized by immunoglobulins and B and T cells, their galectins function as recognition and effector molecules as part of innate immune responses, exerting defense roles through the various mechanisms described below. In vertebrates, however, in addition to innate immune responses similar to those operative in invertebrates, several galectin-mediated key immunoregulatory functions (discussed in Sect. 7.3) triggered by the infectious challenge further contribute to maintain immune homeostasis and to potentially develop an efficient and long-lasting adaptive immunity against the potential pathogen. The recognition of pathogens and parasites by galectins from both invertebrate and vertebrate hosts, however, can also lead to contrasting outcomes, either by promoting effective host defense mechanisms as discussed above, or by facilitating pathogen attachment, entry and infection of the host. In this regard, whether the recognition of pathogens and parasites by host galectins is beneficial to the host, or constitutes a pathogen's host attachment and entry strategies, can be interpreted as the outcome of the host–pathogen co-evolutionary process (Vasta 2009).

1. Blocking of pathogen attachment to the host cell surface: Gal1 can bind to the envelope glycoprotein of influenza A virus (IAV) and reduce infection severity, possibly by hindering viral attachment to the cell surface sialylated glycan receptors (Yang et al. 2011a, b, c) (Fig. 7.3a). However, the detailed mechanisms involved have not been fully elucidated yet (reviewed in Machala et al. 2019). Similarly, Hattori et al. reported that Gal9 bound to the influenza A virus (PR8/H1N1 strain) and blocked virus attachment to the host cells in a lactose-specific manner (Hattori et al. 2013). Furthermore, in the experimentally IAV-infected mice, Gal9 expression was upregulated (Hattori et al. 2013), an observation consistent with the elevated levels of Gal9 observed in patients with IAV infection (Katoh et al. 2014). In contrast, Gal3 can function as an anti-viral galectin not by direct interactions as reported by

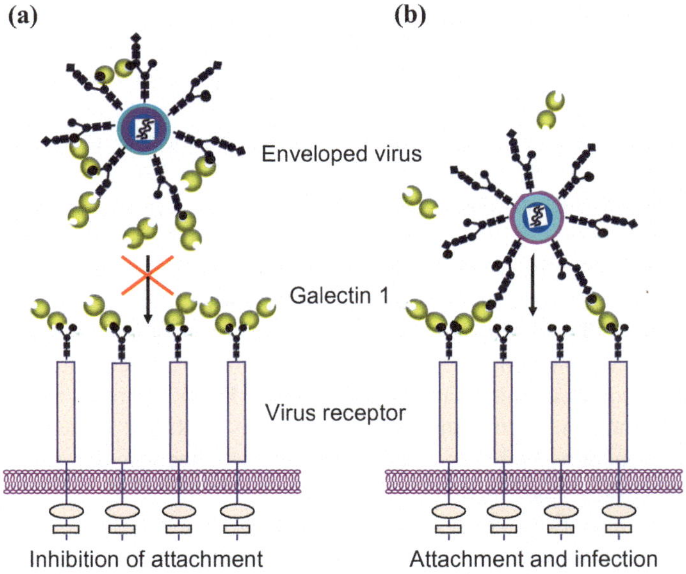

Fig. 7.3 Galectins can inhibit or facilitate attachment of enveloped viruses to the host cell surface. **a** Gal1 can block the attachment of viruses such as dengue by binding to and "coating" the envelope oligosaccharides (influenza A and IHNV), or by hindering access to the viral receptors (IHNV and dengue) on the host cell surface. **b** Gal1 can also cross-link the viral envelope to the host cell surface receptors facilitating attachment (Nipah and HIV)

Gal1 and Gal9, but by activating signaling via the JAK-STAT pathway, leading to an enhanced innate immune response (Jeon et al. 2010). A recent study reported that Gal1 can directly interact with dengue virus (DENV), a mosquito-transmitted enveloped RNA virus that can cause hemorrhagic fever. Gal1 directly binds to DENV and inhibits in vitro viral adhesion and internalization into host cells (Toledo et al. 2014). Prior exposure of the cells to dimeric Gal1, however, resulted in inhibition of viral attachment and infection that was greater than exposure of the virus alone. The role of Gal1 was also examined in vivo using Gal1 knockout mice, and demonstrated that the expression of endogenous Gal1 contributes to resistance against DENV infection (Toledo et al. 2014). During infection by Nipah virus (NiV) Gal1 can cross-link the N glycans displayed in the NiV envelope glycoproteins and reduce cell–cell fusion, thereby attenuating the pathophysiologic effects of NiV infection (Levroney et al. 2005). However, the beneficial effects of Gal1 in NiV infection are conditioned by the timing of the virus–galectin interaction, as during early stages of the viral exposure, Gal1 can enhance viral attachment and entry (Garner et al. 2015).

During the past few years we have used the zebrafish model to examine the roles of galectins in viral adhesion and entry by the infectious hematopoietic necrosis virus (IHNV), which is responsible for significant losses in both farmed and wild salmon and trout populations (Nita-Lazar et al. 2016). IHNV enters the host through the epithelial cells of the skin, the gills, and the gut, but the viral adhesion and entry mechanisms are not fully understood (Harmache et al. 2006). These epithelial cells express all three galectin types. They are secreted to the extracellular space, and are abundant in the mucus that coats the fish external surfaces. Results of the study showed that the zebrafish galectins Drgal1-L2 and Drgal3-L1 interact directly with the glycosylated envelope of IHNV significantly reducing viral attachment (Nita-Lazar et al. 2016). The structure of the complex of Drgal1-L2 with *N*-acetyl-D-lactosamine at 2.0 Å resolution together with models of Drgal3-L1 and the ectodomain of the IHNV glycoprotein provided insight into the mechanisms by which the binding of these galectins to the IHNV glycoprotein hinders the viral attachment (Ghosh et al. 2019). The IHNV envelope in glycoprotein is arranged in a honeycomb-like (hexagonal) arrangement of spikes, decorated by N-linked biantennary oligosaccharides that are also displayed by the host epithelial cells. Drgal1-L2 dimers can cross-link biantennary oligosaccharides from two spikes, thereby occluding two of the six co-receptor attachment sites on the surface of the virus, while the single C-terminal CRD of Drgal3-L1 oligomers can block three sites (Ghosh et al. 2019). Thus, the viral surface coverage by Drgal3-L1 is greater than that for Drgal1-L2 by a factor of 3/2 (Ghosh et al. 2019), which is consistent with the ratio of their inhibitory efficiency of 65%/40% determined in viral attachment experiments (Nita-Lazar et al. 2016). However, because the Drgal1-L2 and Drgal3-L1 secreted to the extracellular space also bind strongly to the fish epithelial cell surface, they can block IHNV cell surface receptors and hinder viral attachment via an alternative mechanism (Nita-Lazar et al. 2016). Furthermore, as the secreted galectins also bind strongly to the skin mucus glycans, a third defense mechanism mediated by these galectins would consist in cross-linking and immobilization of the virus in the mucus matrix, which is sloughed off periodically from the fish skin (Abernathy and Vasta, unpublished). A similar mechanism has been proposed for galectin-4 (Gal4), which is expressed and secreted by gut epithelial cells. Once in the extracellular space, Gal4 would hinder attachment of *Bordetella pertussis* and *Helicobacter pylori* by binding to their gut epithelial cell surface receptors (Danielsen and Hansen 2006; Ideo et al. 2005).

In addition to blocking pathogen attachment to the host cell surface described above, galectins can inhibit interaction of the pathogen's virulence factors with host receptors. For example, the glycolipid-binding galectin Lec-8, which is strongly expressed in sections of the digestive tract of the nematode *Caenorhabditis elegans*, contributes to host defense against bacterial virulence by competitive binding to glycolipid receptors for the pore-forming toxin Cry5B from *Bacillus thuringiensis*, a nematocidal pathogen (Ideo et al. 2009). Interestingly, nematodes can also be susceptible to galectins from their fungal prey. For example, when the nematotoxic galectin CGL2 from the fungus *Coprinopsis cinerea* is ingested by *C. elegans*, it inhibits development and reproduction, and kills the nematode by specifically binding to a trisaccharide (Galβ1,4Fucα1,6GlcNAc) displayed on the nematode intestine, suggesting that fungal galectins constitute a defense mechanism against predator nematodes (Butschi et al. 2010). Similarly, the mammalian Gal2 can also suppress *C. elegans* development by binding to the Galβ1,4Fuc glycotope, a moiety that is also recognized on the parasitic nematodes of humans, such as *Ascaris*, *Nippostrongylus*, and *Brugia* spp., suggesting that it may contribute to anti-parasitic responses (Takeuchi et al. 2019). Gal11 may also contribute to host defense in ruminants against the gastrointestinal nematode parasite, *Haemonchus contortus* (Preston et al. 2015).

2. Direct killing of the pathogen: Some galectins have been reported to not only bind to, but also directly kill the pathogens, possibly by disrupting their cell surface integrity and their normal physiology. For example, the tandem repeat Gal4 and galectin-8 (Gal8), which are expressed in the human intestinal tract, can specifically recognize and kill *Escherichia coli* strains that display B-blood group oligosaccharides (BGB+ *E. coli*), while other *E. coli* strains or bacterial species are not affected (Stowell et al. 2010). The killing activity of both galectins is mediated by their C-terminal domains, and appears to be caused by compromising the integrity of the bacterial cell surface. Mutation of key residues in CRD revealed that the C-CRD mediates recognition of the BGB+ *E. coli* but does not affect its viability, while the N-CRD might be endowed with killing activity (Stowell et al. 2010).

More recently, Park et al. (2016) reported that recombinant Gal3 agglutinated *Helicobacter pylori* and displayed a potent bactericidal effect, as revealed by propidium iodide uptake and drastic morphological changes. The significance of this observation was buttressed by the higher bacterial loads in Gal3-deficient mice than in WT mice that had been experimentally infected with *H. pylori* via gastric tube, supporting the notion that Gal3 plays an important role in innate immunity to infection and gastric colonization by *H. pylori* (Park et al. 2016).

Although galectins can modulate phagocytosis and cytokine responses (IL-17, IL-23, TNFα, and others) to several fungal pathogens, including *Candida albicans* (Linden et al. 2013a, b), *Cryptococcus neoformans* (Almeida et al. 2017), *Histoplasma capsulatum* (Wu et al. 2013), and *Paracoccidioides brasiliensis* (Ruas et al.

2009), the galectin-mediated anti-fungal defense mechanisms may also include binding to and direct killing of the pathogen. The first two, *C. albicans* and *C. neoformans*, are particularly susceptible to direct recognition and fungicidal activity by galectins, although the mechanisms involved appear to be different. Gal3 recognizes and kills *Candida* species that display β1,2-linked oligomannans on the cell surface, but does not bind to *Candida* species or strains, or other fungal species such as *Saccharomyces cerevisiae* that lack these glycans. The binding of Gal3 to *C. albicans* oligomannans is intriguing, as it is well established that like other members of the galectin family, β-galactosyl moieties, particularly LacNAc, are the preferred ligands. Like for the bacteriocidal activity of Gal4 for *E. coli* described above (Stowell et al. 2010), changes in the *C. albicans* cell morphology upon exposure to Gal3 suggested damage to the cell membrane as the basis for the fungicidal activity, although the detailed mechanism has not been elucidated (Kohatsu et al. 2006). In contrast, exposure of *C. neoformans* to Gal3 causes lysis of the fungal extracellular vesicles which contain virulence factors and inhibits fungal growth (Almeida et al. 2017). The disruption of the extracellular vesicles by Gal3 prevents the efficient delivery of their contents to macrophages, and together with its fungistatic activity favors the host's anti-fungal immune response. The cell surface of *C. neoformans* appears to lack β-galactosides, or the β-oligomannosides recognized by Gal3 on *C. albicans*, and the ligand(s) recognized on the *C. neoformans* capsule or vesicles have not been identified so far (Almeida et al. 2017).

3. Opsonization, phagocytosis, encapsulation, and clearance of the pathogen: As mentioned above, galectins can bind to and promote phagocytosis of pathogens, which are killed by intracellular oxidative burst, and cleared from the internal milieu. An example that illustrates the opsonic role of galectins in host defense is the galectin MjGal from the kuruma shrimp, *Marsupenaeus japonicus* (Shi et al. 2014). This galectin is upregulated in circulating phagocytic cells (hemocytes) and hepatopancreas upon bacterial infection, and can bind to both gram-positive and gram-negative bacteria through the recognition of lipoteichoic acid or lipopolysaccharide, respectively. By also binding to the shrimp hemocyte surface, MjGal functions as an opsonin, cross-linking the potentially pathogenic bacteria to the hemocyte surface and promoting their phagocytosis and facilitating their clearance from circulation, as shown in vivo by RNA interference (Shi et al. 2014). There are several reports of galectins from both invertebrates and vertebrates which have been reported as opsonic, or at least promoting phagocytosis by more complex mechanisms that may involve additional receptors on the phagocytic cell surface. Gal3 can recognize surface glycans on the opportunistic fungal pathogens *Candida* spp. and not only exert direct fungicidal activity, as described above, but also promote phagocytosis by neutrophils (Linden et al. 2013a, b). In contrast, phagocytosis of *Candida* spp. by macrophages appears to require TLR2 (Jouault et al. 2006).

In addition to phagocytosis, a defense response to infectious challenge that is typical of invertebrates consists of encapsulation of potential pathogens with multiple layers of cells, particularly when the infectious particle is too large to be phagocytosed by a single cell. The encapsulated, immobilized pathogen can then be killed by diverse mechanisms, such as oxidative stress or melanization (Xia et al. 2018; Vazquez et al.

2009), which have been conserved in vertebrates from fish to mammals. Parasitic nematodes (*Cucullanus* spp.) in the abdominal cavity of the conger eel (*Conger myriaster*) are immobilized and encapsulated by layers of cells with the participation of congerins I and II, which are galectins that can bind to glycans on both the nematode surface as well as the encapsulating cells (Nakamura et al. 2012). In mammals, the formation of granulomas appears as an analogous defense mechanism in which galectins may play a key role for the immobilization of pathogen or parasites. Such is the case of recognition of LacdiNAc (GalNAcβ1,4-GlcNAc) of the eggs and parasite surface of the helminth *Schistosoma mansoni* by the host Gal3, and its potential role on the formation of liver granulomas (van den Berg et al. 2004). Gal3-null mice experimentally infected with *S. mansoni* showed reduced liver granulomas in both the acute and chronic phases, as compared with wild-type mice (Breuilh et al. 2007).

The recognition and clearance of both intra- and extracellular pathogens and parasites, however, can be accomplished by alternative mechanisms that lead to autophagy, and Gal8 has been shown to function as a key participant in this process. Epithelial cells infected with *Salmonella typhi*, *Listeria monocytogenes*, or *Shigella flexneri* exhibit damage to cytoplasmic endosomes and lysosomes (Thurston et al. 2012). It has been proposed that the damaged vacuolar membranes signal to recruit Gal8, which by binding to the exposed glycans activates autophagy (Thurston et al. 2012). A recent report describes a related anti-microbial mechanism by which *Helicobacter pylori* infection causes lysosomal damage in epithelial cells of the gastric epithelium (Li et al. 2019). Lysosome damage exposes membrane luminal *O*-glycans that are recognized by and induce the aggregation of cytoplasmic Gal8, which in turn increases autophagy activity in the *H. pylori*-infected cells (Li et al. 2019).

7.5 "Subversion" of the Galectins' Defense Functions by Pathogens and Parasites

In recent years, mounting experimental evidence has shown that some pathogens and parasites can "subvert" the defense roles of galectins from the host or invertebrate vector, and use galectin-mediated recognition to attach to, or to gain entry to their cells (Vasta 2009). These galectin-mediated interactions are clearly beneficial for the pathogen or parasite and can take place by various mechanisms. One fairly prevalent mechanism consists of the direct cross-linking of the pathogen to the host or vector cells by recognition of similar or different glycans on the cell surfaces, either by multivalent TR galectin or oligomeric proto- or chimera-type galectins. Another mechanism that can result as a downstream from the first consists of the downregulation of the host innate and adaptive immune response by galectin overexpression, secretion, or binding to host cell surfaces upon infectious challenge.

The participation of galectin interactions in the infection mechanisms of HIV (human immunodeficiency virus) has been widely reported (Ouellet et al. 2005;

Mercier et al. 2008). Gal1, which is abundant in organs that represent major reservoirs for HIV-1, such as the thymus and lymph nodes, promotes infection by HIV-1 by cross-linking the LacNAc moieties on the viral glycoprotein gp120 to its cellular glycoprotein receptor CD4 on T cells facilitating viral attachment, increasing viral residence time on the cell surface, and infection efficiency (Sato et al. 2012) (Fig. 7.3b). Additionally, Gal1 would enhance the uptake of the virus by macrophages acting as a soluble scavenger receptor (Sato et al. 2012). In contrast, Gal3, which is upregulated by the HIV Tat protein in several human cell lines, has no effect on HIV-1 attachment or entry (Fogel et al. 1999), but may exert anti-viral activity by inducing apoptosis of HIV-infected cells (Xue et al. 2017). Gal9 can also enhance HIV entry into T cells but via an indirect mechanism based on changes in the redox status of the cell surface (Bi et al. 2011). It is noteworthy that DC-SIGN, a C-type lectin, also facilitates HIV entry into dendritic cells (Sato et al. 2012), illustrating the diversity of protein–carbohydrate interactions that may participate in HIV infection (Ouellet et al. 2005; Mercier et al. 2008). By a mechanism similar to HIV, Gal1 can enhance and stabilize attachment of human T-cell lymphotropic virus (HTLV) to human T cells (Gauthier et al. 2008). Gal1 can also interact with capsid proteins of enteroviruses and promote infection of epithelial cells, although the mechanism(s) involved are not clear (Lee et al. 2015). In contrast, Gal3 would increase cell survival and inhibit apoptosis of the enterovirus-infected cells, facilitating release of the mature viral progeny (Huang et al. 2016). As indicated above for the Nipah virus, if Gal1 is present during the initial phase of virus exposure, it can enhance NiV attachment to the endothelial cell surface by bridging glycans on the viral envelope to host cell glycoproteins (Garner et al. 2015). A similar mechanism has been proposed for the interaction of Gal3 with the herpes simplex virus type 1 (HSV-1). Gal3 enhances HSV-1 attachment to and infection of human corneal keratinocytes, which can be ameliorated by the presence of MUC16, a soluble mucin that is secreted by the corneal cells and is strongly bound by Gal3 (Woodward et al. 2013). In the zebrafish-IHNV infection model described above, different members of the galectin repertoire can display opposite functions: while the proto-type galectin Drgal1-L2 and the chimera-type galectin DrGal3-L1 can inhibit viral adhesion to epithelial cells (Nita-Lazar et al. 2016), the TR galectin DrGal9-L1 can significantly promote viral adhesion and infection (Mancini and Vasta, unpublished).

Galectins can also promote the attachment of bacterial pathogens and facilitate infection, as shown in a murine model for influenza A infection and pneumococcal pneumonia revealed. Neuraminidases from both influenza A virus (IAV) and *Streptococcus pneumoniae* significantly desialylate the airway epithelial surface and modulate expression and release of Gal1 and Gal3 to the bronchoalveolar space (Nita-Lazar et al. 2015a). Studies on the human airway epithelial cell line A549 supported the observations made in the mouse model, and revealed that both Gal1 and Gal3 bind strongly to IAV and to *S. pneumoniae*. Furthermore, exposure of A549 cells to viral neuraminidase or influenza infection significantly increased galectin-mediated *S. pneumoniae* adhesion to the cell surface (Nita-Lazar et al. 2015a). Thus, these observations suggest that upon influenza infection, pneumococcal attachment to the airway epithelial surface is enhanced by the activity of both viral and pneumococcal neuraminidases and the secreted host galectins, and possibly contributes to the greater susceptibility of influenza patients to secondary pneumonia (Nita-Lazar et al. 2015a). In addition, the study revealed that the binding of Gal1 and Gal3 to the epithelial cell surface downregulates the expression of SOCS1 and RIG1, and activation of ERK, AKT, or JAK/STAT1 signaling pathways, leading to overexpression and release of pro-inflammatory cytokines (Nita-Lazar et al. 2015b). These results suggest that upon influenza infection, the binding of secreted Gal3 to the desialylated airway epithelia can severely dysregulate the immune response, leading to the frequently observed "cytokine storm" (Nita-Lazar et al. 2015b).

More recently, the role of Gal1 in infections by *Chlamydia trachomatis*, a highly prevalent sexually transmitted bacterium worldwide, was investigated in detail, revealing that Gal1 enhanced *C. trachomatis* attachment to cervical epithelial cells through recognition of bacterial glycoproteins and N-glycosylated host cell receptors, particularly platelet-derived growth factor receptor (PDGFR)β and β1/αVβ3 integrins (Lujan et al. 2018). Bacterial entry was facilitated by exposure to Gal1, mainly in its dimeric form, which favored interactions among *C. trachomatis*, and between the bacteria and host cells. In vivo studies in mice lacking Gal1 or complex branched *N*-glycans supported the in vitro results (Lujan et al. 2018).

Interactions of galectins with eukaryotic parasites can also promote their attachment and entry into epithelial or phagocytic cells, the latter by opsonic effect. The protozoan parasite *Trichomonas vaginalis* is the causative agent of a sexually transmitted human infection, which can effectively colonize cervical epithelial cells, placenta, endometrial and decidual tissue, as well as prostate (Okumura et al. 2008). Gal1 was identified as the receptor for *T. vaginalis* on epithelial cells: the parasite exhibits abundant lipophosphoglycan (LPG) with galactosyl moieties that are recognized by Gal1 expressed by the epithelial cells, facilitating parasite attachment to the cervix linings (Okumura et al. 2008).

Taking advantage of galectin functions as opsonins, the protozoan parasite *Perkinsus marinus*, a facultative intracellular parasite of the eastern oyster *Crassostrea virginica* is recognized via the oyster's 4-CRD galectins CvGal1 and CvGal2 that recognize and promote phagocytosis of the parasite by the circulating hemocytes (Fig. 7.4a). Both CvGal1 and CvGal2 recognize and bind to microalgae such as *Tetraselmis* spp., and potentially pathogenic bacterial species such as *Aeromonas* spp., *Carnobacterium* spp., *Streptococcus* spp., *Bacillus* spp., and *Vibrio* spp., supporting the notion that during filter-feeding, the oyster galectins contribute to feeding and anti-microbial defense in the gut lumen by promoting phagocytosis of both phytoplankton and bacteria, killing by intracellular respiratory burst, and digesting them in the phagosome compartment by lysosomal enzymes. Therefore, *P. marinus* parasites may have co-evolved with their host to subvert the defense and feeding roles of the oyster galectins. This would have taken place by adaptation of the parasite's glycocalyx to be competitively recognized by the hemocyte galectins over microalgal food and microbial pathogens, and phagocytosed by the oyster hemocytes, where they inhibit respiratory burst and proliferate, eventually causing systemic infection

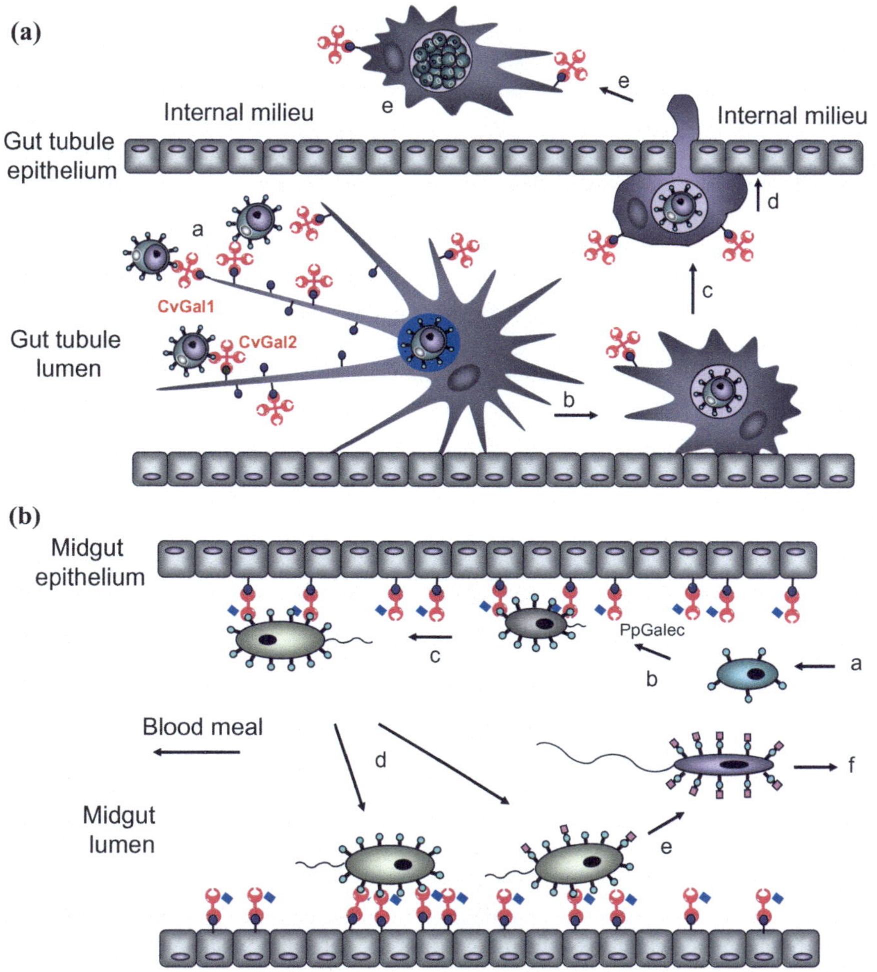

Fig. 7.4 Galectins from the host or vector can facilitate attachment of parasites. **a** The galectins CvGal1 and CvGal2 from the oyster (*Crassostrea virginica*) recognize and opsonize *Perkinsus marinus* trophozoites and promote their phagocytosis (a) by hemocytes (phagocytic cells in the oyster hemolymph and tissues). The phagocytosed *P. marinus* trophozoites avoid intracellular killing by inhibiting the hemocyte oxidative burst (b), and transmigrate through the gut epithelium (c, d) into the internal milieu, where it proliferates (e), causing systemic infection and eventually death of the oyster host. **b** The sandfly (*Phlebotomus papatasi*) TR galectin PpGalec recognizes and binds to poly-Gal(β1–3) side chains (light blue circles) on the lipophosphoglycan (LPG) of *Leishmania major* amastigotes (a) and facilitates attachment of the parasite to the midgut (b), preventing their excretion along with the digested blood meal. The amastigotes mature into promastigotes (c) and undergo numerous divisions (d) before differentiating into infective metacyclics (e, f). During metacyclogenesis, LPG can be capped with arabinose (pink squares), inhibiting recognition by PpGalec and allowing the free-swimming infective metacyclic promastigotes to detach from the sandfly midgut and migrate to the salivary glands for transmission to the mammalian host (Adapted from Vasta 2009)

and death of the oyster host (Tasumi and Vasta 2007; Feng et al. 2013, 2015a, b; Kurz et al. 2013; Vasta et al. 2015).

Eukaryotic parasites also take advantage of the binding properties of galectins not only from the host but also from invertebrate vectors to attach to their cell surfaces. These can be illustrated by interactions of *Leishmania* species with the midgut linings of their insect vectors prior to transmission to the vertebrate hosts (Fig. 7.4b). *Leishmania* amastigotes attach to the insect midgut epithelium via the sandfly galectin PpGalec that binds to the Gal(ß1-3) side chains on the *Leishmania* LPG, to prevent their excretion along with the digested bloodmeal (Kamhawi et al. 2004). During differentiation of the amastigotes into free-swimming infective metacyclics, modifications of their glycocalyx reduce binding by PpGalec, releasing flagellated metacyclics that migrate to the vector's salivary glands, and upon the sandfly's next feeding will infect a new vertebrate host (Kamhawi et al. 2004).

7.6 Conclusion

It is currently well established that by recognizing endogenous (self) cell surface and soluble glycans ligands, galectins participate in early development, tissue repair, regulate immune homeostasis, and contribute to trigger immune responses upon infectious challenge. As discussed throughout this review, as galectins can recognize exogenous (non-self) glycans present on the surface of virus, bacteria, and eukaryotic parasites, they are also considered bona fide pattern recognition receptors (PRRs). The capacity of galectins to recognize structural topologies ("patterns") by a galectin can be illustrated by the recognition with similar binding affinity of the disaccharides LacNAc and TDG by the proto-type galectin from the South American toad *Bufo arenarum* (Bianchet et al. 2000). The crystal structures of the *B. arenarum* galectin complexes with LacNAc and TDG allowed us to rationalize the structural basis of the PRR concept as it concerns the recognition of "self" and "non-self" carbohydrate moieties by galectins (Fig. 7.5). The structures revealed that the non-reducing terminal galactose, shared by both disaccharide ligands, shows identical interactions with the protein. The second moiety (GlcNAc in LacNAc and another galactose in TDG) establishes different contacts in the TDG and in the LacNAc complex, although the same number and quality of H-bonds are present in both cases (two direct and one water mediated to the protein) (Fig. 7.5b, c). This structural evidence reveals that the galectin is exquisitely specific, displaying a surprising structural plasticity to establish well-defined amino acid/sugar interactions for the two chemically distinct disaccharides, LacNAc and TDG, resulting in similar binding affinity for both (Bianchet et al. 2000).

Based on the Medzhitov and Janeway model (Medzhitov and Janeway 2002), however, PRRs such as the mannose-binding lectin (MBL) recognize pathogens via highly conserved microbial surface molecules of wide distribution such as lipopolysaccharide or peptidoglycan (pathogen- or microbe-associated molecular patterns, PAMPs or MAMPs), which are absent in the host (Fig. 7.6a). Although

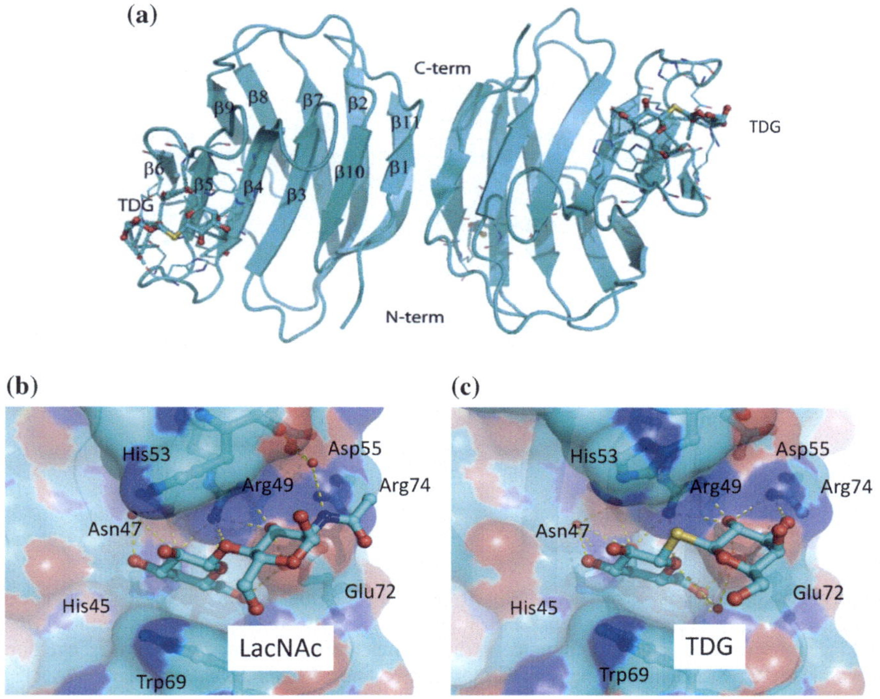

Fig. 7.5 Recognition of topologically similar disaccharides (LacNAc and TDG) by the *Bufo arenarum* proto-type galectin. **a** Structure of the dimer of *B. arenarum* galectin-thiodigalactoside (TDG) complex. Bound TDG disaccharides are shown on the binding clefts of both galectin subunits. **b** N-acetyllactosamine (LacNAc) bound to *B. arenarum* galectin (monomer B). The relevant CRD residues and the hydrogen bonds (dashed lines) to the sugar are shown. Carbon atoms are in white, oxygen in red; nitrogen in blue. **c** Thiodigalactoside (TDG) bound to *B. arenarum* galectin (monomer B). The relevant CRD residues and the hydrogen bonds (dashed lines) to the sugar are shown. Carbon atoms are in white, oxygen in red; sulfur in yellow; nitrogen in blue (Adapted from Bianchet et al. 2000)

this is true for some galectin ligands such as LacDiNAc in the helminth *Schistosoma mansoni*, the *Candida albicans* β1,2-linked oligomannans, and the *Perkinsus marinus* trophozoite surface glycans, most of the galectin carbohydrate ligands on the pathogen or parasite surface are identical or highly similar to those endogenous host ligands (Fig. 7.6b). Therefore, galectins do not rigorously fit the definition of PRRs, as they can recognize carbohydrate ligands that are displayed on both the host and the pathogen cell surface. Because TR galectins display in tandem-arrayed CRDs of similar but distinct specificity in a single polypeptide monomer, the binding and cross-linking of endogenous and exogenous glycans can be rationalized by the distinct properties of their binding sites. For other lectins such as the proto- and chimera-type galectins that display a single binding site per monomer, their capacity to recognize both endogenous and exogenous glycans through the same binding

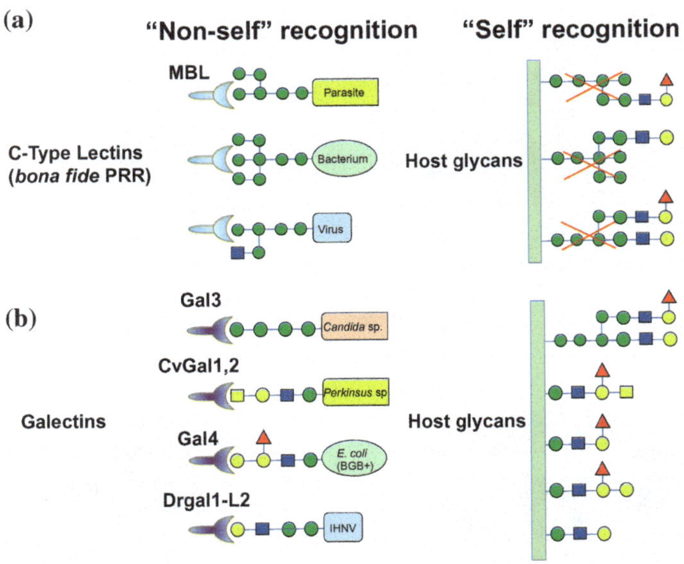

Fig. 7.6 Recognition of exogenous ("non-self") carbohydrate ligands on viruses, bacteria, parasites, and fungi by a *bona fide* pattern recognition receptor (PRR: mannose-binding lectin; MBL) and galectins: **a** The mannose-binding lectin (MBL) functions as a *bona fide* pattern recognition receptor (PRR) by recognizing oligosaccharides (microbe-associated molecular patterns, MAMPs, or pathogen-associated molecular patterns, PAMPs) on the surface of potential pathogens (parasites, bacteria, and viruses) that are either absent or unavailable for binding in the host cell surfaces. **b** Galectin-3 (Gal3) can function as true PRR by recognizing β1,2-linked oligomannans on the surface of *Candida albicans*, which are absent from the host, but can also recognize LacNAc and polylactosamines abundant on the host cell surface. Other galectins such as the oyster CvGal1 and CvGal2, Gal4 and the zebrafish Drgal1-L2 can recognize exogenous ("non-self") carbohydrate ligands on viruses, bacteria, parasites, and fungi that may be topologically similar (*Perkinsus marinus*, recognized by CvGal1 and CvGal2) or chemically identical (IHNV envelope oligosaccharides recognized by Drgal1-L2) to the host's endogenous ligands

site cannot be rigorously explained with the current evidence. This apparent paradox reveals our limited knowledge about the actual diversity in recognition of the host galectin repertoire and the structural and biophysical aspects of ligand binding preference (Dam and Brewer 2010). This lack of detailed information particularly concerns the diverse architectural display of the galectin ligands within the complex carbohydrate moieties of the host cell and the microbial surface, and how these features impact the affinity and avidity of the oligomeric galectins in the extracellular space. Furthermore, multiple factors pertaining to the local galectin concentrations and oligomerization, susceptibility to oxidative inactivation and proteolytic cleavage, and the biophysical properties of the microenvironment(s) in which the aforementioned interactions take place warrant further investigation (Vasta 2009).

From a functional point of view, recognition of glycans on the surface of virus, bacteria, and parasites by host galectins can lead to various and sometimes opposite outcomes that benefit either the host or the pathogen. For example, as discussed above galectins can block pathogen attachment and infection, or directly kill or promote phagocytosis, intracellular killing, and clearance of the pathogen from the host, which are defense functions that seem to have arisen early in evolution. In contrast, galectins can also promote attachment of pathogens and parasites to the host cells, and facilitate infection. From an evolutionary standpoint, given that host galectins play key roles in early development, tissue repair, and regulation of immune homeostasis via recognition of "self" carbohydrate moieties, the substantial conservation of this lectin family from protistans, and fungi, to invertebrates and vertebrates supports the notion that strong functional constraints would prevent any dramatic evolutionary changes in galectin structure and carbohydrate specificity that would be detrimental to the host. Together with the well-recognized evolutionary plasticity of pathogens for colonization of host tissues, it seems plausible that pathogens would have rather evolved their surface glycosomes to mimic their hosts, and subvert the roles of host galectins by taking advantage of their carbohydrate-binding properties for attachment and entry into the host cells in a "Trojan horse" model (Vasta 2009).

Acknowledgements In the text, information is reviewed from studies in the author's laboratory that were supported by the National Institutes of Health, National Science Foundation, Maryland Sea Grant Program, National Oceanic and Atmospheric Administration, and the Mizutani Foundation for Glycoscience (Japan).

References

Ahmed H, Vasta GR (2008) Unlike mammalian GRIFIN, the zebrafish homologue (DrGRIFIN) represents a functional carbohydrate-binding galectin. Biochem Biophys Res Commun 371:350–355

Ahmed H, Bianchet MA, Amzel LM, Hirabayashi J, Kasai K, Giga-Hama Y, Tohda H, Vasta GR (2002) Novel carbohydrate specificity of the 16- kDa galectin from *Caenorhabditis elegans*: binding to blood group precursor oligosaccharides (type 1, type 2, Talpha, and Tbeta) and gangliosides. Glycobiology 12:451–461

Ahmed H, Cappello F, Rodolico V, Vasta GR (2009a) Evidence of heavy methylation in the galectin 3 promoter in early stages of prostate adenocarcinoma: development and validation of a methylated marker for early diagnosis of prostate cancer. Transl Oncol 2:146–156

Ahmed H, Du SJ, Vasta GR (2009b) Knockdown of a galectin-1-like protein in zebrafish (*Danio rerio*) causes defects in skeletal muscle development. Glycoconj J 26(3):277–283

Ahmed H, Al Sadek DM (2015) Galectin-3 as a potential target to prevent cancer metastasis. Clin Med Insights Oncol 25(9):113–121

Almeida F, Wolf JM, da Silva TA, DeLeon-Rodriguez CM, Rezende CP, Pessoni AM, Fernandes FF, Silva-Rocha R, Martinez R, Rodrigues ML, Roque-Barreira MC, Casadevall A (2017) Galectin-3 impacts *Cryptococcus neoformans* infection through direct antifungal effects. Nat Commun 8(1):1968

Bi S, Hong PW, Lee B, Baum LG (2011) Galectin-9 binding to cell surface protein disulfide isomerase regulates the redox environment to enhance T-cell migration and HIV entry. Proc Natl Acad Sci USA 108:10650–10655

Bianchet MA, Ahmed H, Vasta GR, Amzel LM (2000) Soluble beta-galactosyl-binding lectin (galectin) from toad ovary: crystallographic studies of two protein-sugar complexes. Proteins 40(3):378–388 PMID: 10861929

Blidner AG, Mendez-Huergo SP, Cagnoni AJ, Rabinovich GA (2015) Re-wiring regulatory cell networks in immunity by galectin-glycan interactions. FEBS Lett 589:3407–3418

Blois SM, Ilarregui JM, Tometten M, Garcia M, Orsal AS, Cordo-Russo R, Toscano MA, Bianco GA, Kobelt P, Handjiski B, Tirado I, Markert UR, Klapp BF, Poirier F, Szekeres-Bartho J, Rabinovich GA, Arck PC (2007) A pivotal role for galectin-1 in fetomaternal tolerance. Nat Med 13:1450–1457

Blois SM, Dveksler G, Vasta GR, Freitag N, Blanchard V, Barrientos G (2019) Pregnancy galectinology: insights into a complex network of glycan binding proteins. Front Immunol 10:1166

Bourne Y, Bolgiano B, Liao DI, Strecker G, Cantau P, Herzberg O, Feizi T, Cambillau C (1994) Crosslinking of mammalian lectin (galectin-1) by complex biantennary saccharides. Nat Struct Biol 1(12):863–870

Breuilh L, Vanhoutte F, Fontaine J, van Stijn CM, Tillie-Leblond I, Capron M, Faveeuw C, Jouault T, van Die I, Gosset P, Trottein F (2007) Galectin-3 modulates immune and inflammatory responses during helminthic infection: impact of galectin-3 deficiency on the functions of dendritic cells. Infect Immun 75(11):5148–5157

Butschi A, Titz A, Wälti MA, Olieric V, Paschinger K, Nöbauer K, Guo X, Seeberger PH, Wilson IB, Aebi M, Hengartner MO, Künzler M (2010) *Caenorhabditis elegans* N-glycan core beta-galactoside confers sensitivity towards nematotoxic fungal galectin CGL2. PLoS Pathog 6(1):e1000717

Caldararu O, Manzoni F, Oksanen E, Logan DT, Ryde U (2019) Refinement of protein structures using a combination of quantum-mechanical calculations with neutron and X-ray crystallographic data. Acta Crystallogr D Struct Biol 75(Pt 4):368–380

Carlsson S, Oberg CT, Carlsson MC, Sundin A, Nilsson UJ, Smith D, Cummings RD, Almkvist J, Karlsson A, Leffler H (2007) Affinity of galectin-8 and its carbohydrate recognition domains for ligands in solution and at the cell surface. Glycobiology 17:663–676

Cho M, Cummings RD (1995a) Galectin-1, a β-galactoside-binding lectin in Chinese hamster ovary cells. II. Localization and biosynthesis. J Biol Chem 270:5207–5212

Cho M, Cummings RD (1995b) Galectin-1, a β-galactoside-binding lectin in Chinese hamster ovary cells. I. Physical and chemical characterization. J Biol Chem 270:5198–5206

Cleves AE, Cooper DN, Barondes SH, Kelly RB (1996) A new pathway for protein export in *Saccharomyces cerevisiae*. J Cell Biol 133:1017–1026

Colnot C, Ripoche M, Fowlis D, Cannon V, Scaerou F, Cooper DNW, Poirier F (1997) The role of galectins in mouse development. Trends Glycosci Glycotechnol 9:31–40

Colnot C, Sidhu SS, Balmain N, Poirier F (2001) Uncoupling of chondrocyte death and vascular invasion in mouse galectin 3 null mutant bones. Dev Biol 229:203–214

Cooper DN (2002) Galectinomics: finding themes in complexity. Biochim Biophys Acta 1572:209–231

Craig SE, Thummel R, Ahmed H, Vasta GR, Hyde DR, Hitchcock PF (2010) The zebrafish galectin Drgal1-l2 is expressed by proliferating Müller glia and photoreceptor progenitors and regulates the regeneration of rod photoreceptors. Invest Ophthalmol Vis Sci 51(6):3244–3252

Croci DO, Cerliani JP, Dalotto-Moreno T, Mendez-Huergo SP, Mascanfroni ID, Dergan-Dylon S, Toscano MA, Caramelo JJ, Garcia-Vallejo JJ, Ouyang J, Mesri EA, Junttila MR, Bais C, Shipp MA, Salatino M, Rabinovich GA (2014) Glycosylation-dependent lectin- receptor interactions preserve angiogenesis in anti-VEGF refractory tumors. Cell 156:744–758

Cummings RD, Liu FT, Vasta GR (2015–2017) Galectins. In: Varki A, Cummings RD, Esko JD, Stanley P, Hart GW, Aebi M, Darvill AG, Kinoshita T, Packer NH, Prestegard JH, Schnaar RL, Seeberger PH (eds) Essentials of glycobiology, 3rd edn. Cold Spring Harbor Laboratory Press, Cold Spring Harbor (NY) (Chapter 36)

Dam TK, Brewer CF (2008) Effects of clustered epitopes in multivalent ligand-receptor interactions. Biochemistry 47:8470–8476

Dam TK, Brewer CF (2010) Maintenance of cell surface glycan density by lectin-glycan interactions: a homeostatic and innate immune regulatory mechanism. Glycobiology 20:1061–1064

Danielsen EM, Hansen GH (2006) Lipid raft organization and function in brush borders of epithelial cells. Mol Membr Biol 23:71–79

Di Lella S, Sundblad V, Cerliani JP, Guardia CM, Estrin DA, Vasta GR, Rabinovich GA (2011) When galectins recognize glycans: from biochemistry to physiology and back again. Biochemistry 50:7842–7857

DiLella S, Marti MA, Croci DO, Guardia CMA, Dıáz Ricci JC, Rabinovich GA, Caramelo JJ, Estrin DA (2010) Linking the structure and thermal stability of β-galactoside-binding protein galectin-1 to ligand binding and dimerization equilibria. Biochemistry 49(35):7652–7658

Eastlake K, Heywood WE, Tracey-White D, Aquino E, Bliss E, Vasta GR, Mills K, Khaw PT, Moosajee M, Limb GA (2017) Comparison of proteomic profiles in the zebrafish retina during experimental degeneration and regeneration. Sci Rep 16(7):44601

Esteban A, Popp MW, Vyas VK, Strijbis K, Ploegh HL, Fink GR (2011) Fungal recognition is mediated by the association of dectin-1 and galectin-3 in macrophages. Proc Natl Acad Sci U S A 108:14270–14275. https://doi.org/10.1073/pnas.1111415108

Feng C, Ghosh A, Amin MN, Giomarelli B, Shridhar S, Banerjee A, Fernandez-Robledo JA, Bianchet MA, Wang LX, Wilson IB, Vasta GR (2013) The galectin CvGal1 from the eastern oyster (*Crassostrea virginica*) binds to blood group A oligosaccharides on the hemocyte surface. J Biol Chem 288:24394–24409

Feng C, Nita-Lazar M, González-Montalbán N, Wang J, Mancini J, Ravindran C, Ahmed H, Vasta GR (2015a) Manipulating galectin expression in zebrafish (*Danio rerio*). Methods Mol Biol 1207:327–341

Feng C, Ghosh A, Amin MN, Bachvaroff TR, Tasumi S, Pasek M, Banerjee A, Shridhar S, Wang LX, Bianchet MA, Vasta GR (2015b) Galectin CvGal2 from the Eastern Oyster (*Crassostrea virginica*) displays unique specificity for ABH blood group Oligosaccharides and differentially recognizes sympatric *Perkinsus* species. Biochemistry 54:4711–4730

Fogel S, Guittaut M, Legrand A, Monsigny M, Hebert E (1999) The tat protein of HIV-1 induces galectin-3 expression. Glycobiology 9:383–387

Fortuna-Costa A, Gomes AM, Kozlowski EO, Stelling MP, Pavao MS (2014) Extracellular galectin-3 in tumor progression and metastasis. Front Oncol 4:138

Gabius HJ (1997) Animal lectins. Eur J Biochem FEBS 243:543–576

García Caballero G, Kaltner H, Michalak M, Shilova N, Yegres M, André S, Ludwig AK, Manning JC, Schmidt S, Schnölzer M, Bovin NV, Reusch D, Kopitz J, Gabius HJ (2016) Chicken GRIFIN: a homodimeric member of the galectin network with canonical properties and a unique expression profile. Biochimie 128–129:34–47

Garner OB, Baum LG (2008) Galectin-glycan lattices regulate cell-surface glycoprotein organization and signalling. Biochem Soc Trans 36(Pt 6):1472–1477

Garner OB, Yun T, Pernet O, Aguilar HC, Park A, Bowden TA, Freiberg AN, Lee B, Baum LG (2015) Timing of galectin-1 exposure differentially modulates Nipah virus entry and syncytium formation in endothelial cells. J Virol 89:2520–2529

Gauthier S, Pelletier I, Ouellet M, Vargas A, Tremblay MJ et al (2008) Induction of galectin-1 expression by HTLV-I Tax and its impact on HTLV-I infectivity. Retrovirology 5:105

Georgiadis V, Stewart HJ, Pollard HJ, Tavsanoglu Y, Prasad R, Horwood J, Deltour L, Goldring K, Poirier F, Lawrence-Watt DJ (2007) Lack of galectin-1 results in defects in myoblast fusion and muscle regeneration. Dev Dyn 236:1014–1024

Gorski JP et al (2002) New alternatively spliced form of galectin-3, a member of the beta-galactoside-binding animal lectin family, contains a predicted transmembrane-spanning domain and a leucine zipper motif. J Biol Chem 277:18840–18848

Ghosh A, Banerjee A, Amzel LM, Vasta GR, Bianchet MA (2019) Structure of the zebrafish galectin-1-L2 and model of its interaction with the infectious hematopoietic necrosis virus (IHNV) envelope glycoprotein. Glycobiology 29(5):419–430

Guha P, Kaptan E, Bandyopadhyaya G, Kaczanowska S, Davila E, Thompson K, Martin SS, Kalvakolanu DV, Vasta GR, Ahmed H (2013) Cod glycopeptide with picomolar affinity to galectin-3 suppresses T-cell apoptosis and prostate cancer metastasis. Proc Natl Acad Sci U S A 110:5052–5057

Guzman-Aranguez A, Mantelli F, Argueso P (2009) Mucin-type O-glycans in tears of normal subjects and patients with non-Sjogren's dry eye. Invest Ophthalmol Vis Sci 50:4581–4587

Harmache A, LeBerre M, Droineau S, Giovannini M, Bremont M (2006) Bioluminescence imaging of live infected salmonids reveals that the fin bases are the major portal of entry for Novirhabdovirus. J Virol 80:3655–3659

Hattori T, Arikawa T, Fujioka Y, Maruyama J, Nakayama Y, Ohba Y, Niki T, Miyazaki T, Hirashima M, Kida H (2013) Inhibition of influenza A virus infection by Galectin-9. Jpn J Vet Res 61(1&2):5–18

Hill M, Mazal D, Biron VA, Pereira L, Ubillos L, Berriel E, Ahmed H, Freire T, Rondan M, Vasta GR, Liu FT, Iglesias MM, Osinaga E (2010) A novel clinically relevant animal model for studying galectin-3 and its ligands during colon carcinogenesis. J Histochem Cytochem 58:553–565

Hirabayashi J, Kasai K (1993) The family of metazoan metal-independent beta-galactoside-binding lectins: structure, function and molecular evolution. Glycobiology 3:297–304

Hirabayashi J, Hashidate T, Arata Y, Nishi N, Nakamura T, Hirashima M, Urashima T, Oka T, Futai M, Muller WE, Yagi F, Kasai K (2002) Oligosaccharide specificity of galectins: a search by frontal affinity chromatography. Biochim Biophys Acta 1572:232–254

Hirashima M, Kashio Y, Nishi N, Yamauchi A, Imaizumi TA, Kageshita T, Saita N, Nakamura T (2002) Galectin-9 in physiological and pathological conditions. Glycoconj J 19(7–9):593–600

Houzelstein D, Gonçalves IR, Fadden AJ, Sidhu SS, Cooper DNW, Drickamer K, Leffler H, Poirier F (2004) Phylogenetic analysis of the vertebrate galectin family. Mol Biol Evol 21(7):1177–1187

Hsu DK, Liu F-T (2008) Regulation of immune responses by Galectin-3. In: Vasta GR, Ahmed H (eds) Animals lectins: a functional view. CRC Press, Boca Raton, FL

Hsu YA, Kuo YH, Chen CS, Chen YC, Huang CC, Chang CY, Lin CJ, Lin CW, Lin HJ, Liu FT, Wan L (2018) Galectin-12 is involved in corn silk-induced anti-adipogenesis and anti-obesity effects. Am J Chin Med 5:1–19

Huang WC, Chen HL, Chen HY, Peng KP, Lee Y et al (2016) Galectin-3 and its genetic variation rs4644 modulate enterovirus 71 infection. PLoS ONE 11(12):e0168627

Ideo H, Seko A, Ishizuka I, Yamashita K (2003) The N-terminal carbohydrate recognition domain of galectin-8 recognizes specific glycosphingolipids with high affinity. Glycobiology 13(10):713–723

Ideo H, Seko A, Yamashita K (2005) Galectin-4 binds to sulfated glycosphingolipids and carcinoembryonic antigen in patches on the cell surface of human colon adenocarcinoma cells. J Biol Chem 280:4730–4737

Ideo H, Fukushima K, Gengyo-Ando K, Mitani S, Dejima K, Nomura K, Yamashita K (2009) A *Caenorhabditis elegans* glycolipid-binding galectin functions in host defense against bacterial infection. J Biol Chem 284(39):26493–26501

Ideo H, Matsuzaka T, Nonaka T, Seko A, Yamashita K (2011) Galectin-8-N-domain recognition mechanism for sialylated and sulfated glycans. J Biol Chem 286(13):11346–11355

Liu F-T, Yang R-Y, Hsu DK (2012) Galectins in acute and chronic inflammation. Ann N Y Acad Sci 1253(1):80–91

Jeon S-B, Yoon HJ, Chang CY, Koh HS, Jeon S-H et al (2010) Galectin-3 exerts cytokine-like regulatory actions through the JAK-STAT pathway. J Immunol 185(11):7037–7046

Jouault T, ElAbed-ElBehi M, Martínez-Esparza M, Breuilh L, Trinel PA, Chamaillard M et al (2006) Specific recognition of Candida albicans by macrophages requires galectin-3 to discriminate *Saccharomyces cerevisiae* and needs association with TLR2 for signaling. J Immunol 177:4679–4687

Kamhawi S, Ramalho-Ortigao M, Pham VM, Kumar S, Lawyer PG, Turco SJ, Barillas-Mury C, Sacks DL, Valenzuela JG (2004) A role for insect galectins in parasite survival. Cell 119:329–341

Katoh S, Ikeda M, Shimizu H, Mouri K, Obase Y, Kobashi Y, Fukushima K, Hirashima M, Oka M (2014) Increased levels of plasma galectin-9 in patients with influenza virus infection. Tohoku J Exp Med 232(4):263–267

Kohatsu L, Hsu DK, Jegalian AG, Liu FT, Baum LG (2006) Galectin-3 induces death of Candida species expressing specific beta-1,2-linked mannans. J Immunol 177(7):4718–4726

Krejcirikova V, Pachl P, Fabry M, Maly P, Rezacova P, Brynda J (2011) Structure of the mouse galectin-4N-terminal carbohydrate-recognition domain reveals the mechanism of oligosaccharide recognition. Acta Crystallogr D Biol Crystallogr 67:204–211

Krzeminski M, Singh T, André S, Lensch M, Wu AM, Bonvin AM, Gabius HJ (2011) Human galectin-3 (Mac-2 antigen): defining molecular switches of affinity to natural glycoproteins, structural and dynamic aspects of glycan binding by flexible ligand docking and putative regulatory sequences in the proximal promoter region. Biochim Biophys Acta 1810:150–161

Kurz S, Jin C, Hykollari A, Gregorich D, Giomarelli B, Vasta GR, Wilson IB, Paschinger K (2013) Hemocytes and plasma of the eastern oyster (*Crassostrea virginica*) display a diverse repertoire of sulfated and blood group A-modified N-glycans. J Biol Chem 288:24410–24428

Kutzner TJ, Gabba A, Fitzgerald FG, Shilova NV, García Caballero G, Ludwig AK, Manning JC, Knospe C, Kaltner H, Sinowatz F, Murphy PV, Cudic M, Bovin NV, Gabius HJ (2019) How altering the modular architecture affects aspects of lectin activity: case study on human galectin-1. Glycobiology cwz034. https://doi.org/10.1093/glycob/cwz034 (Epub ahead of print)

Lee PH, Liu CM, Ho TS, Tsai YC, Lin CC et al (2015) Enterovirus 71 virion-associated galectin-1 facilitates viral replication and stability. PLoS ONE 10(2):e0116278

Levroney EL, Aguilar HC, Fulcher JA, Kohatsu L, Pace KE, Pang M, Gurney KB, Baum LG, Lee B (2005) Novel innate immune functions for galectin-1: galectin-1 inhibits cell fusion by Nipah virus envelope glycoproteins and augments dendritic cell secretion of proinflammatory cytokines. J Immunol 175:413–420

Li FY, Weng IC, Lin CH, Kao MC, Wu MS, Chen HY, Liu FT (2019) Helicobacter pylori induces intracellular galectin-8 aggregation around damaged lysosomes within gastric epithelial cells in a host O-glycan-dependent manner. Glycobiology 29(2):151–162

Liao D-I, Kapadia G, Ahmed H, Vasta GR, Herzberg O (1994) Structure of S-lectin, a developmentally regulated vertebrate b-galactoside-binding protein. Proc Natl Acad Sci 91:1428–1432

Linden JR, De Paepe ME, Laforce-Nesbitt SS, Bliss JM (2013a) Galectin-3 plays an important role in protection against disseminated candidiasis. Med Mycol 51:641–651

Linden JR, Kunkel D, Laforce-Nesbitt SS, Bliss JM (2013b) The role of galectin-3 in phagocytosis of *Candida albicans* and *Candida parapsilosis* by human neutrophils. Cell Microbiol 15:1127–1142

Lipkowitz MS, Leal-Pinto E, Cohen BE, Abramson RG (2004) Galectin 9 is the sugar-regulated urate transporter/channel UAT. Glycoconj J 19:491–498

Lobsanov YD, Gitt MA, Leffler H, Barondes SH, Rini JM (1993) X-ray crystal structure of the human dimeric S-Lac lectin, L-14-II, in complex with lactose at 2.9-Å resolution. J Biol Chem 268(36):27034–27038

Lopez-Lucendo MF, Solis D, Andre S, Hirabayashi J, Kasai K, Kaltner H, Gabius HJ, Romero A (2004) Growth-regulatory human galectin-1: crystallographic characterization of the structural changes induced by single-site mutations and their impact on the thermodynamics of ligand binding. J Mol Biol 343(4):957–970

Lujan AL, Croci DO, Gambarte Tudela JA, Losinno AD, Cagnoni AJ, Mariño KV, Damiani MT, Rabinovich GA (2018) Glycosylation-dependent galectin-receptor interactions promote *Chlamydia trachomatis* infection. Proc Natl Acad Sci USA 115(26):E6000–E6009

Machala EA, McSharry BP, Rouse BT, Abendroth A, Slobedman B (2019) Gal power: the diverse roles of galectins in regulating viral infections. J Gen Virol 100(3):333–349

Medzhitov R, Janeway CA Jr (2002) Decoding the patterns of self and nonself by the innate immune system. Science 296:298–300

Méndez-Huergo SP, Blidner AG, Rabinovich GA (2017) Galectins: emerging regulatory checkpoints linking tumor immunity and angiogenesis. Curr Opin Immunol 45:8–15

Mercier S, St-Pierre C, Pelletier I, Ouellet M, Tremblay MJ, Sato S (2008) Galectin-1 promotes HIV-1 infectivity in macrophages through stabilization of viral adsorption. Virology 371:121–129

Morris S, Ahmad N, Andre S, Kaltner H, Gabius HJ, Brenowitz M, Brewer F (2004) Quaternary solution structures of galectins-1, -3, and -7. Glycobiology 14:293–300

Nabi IR, Shankar J, Dennis JW (2015) The galectin lattice at a glance. J Cell Sci 128(13):2213–2219
Nagae M, Nishi N, Murata T, Usui T, Nakamura T, Wakatsuki S, Kato R (2009) Structural analysis of the recognition mechanism of poly-N-acetyllactosamine by the human galectin-9 N-terminal carbohydrate recognition domain. Glycobiology 19:112–117
Nakamura O, Watanabe M, Ogawa T, Muramoto K, Ogawa K, Tsutsui S, Kamiya H (2012) Galectins in the abdominal cavity of the conger eel *Conger myriaster* participate in the cellular encapsulation of parasitic nematodes by host cells. Fish Shellfish Immunol 33(4):780–787
Nemoto-Sasaki Y, Hayama K, Ohya H, Arata Y, Kaneko MK, Saitou N, Hirabayashi J, Kasai K (2008) *Caenorhabditis elegans* galectins LEC-1-LEC-11: structural features and sugar-binding properties. Biochim Biophys Acta 1780(10):1131–1142
Nita-Lazar M, Banerjee A, Feng C, Amin MN, Frieman MB, Chen WH, Cross AS, Wang LX, Vasta GR (2015a) Desialylation of airway epithelial cells during influenza virus infection enhances pneumococcal adhesion via galectin binding. Mol Immunol 65:1–16
Nita-Lazar M, Banerjee A, Feng C, Vasta GR (2015b) Galectins regulate the inflammatory response in airway epithelial cells exposed to microbial neuraminidase by modulating the expression of SOCS1 and RIG1. Mol Immunol 68:194–202
Nita-Lazar M, Mancini J, Feng C, Gonzalez-Montalban N, Ravindran C, Jackson S, Heras-Sanchez Ade L, Giomarelli B, Ahmed H, Haslam SM, Wu G, Dell A, Ammayappan A, Vakharia VN, Vasta GR (2016) The zebrafish galectins Drgal1-L2 and Drgal3-L1 bind in vitro to the infectious hematopoietic necrosis virus (IHNV) glycoprotein and reduce viral adhesion to fish epithelial cells. Dev Comp Immunol 55:241–252
Ogden AT, Nunes I, Ko K, Wu S, Hines CS, Wang AF et al (1998) GRIFIN, a novel lens-specific protein related to the galectin family. J Biol Chem 273:28889–28896
Okumura CY, Baum LG, Johnson PJ (2008) Galectin-1 on cervical epithelial cells is a receptor for the sexually transmitted human parasite *Trichomonas vaginalis*. Cell Microbiol 10:2078–2090
Ouellet M, Mercier S, Pelletier I, Bounou S, Roy J, Hirabayashi J, Sato S, Tremblay MJ (2005) Galectin-1 acts as a soluble host factor that promotes HIV-1 infectivity through stabilization of virus attachment to host cells. J Immunol 174:4120–4126
Pang J, Rhodes DH, Pini M, Akasheh RT, Castellanos KJ, Cabay RJ, Cooper D, Perretti M, Fantuzzi G (2013) Increased adiposity, dysregulated glucose metabolism and systemic inflammation in Galectin-3 KO mice. PloS ONE 8:e57915
Pace KE, Lebestky T, Hummel T, Arnoux P, Kwan K, Baum LG (2002) Characterization of a novel *Drosophila melanogaster* galectin. J Biol Chem 277:13091–13098
Park AM, Hagiwara S, Hsu DK, Liu FT, Yoshie O (2016) Galectin-3 plays an important role in innate immunity to gastric infection by *Helicobacter pylori*. Infect Immun 84(4):1184–1193
Pejnovic NN, Pantic JM, Jovanovic IP, Radosavljevic GD, Djukic A, Arsenijevic NN, Lukic ML (2013) Galectin-3 is a regulator of metaflammation in adipose tissue and pancreatic islets. Adipocyte 2:266–271
Preston SJ, Beddoe T, Walkden-Brown S, Meeusen E, Piedrafita D (2015) Galectin-11: a novel host mediator targeting specific stages of the gastrointestinal nematode parasite, *Haemonchus contortus*. Int J Parasitol 45(12):791–796
Rabinovich GA, Toscano MA (2009) Turning 'sweet' on immunity: galectin-glycan interactions in immune tolerance and inflammation. Nat Rev Immunol 9:338–352
Rabinovich G, Toscano MA, Jackson SS, Vasta GR (2007a) Functions of cell surface galectin-glycoprotein lattices. Curr Opin Struct Biol 17:513–520
Rabinovich GA, Liu FT, Hirashima M, Anderson A (2007b) An emerging role for galectins in tuning the immune response: lessons from experimental models of inflammatory disease, autoimmunity and cancer. Scand J Immunol 66:143–158
Rabinovich GA, van Kooyk Y, Cobb BA (2012) Glycobiology of immune responses. Ann N Y Acad Sci 1253:1–15
Ramaswamy S, Sleiman MH, Masuyer G, Arbez-Gindre C, Micha-Screttas M, Calogeropoulou T, Steele BR, Acharya KR (2015) Structural basis of multivalent galactose-based dendrimer recognition by human galectin-7. FEBS J 282(2):372–387

Rossi B, Espeli M, Schiff C, Gauthier L (2006) Clustering of pre-B cell integrins induces galectin-1-dependent pre-B cell receptor relocalization and activation. J Immunol 177:796–803

Ruas LP, Bernardes ES, Fermino ML, de Oliveira LL, Hsu DK, Liu FT, Chammas R, Roque-Barreira MC (2009) Lack of galectin-3 drives response to *Paracoccidioides brasiliensis* toward a Th2-biased immunity. PLoS ONE 4:e4519

Sato S, Ouellet M, St-Pierre C, Tremblay MJ (2012) Glycans, galectins, and HIV-1 infection. Ann N Y Acad Sci 1253:133–148

Schwarz FP, Ahmed H, Bianchet MA et al (1998) Thermodynamics of bovine spleen galectin-1 binding to disaccharides: correlation with structure and its effect on oligomerization at the denaturation temperature. Biochemistry 37:5867–5877

Seetharaman J, Kanigsberg A, Slaaby R et al (1998) X-ray crystal structure of the human galectin-3 carbohydrate recognition domain at 2.1-Å resolution. J Biol Chem 273:13047–13052

Shi X-Z, Wang L, Xu S, Zhang X-W, Zhao X-F, Vasta GR, Wang J-X (2014) A galectin from the kuruma shrimp (*Marsupenaeus japonicus*) functions as an opsonin and promotes bacterial clearance from hemolymph. PLoS ONE 9:e91794

Si Y, Feng S, Gao J, Wang Y, Zhang Z, Meng Y, Zhou Y, Tai G, Su J (2016) Human galectin-2 interacts with carbohydrates and peptides non-classically: new insight from X-ray crystallography and hemagglutination. Acta Biochim Biophys Sin (Shanghai) 48(10):939–947

Stowell SR, Qian Y, Karmakar S, Koyama NS, Dias-Baruffi M, Leffler H, McEver RP, Cummings RD (2008) Differential roles of galectin-1 and galectin-3 in regulating leukocyte viability and cytokine secretion. J Immunol 180:3091–3102

Stowell SR, Arthur CM, Dias-Baruffi M, Rodrigues LC, Gourdine J-P, Heimburg-Molinaro J, Ju T, Molinaro RJ, Rivera-Marrero C, Xia B, Smith DF, Cummings RD (2010) Innate immune lectins kill bacteria expressing blood group antigen. Nat Med 16:295–301

Stowell SR, Cho M, Feasley CL, Arthur CM, Song X, Colucci JK, Karmakar S, Mehta P, Dias-Baruffi M, McEver RP, Cummings RD (2009) Ligand reduces galectin-1 sensitivity to oxidative inactivation by enhancing dimer formation. J Biol Chem 284(8):4989–4999

Su J, Gao J, Si Y, Cui L, Song C, Wang Y, Wu R, Tai G, Zhou Y (2018) Galectin-10: a new structural type of prototype galectin dimer and effects on saccharide ligand binding. Glycobiology 28(3):159–168

Takeuchi T, Tamura M, Ishiwata K, Hamasaki M, Hamano S, Arata Y (2019) Hatanaka T Galectin-2 suppresses nematode development by binding to the invertebrate-specific galactoseβ1-4fucose glyco-epitope. Glycobiology 29(6):504–512

Tasumi S, Vasta GR (2007) A galectin of unique domain organization from hemocytes of the Eastern oyster (*Crassostrea virginica*) is a receptor for the protistan parasite *Perkinsus marinus*. J Immunol 179:3086–3098

Than NG, Sumegi B, Than GN, Berente Z, Bohn H (1999) Isolation and sequence analysis of a cDNA encoding human placental tissue protein 13 (PP13), a new lysophospholipase, homolog of human eosinophil Charcot-Leyden Crystal protein. Placenta 20:703–710

Than NG, Pick E, Bellyei S, Szigeti A, Burger O, Berente Z et al (2004) Functional analyses of placental protein 13/galectin-13. Eur J Biochem 271:1065–1078

Thurston TL, Wandel MP, von Muhlinen N, Foeglein A, Randow F (2012) Galectin8 targets damaged vesicles for autophagy to defend cells against bacterial invasion. Nature 482:414–418

Toledo KA, Fermino ML, Andrade Cdel C, Riul TB, Alves RT, Muller VD, Russo RR, Stowell SR, Cummings RD, Aquino VH, Dias- Baruffi M (2014) Galectin-1 exerts inhibitory effects during DENV-1 infection. PLoS ONE 9:e112474

Tomizawa T et al (2005) Solution structure of the C-terminal gal-bind lectin domain from human galectin-4. Structural Genomics and Proteomics Initiative (RSGI), Riken

Tsay YG, Lin NY, Voss PG, Patterson RJ, Wang JL (1999) Export of galectin-3 from nuclei of digitonin-permeabilized mouse 3T3 fibroblasts. Exp Cell Res 252:250–261

van den Berg TK, Honing H, Franke N, van Remoortere A, Schiphorst WE, Liu FT et al (2004) LacdiNAc glycans constitute a parasite pattern for galectin-3-mediated immune recognition. J Immunol 173:1902–1907

Vasta GR (2009) Roles of galectins in infection. Nat Rev Microbiol 7:424–438
Vasta GR, Ahmed H (2008) Animal lectins: a functional view. CRC Press, New York
Vasta GR, Ahmed H, Du S-J, Henrikson D (2004) Galectins in teleost fish: Zebrafish (*Danio rerio*) as a model species to address their biological roles in development and innate immunity. Glycoconj J 21:503–521
Vasta GR, Ahmed H, Bianchet MA, Fernández-Robledo JA, Amzel LM (2012) Diversity in recognition of glycans by F-type lectins and galectins: molecular, structural, and biophysical aspects. Ann N Y Acad Sci 1253(1):14–26
Vasta GR, Feng C, Bianchet MA, Bachvaroff TR, Tasumi S (2015) Structural, functional, and evolutionary aspects of galectins in aquatic mollusks: from a sweet tooth to the Trojan horse. Fish Shellfish Immunol 46(1):94–106
Vasta GR, Amzel LM, Bianchet MA, Cammarata M, Feng C, Saito K (2017) F-type lectins: a highly diversified family of fucose-binding proteins with a unique sequence motif and structural fold, involved in self/non-self-recognition. Front Immunol 8:1648
Vazquez L, Alpuche J, Maldonado G, Agundis C, Pereyra-Morales A, Zenteno E (2009) Review: immunity mechanisms in crustaceans. Innate Immun 15(3):179–188
Woodward AM, Mauris J, Argueso P (2013) Binding of transmembrane mucins to galectin-3 limits herpesvirus 1 infection of human corneal keratinocytes. J Virol 87(10):5841–5847
Wu AM et al (2002) Fine specificity of domain-I of recombinant tandem-repeat-type galectin-4 from rat gastrointestinal tract (G4-N). Biochem J 367:653–664
Wu SY, Yu JS, Liu FT, Miaw SC, Wu-Hsieh BA (2013) Galectin-3 negatively regulates dendritic cell production of IL-23/IL-17-axis cytokines in infection by *Histoplasma capsulatum*. J Immunol 190:3427–3437
Xia X, You M, Rao XJ, Yu XQ (2018) Insect C-type lectins in innate immunity. Dev Comp Immunol 83:70–79
Xue J, Fu C, Cong Z, Peng L, Peng Z et al (2017) Galectin-3 promotes caspase-independent cell death of HIV-1-infected macrophages. FEBS J 284(1):97–113
Yang RY, Yu L, Graham JL, Hsu DK, Lloyd KC, Havel PJ, Liu FT (2011a) Ablation of a galectin preferentially expressed in adipocytes increases lipolysis, reduces adiposity, and improves insulin sensitivity in mice. Proc Natl Acad Sci U S A 108:18696–18701
Yang M-L, Chen Y-H, Wang S-W, Huang Y-J, Leu C-H et al (2011b) Galectin-1 binds to influenza virus and ameliorates influenza virus pathogenesis. J Virol 85(19):10010–10020
Yang ML, Chen YH, Wang SW, Huang YJ, Leu CH, Yeh NC, Chu CY, Lin CC, Shieh GS, Chen YL, Wang JR, Wang CH, Wu CL, Shiau AL (2011c) Galectin-1 binds to influenza virus and ameliorates influenza virus pathogenesis. J Virol 85:10010–10020
Zelensky AN, Gready JE (2005) The C-type lectin-like domain superfamily. FEBS J 272:6179–6217
Zhou D, Ge H, Sun J, Gao Y, Teng M, Niu L (2008) Crystal structure of the C-terminal conserved domain of human GRP, a galectin-related protein, reveals a function mode different from those of galectins. Proteins 15;71(3):1582–1588
Zhu C, Anderson AC, Schubart A, Xiong H, Imitola J, Khoury SJ, Zheng XX, Strom TB, Kuchroo VK (2005) The Tim-3 ligand galectin-9 negatively regulates T helper type 1 immunity. Nat Immunol 6:1245–1252

Chapter 8
Siglecs at the Host–Pathogen Interface

Yung-Chi Chang and Victor Nizet

Abstract Siglecs are sialic acid (Sia) recognizing immunoglobulin-like receptors expressed on the surface of all the major leukocyte lineages in mammals. Siglecs recognize ubiquitous Sia epitopes on various glycoconjugates in the cell glycocalyx and transduce signals to regulate immunological and inflammatory activities of these cells. The subset known as CD33-related Siglecs is principally inhibitory receptors that suppress leukocyte activation, and recent research has shown that a number of bacterial pathogens use Sia mimicry to engage these Siglecs as an immune evasion strategy. Conversely, Siglec-1 is a macrophage phagocytic receptor that engages GBS and other sialylated bacteria to promote effective phagocytosis and antigen presentation for the adaptive immune response, whereas certain viruses and parasites use Siglec-1 to gain entry to immune cells as a proximal step in the infectious process. Siglecs are positioned in crosstalk with other host innate immune sensing pathways to modulate the immune response to infection in complex ways. This chapter summarizes the current understanding of Siglecs at the host–pathogen interface, a field of study expanding in breadth and medical importance, and which provides potential targets for immune-based anti-infective strategies.

Keywords Sialic acid · Streptococcus · Pattern-recognition receptor · Trans-infection

Y.-C. Chang (✉)
Graduate Institute of Microbiology, College of Medicine, National Taiwan University, No. 1, Sec. 1, Jen-Ai Rd., Taipei 10051, Taiwan
e-mail: yungchiychang@ntu.edu.tw

V. Nizet
Division of Host-Microbe Systems and Therapeutics, Department of Pediatrics, and Skaggs School of Pharmacy and Pharmaceutical Sciences, UC San Diego, 9500 Gilman Drive Mail Code 0760, La Jolla, CA 92093, USA
e-mail: vnizet@ucsd.edu

S.-L. Hsieh (ed.), *Lectin in Host Defense Against Microbial Infections*,
Advances in Experimental Medicine and Biology 1204,
https://doi.org/10.1007/978-981-15-1580-4_8

8.1 Introduction

Sialic acid-binding immunoglobulin-type lectins (Siglecs) are cell surface receptors belonging to the immunoglobulin (Ig) superfamily. The extracellular domains of these receptors comprise one ligand-binding V-set domain and a variable number of C2-set domains, with high-sequence similarities to the variable and constant region of antibodies, respectively. Siglecs are mostly located on the cell surface of hematopoietic cells, with the exception of Siglec-4 (Schwann cells) and Siglec-6 (epithelial cells). Until now, 14 human Siglecs and 9 murine Siglecs have been identified with different preferences for binding to terminal sialic acids (Sia) on glycan structures in a linkage-sensitive fashion. The arginine residue in the Siglec V-set domain is critical for contacting the carboxyl group of the Sia to form a salt bridge that stabilizes the binding interaction.

Phylogenetically, Siglec family members can be subdivided into two groups: First, there are Siglecs that are conserved among different species, but showing low-sequence identity to one another, including Siglec-1, Siglec-2, Siglec-4, and Siglec-15. The other group comprises the so-called CD33 (Siglec-3)-related Siglecs (CD33rSiglecs), which show low gene conservation but possess higher degrees of sequence identity among the subfamily members. Both conserved Siglecs and CD33rSiglecs include activating and inhibitory receptors. The inhibitory Siglecs contain an immunoreceptor-based inhibition motif (ITIM) in their intracellular domain, conferring the ability to antagonize immune signaling pathways through the recruitment of the SHP phosphatases. Conversely, activating Siglecs have an aspartic acid residue in their trans-membrane domain that associates with immunoreceptor-based activation motif (ITAM)-containing adaptor DAP12 (DNAX-activation protein of 12 kDa) to promote signaling and immune responses. Lastly, Siglec-1 does not possess a functional intracellular domain and is not known to signal directly, but rather plays a role in cell–cell and cell–microbe interactions.

8.2 Siglecs in Bacterial Infection

Glycans are ubiquitous on eukaryotic cell surfaces via their incorporation in glycoproteins, glycosphingolipids, and glycerophosphatides. In mammalian cells, Sia is usually the outmost sugar residue on the oligosaccharide chains of cell surface or serum glycoconjugates, where it functions in recognition and anti-recognition phenomena ranging from the regulation of complement activation to the control of cell–cell apposition (Varki 1993). Like their host cells, bacteria have also evolved complex biosynthetic pathways to produce a diverse array of carbohydrates that form the building blocks of capsular polysaccharides (CPS), lipopolysaccharides (LPS), lipooligosaccharides (LOS), and peptidoglycans. These specialized bacterial glycans play pivotal roles in a number of biological processes, particularly mediating microbe–host interactions during the onset and development of infectious disease.

It has been long known that the bacterial CPS represents a key virulence factor for most encapsulated bacterial pathogens by protecting them from immune clearance within the host. Some bacterial capsules interfere with the binding and activation of complement factors on the bacterial surface by inhibiting the C3b convertase or recruiting inhibitory complement factor H (Cross and Kelly 1990). Bacterial capsules exerting this immune resistance mechanism usually contain sialic or polysialic acids or hyaluronic acid (HA), structurally identical or similar to the polysaccharides found in mammalian tissues, as exemplified by *Escherichia coli* K1, *Neisseria meningitidis* types B and C, or group A or group B *Streptococcus* (Stevens et al. 1978; Wessels et al. 1989; Foley and Wood 1959; Dale et al. 1996). Envelopment in these host-like capsular structures confers resistance to complement-mediated killing and phagocytic uptake by neutrophils and macrophages, increasing the chance of bacterial survival and dissemination into host tissues. In addition to these passive mechanisms of protection, an emerging hypothesis has been raised based upon the discovery of inhibitory members of the Siglec protein family on immune cells. Researchers have explored whether these host-mimicking pathogens can blunt activation of antimicrobial responses via engagement of inhibitory Siglecs using the Sia in their CPS. In parallel, studies have asked how hosts have evolved to better recognize and destroy such camouflaged pathogens. The interplay between a particular sialylated human pathogen, group B *Streptococcus* (GBS), and host immune responses serves as a good first example to illustrate the role of Siglecs and bacterial expression of Sia in the pathogenesis of infection.

GBS is a leading cause of neonatal pneumonia, sepsis, and meningitis and is increasingly recognized as a pathogen in elderly and immunocompromised adult populations (Heath and Schuchat 2007; Thigpen et al. 2011; Skoff et al. 2009). GBS can be classified into ten serotypes varying in structural and antigenic features, but which share in common the presence of a terminal Sia residue that closely resembles the one presented throughout the abundant surface glycocalyx of all human cells (Cieslewicz et al. 2005). Sialylated CPS from different GBS serotypes interacts with several human Siglecs in a Sia- and serotype-specific manner (Carlin et al. 2007). Through this Sia molecular mimicry, GBS engages inhibitory human Siglec-9 receptors on neutrophils, resulting in reduced production of reactive oxidative species (ROS) and neutrophil extracellular traps (NETs), which together impair bactericidal activity (Carlin et al. 2009b). Recently, GBS Sia engagement of Siglec-9 on human platelets to suppress their activation and release of antimicrobial peptides (AMPs) was shown to contribute to GBS resistance to platelet killing (Uchiyama et al. 2019). Corroborating these findings, transgenic mice expressing a soluble Siglec-9 receptor, which acts as a decoy to prevent neutrophil suppression, were more resistant to GBS infection in vivo (Saito et al. 2016). Linkages to the underlying sugar chain as well as the substitution at certain carbon positions of Sias are critical in determining their binding specificity for Siglecs. One study used isogenic mutants with different *O*-acetylation phenotypes to show that Sia *O*-acetylation protects GBS CPS from enzymatic removal by microbial sialidases produced by commensal microbes occupying in the same mucosal niche as GBS (Weiman et al. 2009). However, the same modification markedly reduced GBS binding to human Siglec-9, such that highly *O*-acetylated GBS mutants were less able to accomplish Sia-mediated neutrophil suppression and showed reduced virulence in vivo (Weiman et al. 2009, 2010). These observations suggest that GBS must balance competing evolutionary selective pressures to fine tune the *O*-acetylation level of CPS Sias to maximize its survival advantage in the host.

The role of GBS CPS Sia engagement of inhibitory Siglecs in the context of in vivo infection was first addressed in Siglec-E deficient mice due to the similarity of function and cellular distribution between human Siglec-9 and murine Siglec-E. Upon lower dose GBS challenge intranasally or intravenously, Siglec-E deficient mice showed increased production of several inflammatory cytokines and had reduced dissemination of the pathogen to the brain (GBS meningitis). However, exaggerated inflammatory mediators and reduced anti-inflammatory cytokine IL-10 production were observed in the Siglec-E deficient mice during a high-dose lethal challenge with GBS (Chang et al. 2014a). Thus, the sum consequence of the GBS molecular mimicry and inhibitory Siglec engagement is likely to vary based upon the site, stage, and magnitude of infection. Importantly, in addition to Sia-dependent Siglec engagement, certain GBS strains can use the surface-anchored β protein to bind human Siglec-5, another inhibitory Siglec preferentially expressed on macrophages and neutrophils. This protein-mediated engagement increased bacterial attachment to the macrophage surface, but simultaneously paralyzed macrophage killing functions, leading to a net reduction of phagocytosis, ROS production, NET formation, and bactericidal activity (Carlin et al. 2009a).

To counteract inhibitory Siglec hijacking strategies of bacterial pathogens, host Siglec-1 (sialoadhesin, CD169) plays a crucial role in limiting bacterial dissemination. This receptor, uniquely expressed in the marginal metallophilic and subcapsular sinus macrophages, recognizing the same key Sia epitope on the GBS surface, however lacks an intracellular ITM motif (Crocker and Gordon 1989; Crocker et al. 1994). Siglec-1 binding to GBS CPS Sia promoted phagocytic and bactericidal activity of macrophages in vitro and restricts GBS dissemination in vivo (Chang et al. 2014b). Loss of Siglec-1 expression not only affected the macrophage sampling and trapping capabilities but also the production of anti-GBS antibodies, suggesting a key role in optimization of antigen presentation and subsequent adaptive immune response against sialylated pathogens (Chang et al. 2014b). Another evolutionary adaption of the host to defeat pathogen Siglec hijacking is the emergence of activating Siglecs with the potential to counteract inhibitory Siglec-mediated immune suppression. In 2006, Angata et al. discovered Siglec-14, which possesses nearly identical Sia-binding domain as inhibitory Siglec-5 through gene conversion, but is coupled with DAP12, an ITAM motif bearing adaptor (Angata et al. 2006). Thus Siglec-5 and -14 represent paired receptors with opposite signaling effects. For example, on neutrophils and amniotic epithelium, β protein-expressing GBS can bind to Siglec-5 and Siglec-14, with the latter engagement stimulating p38 MAP kinase and AKT signaling to promote more efficient bacterial clearance. Notably, a *SIGLEC14*-null polymorphism is present in some humans, caused by fusion between *SIGLEC14* and *SIGLEC5* genes, resulting in functional deletion of Siglec-14 expression (Yamanaka et al. 2009). A genetic survey of the *SIGLEC14*-null polymorphism and GBS colonization and premature delivery found that *SIGLEC14*-null allele is associated with higher GBS colonization in mothers and more frequent premature birth of infants from GBS-positive pregnancies (Ali et al. 2014).

In addition to GBS, the pathogens *Escherichia coli* K1 and *Neisseria meningitidis* serotype B, important agents of bacterial meningitis in infants and children, possess sialylated capsules. The CPS produced by these two bacteria resembles the same poly-α2,8-Sia (PSA) structure abundantly expressed on neurons and the glia cells of the developing central nervous systems (CNS) is critical for neuron development and function (Finne 1982; Devi et al. 1991; Rutishauser 2008). The human-specific Siglec-11 shows a unique expression pattern in human microglia and selective binding preference to α2,8-linked Sias (Angata et al. 2002; Hayakawa et al. 2005). Engagement of Siglec-11 by endogenous host PSA exerts protective effects against inflammation-mediated neurotoxicity (Wang and Neumann 2010). The mimicry of mammalian PSA by *E. coli* K1 and *N. meningitidis* type B CPS may, therefore, target Siglec-11 expression on microglia to blunt immune responsiveness and facilitate neuroinvasion, consistent with the high rates of mortality and serious neurological sequelae seen with these pathogens (Robbins et al. 1974; Kaper et al. 2004). Interestingly, the activating Siglec-16, which shares over 99% of sequence identity to the first two Ig-like domain of Siglec-11 but is coupled with ITAM-containing DAP12, was later discovered to be a paired receptor of Siglec-11 allowing fine-tuning of immune responses (Cao et al. 2008). Pathogenic *E. coli* K1 engages inhibitory Siglec-11 through their PSA capsule to inhibit macrophage anti-bacterial functions. In contrast, activating Siglec-16 expressing macrophages recognizes the same epitope to promote elimination of pathogen (Schwarz et al. 2017). In sum, macrophage engagement of GBS β protein or *E. coli* K1 PSA capsule illustrates an evolutionary dynamic in which the activating member of the Siglec receptor pair may override the immune suppressive responses generated by pathogen mimicry of the inhibitory Siglec member.

In addition to the CPS, sialylation of LPS and LOS is also a prominent binding target of Siglecs. For example, Siglec-1 and Siglec-5 bind to the sialylated LPS of *N. meningitidis*, and cells expressing either Siglec internalized meningococci in a Siglec- and Sia-dependent manner (Jones et al. 2003). Strains of *Campylobacter jejuni* express various monosialylated and disialylated LOS with α2,3- or α2,3/2,8-linked Sia residues, respectively, which perfectly mimic host neural gangliosides GM1, GD1a, GD3, or GT1a. Colonization with *C. jejuni* strains possessing sialylated LOS is epidemiologically associated with higher risk of developing a postinfectious autoimmune neuropathy termed Guillain-Barré syndrome (Jacobs et al. 1996; Willison et al. 2016; Willison and Yuki 2002). Several human Siglecs recognize these particular sialylated LOS structures, although the functional consequences of this interaction require further investigation. Siglec-7, which is expressed on natural killer (NK) cells, monocytes, and dendritic cells (DCs), can mediate specific Sia-dependent interaction with *C. jejuni* LOS (Avril et al. 2006). *C. jejuni* strains recognized by Siglec-7 express terminal disialylated residues mimicking host GQ1b-like epitopes, including GD1c and GD3. The Siglec-7 binding signal of *C. jejuni* correlates with its ability to elicit anti-GQ1b antibodies, and strains recognized by Siglec-7 were particularly associated with oculomotor weakness in Guillain-Barré syndrome and its so-called Miller-Fisher variant (Heikema et al. 2013). On the other hand, *C. jejuni* with α2,3-linked Sia on the LOS chain showed strong interaction with Siglec-1 (Heikema et al. 2010), which facilitated bacterial uptake and induce higher macrophage production of the pro-inflammatory cytokine IL-6 (Heikema et al. 2013). A key role of Siglec-1 in recognition of sialylated *C. jejuni* LOS was confirmed in Siglec-1 deficient animals. Bone-marrow-derived macrophages from mice lacking Siglec-1 showed greatly reduced phagocytosis of sialylated *C. jejuni*, coupled with reduced production of proinflammatory cytokines and type I interferon responses (Klaas and Crocker 2012).

Sialylated LOS is important in DC and macrophage activation as well as subsequent T cell polarization and B cell activation. Removal of Sia from the *C. jejuni* LOS causes reduced myeloid cell activation and subsequent B cell responses (Kuijf et al. 2010; Huizinga et al. 2012, 2013). Varied sialylated LOS structures differentially modulate DC-mediated T cell polarization in a Siglec-dependent manner, wherein the GD1a/GM1a mimic induced a more pronounced Th2 skewing, while the GD1c mimic preferentially stimulated Th1 responses (Bax et al. 2011). These observations suggest that targeting distinct DC-expressed Siglecs may represent a potential strategy for manipulating Th cell differentiation programs and forestalling autoimmune disease post *C. jejuni* infection. In addition to modulating host DC functions via its sialylated LOS, *C. jejuni* triggers IL-10 production of DCs via engaging Siglec-10 through the pseudaminic acid residues on its flagella (Stephenson et al. 2014). These abundant Sia and Sia-like *C. jejuni* surface structures (e.g. pseudaminic acid) help mediate a complicated interaction with host immune cells that may impact *C. jejuni* human disease associations, ranging from autoimmune neuropathies to asymptomatic colonization in individuals with repeated exposure to *Campylobacter* spp.

In addition to the well-studied microorganisms mentioned above, a growing list of bacterial species have been discovered to display sialoglycoconjugates on their surfaces, such as *Pseudomonas aeruginosa* (Khatua et al. 2010, 2012), *Klebsiella pneumoniae* (Lee et al. 2014), and nontypeable *Haemophilus influenzae* (NTHi) (Kalograiaki et al. 2016). *P. aeruginosa* recruits host sialoglycoproteins and displays them on the bacterial surface, and these absorbed Sias can enhance bacterial survival by reducing complement deposition and engaging inhibitory Siglec-9 to suppress neutrophil bactericidal machinery (Khatua et al. 2010, 2012). *K. pneumoniae* exhibiting a hypermucoviscosity phenotype possess abundant Sia CPS; blocking of the pathogen's ability to engage inhibitory Siglec-9 enhances neutrophil bactericidal activity (Lee et al. 2014). NTHi LOS also contains a terminal Sia residue that interacts with Siglec-14, which enhances inflammatory cytokine production in Siglec-14-expressing macrophages correlating to COPD exacerbation in human patients (Angata et al. 2013). All these observations suggest that Sia molecular mimicry by many medical important bacteria can interplay with various Siglecs to affect infectious risk and clinical disease manifestations.

Glycosaminoglycans (GAGs) are another family of complex carbohydrates ubiquitously present on mammalian host cell surfaces and in the extracellular matrix, regulating a wide range of biological functions from cell adhesion and cell migration to tissue repair and immune responses (Linhardt and Toida 2004). In another instance of molecular mimicry, GAG structures have been discovered in several gram-positive and gram-negative bacterial capsules (Wessels et al. 1994; Jann and Jann 1992; DeAngelis 2002). The GAG HA is structurally identical in animals and in the capsule of group A *Streptococcus* (GAS) where it is serving as a molecular camouflage to evade host immune responses (Wessels et al. 1994; Dale et al. 1996). Recently, GAS was shown to use its HA capsule to target Siglec-9 on neutrophils, thereby blocking NET formation and oxidative burst to inhibit bactericidal function (Secundino et al. 2016). It is an interesting example of convergent evolution that two structurally unrelated carbohydrates, Sia of GBS and HA of GAS, each target distinct epitopes in the V-set domain of inhibitory Siglec-9 to dampen immune responses and promote pathogen survival in the human host (Secundino et al. 2016).

8.3 Siglecs in Viral Infection

Although fundamentally distinct from the Sia molecular mimicry strategies employed by bacterial pathogens to engage inhibitory Siglecs for immune evasion, several lines of evidence indicate viruses can themselves take advantage of Sia-Siglec interactions for cell targeting, spreading, and trans-infection.

Human immunodeficiency virus-1 (HIV-1) exploits a Sia-Siglec axis to facilitate its entry into myeloid cells and *trans*-infection into $CD4^+$ T cells. Notably, Siglec-1 possesses a unique extended 17 Ig-like extracellular domain structure extending out from the surface glycocalyx in an un-masked state which makes it an ideal surface entry target. Moreover, in contrast to infected $CD4^+$ T cells, myeloid cells are relatively resistant to HIV-1-induced cytopathic effects and there is no obvious depletion of myeloid cells in HIV-1-infected patients. To escape from host immune surveillance, HIV-1 may have evolved to hijack this Siglec-1-mediated cellular recognition pathway for its own benefit, to hide within infected macrophages and *trans*-infect $CD4^+$ T cells for efficient viral spread.

In the manner of many enveloped retroviruses, the HIV-1 envelope glycoprotein gp120 is heavily glycosylated, and gp120 mutations that remove N-linked glycan sites on HIV (and simian immunodeficiency virus SIV) impair virus attachment and entry (Auwerx et al. 2008). A direct interaction of HIV-1 and Siglec was first reported by Rempel et al. in 2008, where expression levels of Siglec-1 were correlated to HIV-1 viral load, and Siglec-1 was proven to bind HIV-1 in a Sia-dependent manner and facilitate infection to a permissive reporter cell line (Rempel et al. 2008). Later gp120 was proven to serve as viral ligand for Siglec-1 and for several CD33rSigecs, including Siglec-3, -5, -7, and -9, with varying avidity and HIV-1 strain dependency. Moreover, this Siglec-gp120 interaction facilitates virus infectivity to macrophages and T cells (Zou et al. 2011; Varchetta et al. 2013). Mature DCs (mDCs) capture HIV-1 and viral membrane gangliosides, then transfer the virus to bystander CD4$^+$ T cells via the established immunological synapses between these two immune cell types. Host cell-derived α2,3 sialylated glycosphingolipid (GSL) on the HIV-1 particle membrane is the ligand for Siglec-1 on mDC required for triggering mDC-mediated *trans*-infection of CD4$^+$ T cells in lymphoid organs (Yu et al. 2014; Izquierdo-Useros et al. 2012; Puryear et al. 2013). A population of cervical DCs at the lamina propria of the ectocervix and the endocervix that express Siglec-1 may be particularly important, and ex vivo studies suggest that CD4$^+$ T cell *trans*-infection from these cells can be blocked by addition of anti-Siglec-1 antibodies (Perez-Zsolt et al. 2019a).

A loss-of-function variant (Glu88Ter) of *SIGLEC1* gene was recently identified from the Exome Aggregation Consortium genetic database to be present in 1% of the human population. Monocytes isolated from individuals with this specific variant completely lack Siglec-1 expression and have reduced HIV capture and *trans*-infection phenotypes ex vivo. However, individuals carrying this truncated Siglec-1 protein do not show marked difference in HIV-1 acquisition and AIDS outcomes in vivo, which suggests an indispensable (and perhaps dominant) role of the classical HIV-1 infectious routes in the HIV-1 dissemination within the infected individuals (Martinez-Picado et al. 2016).

The importance of Siglec-1-mediated viral spreading has also been confirmed in vivo in murine retroviral infection models. Murine leukemia virus (MLV) is recognized by murine Siglec-1 through its sialylated gangliosides (Erikson et al. 2015; Sewald et al. 2015). MLV captured by Siglec-1-expressing sinus-lining macrophages and subsequent *trans*-infection into B-1 cells through their synaptic contacts was directly demonstrated by time-lapse intravital 2-photon laser scanning microscopy (Sewald et al. 2015). Virus capture and efficient MLV infection at the lymph node and spleen were significantly reduced by Siglec-1-targeting antibodies as well as in *Siglec1*$^{-/-}$ mice (Sewald et al. 2015). These data confirm a pivotal role of Siglec-1 in initial retroviral capture and suggest that Siglec-1 could represent a therapeutic target to reduce viral laden macrophage reservoirs and to prevent *trans*-infection. However, in a murine model of the splenomegaly-inducing retrovirus Friend virus complex (FVC) infection, Siglec-1-expressing macrophages capture of incoming blood-borne retroviruses limited their spread to erythroblasts in the red pulp where FVC manifests its pathogenesis (Uchil et al. 2019). In this case, Siglec-1-mediated FVC capture was beneficial, and further activated DCs and promoted cytotoxic CD8$^+$ T cell responses, promoting efficient clearing of FVC-infected cells (Uchil et al. 2019).

Another well-documented example of Siglec exploitation for viral entry and immune modulation is the example of porcine reproductive and respiratory syndrome virus (PRRSV). PRRSV infection is one of the most economically devastating diseases in the global pork industry. PRRSV has a narrow cell tropism, primarily targeting cells of the monocyte/macrophage lineage (Duan et al. 1997). Two cellular proteins, CD163 and Siglec-1 have been identified as the primary targets for PRRSV binding and internalization (Duan et al. 1998; Vanderheijden et al. 2003). α2-3 and, to a lesser extent, α2-6-linked Sias on the PRRSV virion mediate viral attachment and infection alveolar macrophages, with Siglec-1 serving as the primary entry receptor responsible for Sia-dependent binding through its N-terminal V-set domain (Delputte and Nauwynck 2004; Delputte et al. 2007). Siglec-1 neutralization blocks PRRSV infection in a dose-dependent manner, and overexpression of Siglec-1 in non-permissive cells enhances PRRSV cell attachment and internalization (Duan et al. 1998; Vanderheijden et al. 2003). In subsequent studies, Sias on the viral envelope structural protein M/GP5 heterodimer were identified as the binding target of Siglec-1 (Van Breedam et al. 2010), and the interaction of PRRSV with Siglec-1 interferes with macrophage phagocytic activity, which may increase the incidence and severity of secondary bacterial infections complicating primary PRRSV disease (De Baere et al. 2012).

Since Siglec-1 and CD163 are the two key receptors for PRRSV entry and internalization into porcine alveolar macrophages, neutralization of these two viral receptors has been explored as a potential target to control PRRSV infection in pigs. Two recent reports support this therapeutic concept. Neutralization of PRRSV infectivity for porcine alveolar macrophages was achieved by addition of soluble Siglec-1 and CD163 Fc fusion proteins, or by recombinant adenovirus- or exosome-delivered microRNA that specifically targeted CD163 and Siglec-1 (Chen et al. 2014b; Zhu et al. 2014). In addition, pigs that received recombinant adenovirus-delivered soluble CD163 plus Siglec-1 had reduced PPRSV viral loads and fecal viral emission, concurrent with improved clinical scores and higher survival rates from the contagious infection (Xia et al. 2018). It is notable that an in vivo study conducted in *SIGLEC1* knockout pigs found that the absence of Siglec-1 expression does not alter to the clinical course and histopathology of PRRSV infection (Prather et al. 2013), implying a potential redundant function of CD163 and Siglec-1 in PRRSV infection and the importance of dual targeting for potential pharmacological interventions.

Although Siglec-1 and CD163 are general recognized as the key entry mediators for PRRSV infection, a recent report demonstrated that certain PRRSV strains can infect Siglec-1-deficient cells by exploiting Siglec-10 in the presence of CD163 (Frydas and Nauwynck 2016; Xie et al. 2017). Siglec-10 showed a higher affinity to type 2 PRRSV vs. type 1 PRRSV and mediates attachment and endocytosis of PRRSV in a Sia-dependent manner, while transfection of Siglec-10 into non-permissive cells also restored the PRRSV infectivity. These findings indicate that PRRSV can use several Siglecs to enter macrophages and may influence strain differences in pathogenesis (Xie et al. 2017). Of note, PRRSV lung infection induces minimal production of type I interferon and inflammatory cytokines compared to infections caused by swine influenza virus and porcine respiratory coronavirus (Van Reeth et al. 1999). Like other inhibitory Siglecs, porcine Siglec-10 contains one ITIM and one ITIM-like motif possibly, likely counteracting key immune activation pathways induced by viral infection (Crocker et al. 2007).

Budding from the GSL-enriched domain is a conserved feature of other enveloped viruses. Incorporation of host GSL into the viral envelope is found in Hendra and Nipah viruses, two enveloped RNA viruses of the family *Paramyxoviridae*, and these viruses can utilize the Sia-Siglec-1 axis to potentiate mDC-dependent capture and *trans*-infection of T cells (Akiyama et al. 2014). Ebola viruses cause lethal hemorrhagic fever in humans, with many recent outbreaks on the African continent. Recently, Ebola virus entry into activated DCs was shown to involve Siglec-1 recognition of sialylated gangliosides anchored to Ebola virus membranes (Perez-Zsolt et al. 2019b). Blockade of Siglec-1 by specific monoclonal antibodies interrupted Ebola viral uptake and cytoplasmic entry, providing cross-protection against other GSL-containing viruses including HIV-1 (Perez-Zsolt et al. 2019b). This finding suggests that incorporation of GSLs in virus particle membranes facilitates Siglec-1 binding to mDC and may be a conserved mechanism for enveloped RNA viruses to exploit mDC for systemic virus dissemination.

Finally, Siglec-1 modulates T cell-mediated anti-viral responses during human respiratory syncytial virus (RSV) infection, with key differences between newborns and adults. Upon RSV infection, expression of Siglec-1 is upregulated on monocytes from both newborns and adults, whereas expression of Siglec-1 ligand, CD43, is only highly upregulated on adult $CD4^+$ T cells (van den Berg et al. 2001; Jans et al. 2018). This finding is consistent with observations that Siglec-1 inhibits IFN-γ production by adult $CD4^+$ T cells but not newborn $CD4^+$ T cells, although the detailed mechanism remains unknown (Jans et al. 2018).

8.4 Siglecs in Parasitic and Fungal Infections

Unlike many bacterial and viral infections, parasitic diseases caused by protozoa and helminths are often chronic, lasting months to years or even lifetimes. Animal parasites have developed remarkable strategies to interfere with or avoid immune clearance mechanisms to establish a chronic infection in their vertebrate hosts, including several examples of Sia-Siglec-mediated interactions that help achieve this immune evasion phenotype.

The protozoan parasite *Trypanosoma cruzi*, the causative agent of Chagas disease, illustrates well how Sia can be exploited by animal parasites to escape host immune surveillance, from the very first encounter with innate immune cells to the late-mounted adaptive immune responses. Although *T. cruzi* cannot synthesize Sias de novo, it uses its *trans*-sialidase (TcTS) enzyme to transfer Sias from host sialylglycoconjugates to its own surface mucins (Pereira 1983; Pereira et al. 1980). These highly sialylated mucin-like coats generated on the *T. cruzi* surface confer resistance to complement-mediated killing by impeding C3 convertase assembly, and block potential lytic effects of host anti-α-Gal antibodies that would recognize the otherwise exposed α-galactosyl epitopes on the mucin-like glycoproteins (Pereira-Chioccola et al. 2000; Tomlinson et al. 1994). Moreover, *T. cruzi* sialyl-glycoconjugates facilitate its binding to Siglec-1 on host macrophages, and this interaction may be involved in the initiation of trypomastigote infection (Monteiro et al. 2005). *T. cruzi* can also use its surface sialyl-glycoconjugates to engage Siglec-E on DCs and actively suppress their production of proinflammatory cytokine IL-12, impairing generation of protective Th1 responses (Erdmann et al. 2009).

Presence of α2-3 and α2-6 linked sialoglycans has also been discovered on another flagellated protozoa, *Leishmania donovani*, the causative organism of Indian visceral leishmaniasis (Chatterjee et al. 2003), such that high Sia-containing virulent strains bind both Siglec-1 and Siglec-5. The Sia-Siglec-1 interaction promotes macrophage uptake of the parasite promoting dissemination to other sites within the body, whereas the Sia-Siglec-5 binding suppresses ROS, NO, and Th2-dominant cytokine responses in infected macrophages by counteracting MAPK and PI3K/Akt signaling pathways (Roy and Mandal 2016).

Lastly, vaginitis caused by the fungal pathogen *Candida* is common and frequently recurrent condition in women's health. Recently, a combined global genetic and immune profiling study of clinical populations identified polymorphisms in *SIGLEC15* as candidate genetic predispositions involved in *Candida* vaginitis susceptibility (Jaeger et al. 2019). A particular SIGLEC15 polymorphism was associated with great inflammasome signaling and IL-1β production in response to *Candida* in vitro, and in vivo silencing of Siglec15 in a murine vaginitis model was associated with increased fungal burden and neutrophilic inflammation (Jaeger et al. 2019).

8.5 Crosstalk Between Siglec and Pattern-Recognition Receptors in Host Defense Mechanism

Recent evidence indicates that Siglecs functionally intersect with other host innate immune sensing pathways to modulate the response to viral and bacterial infection in important ways. For example, the RNA virus vesicular stomatitis virus (VSV) upregulates Siglec-G expression in macrophages through intracellular nucleic acid sensor RIG-I- or NF-κB-dependent mechanisms (Chen et al. 2013). Subsequent recruitment of SHP-2 phosphatase to the Siglec-G intracellular domain initiates a pathway for RIG-I degradation, thus suppressing anti-viral immunity. In corroboration, inactivation of Siglec-G protects mice against lethal VSV infection (Chen et al. 2013). VSV infection also results in upregulation of Siglec-1 in macrophages, which triggers a negative regulation pathway for TBK1 degradation via the ubiquitin ligase TRIM27, suppressing type I interferon production and allowing the virus to escape immune elimination (Zheng et al. 2015). Siglec-G expression is low in CD8α^+ DCs, and Siglec-G deficient mice generate more antigen-specific cytotoxic T cell responses to inhibit intracellular bacterial infection (Ding et al. 2016).

One line of evidence developed in recent years suggests a potential for direct interactions between Siglec receptors, e.g. human Siglec-5 or -9 and murine Siglec-E and -F, and various toll-like receptors (TLRs) to downregulate the inflammatory pathways downstream of their pattern recognition (Chen et al. 2014a). Consequently, Siglec-E deletion was seen to boost DC responses to a broad range of microbial TLR ligands. In this model, activation of the mammalian sialidase Neu1 to the cell surface disrupts Siglec-E-TLR4 interaction, suggesting it can serve to derepress and allow positive feedback of TLR activation during infection (Chen et al. 2014a).

Caution is warranted, however, by another study applied quantitative proteomics to three different strains of Siglec-E-deficient mice. Quantitative proteomics found no consistent differences in TLR4 signaling or TLR4 endocytosis in response to LPS, nor did macrophages from the Siglec-E-deficient mice exhibit significant differences in uptake or killing of *Salmonella enterica* in vitro (Nagala et al. 2017).

8.6 Future Perspectives

Siglecs are important and broadly distributed lectin receptors of leukocytes uniquely poised to detect Sias and its perturbation during homeostasis and in disease states. The importance of Siglecs in immunopathology was proposed as early as the identification of Siglec-3 (CD33) and Siglec-2 (CD22) as biological markers of myeloid leukemias and B cell lymphomas, respectively. Mounting evidence supports the concept that Sia molecular mimicry serves as a virulence mechanism to subvert host innate immunity or to infect permissive target cells through an interplay with various Siglecs. As our understanding of Siglec influences on glycans-mediated host–pathogen interactions is now rapidly expanding and deepening, there is considerable interest in exploiting Siglecs for immunotherapy and disease prevention. Design of glycan-based therapeutics and their requirements for potency and specificity will likely provide a new biotechnological approach to effectively intervene in immunological processes to reduce the incidence and severity infectious diseases.

References

Akiyama H, Miller C, Patel HV, Hatch SC, Archer J, Ramirez NG, Gummuluru S (2014) Virus particle release from glycosphingolipid-enriched microdomains is essential for dendritic cell-mediated capture and transfer of HIV-1 and henipavirus. J Virol 88(16):8813–8825. https://doi.org/10.1128/JVI.00992-14

Ali SR, Fong JJ, Carlin AF, Busch TD, Linden R, Angata T, Areschoug T, Parast M, Varki N, Murray J, Nizet V, Varki A (2014) Siglec-5 and Siglec-14 are polymorphic paired receptors that modulate neutrophil and amnion signaling responses to group B Streptococcus. J Exp Med. https://doi.org/10.1084/jem.20131853

Angata T, Kerr SC, Greaves DR, Varki NM, Crocker PR, Varki A (2002) Cloning and characterization of human Siglec-11. A recently evolved signaling molecule that can interact with SHP-1 and SHP-2 and is expressed by tissue macrophages, including brain microglia. J Biol Chem 277(27):24466–24474

Angata T, Hayakawa T, Yamanaka M, Varki A, Nakamura M (2006) Discovery of Siglec-14, a novel sialic acid receptor undergoing concerted evolution with Siglec-5 in primates. Faseb J 20(12):1964–1973

Angata T, Ishii T, Motegi T, Oka R, Taylor RE, Soto PC, Chang YC, Secundino I, Gao CX, Ohtsubo K, Kitazume S, Nizet V, Varki A, Gemma A, Kida K, Taniguchi N (2013) Loss of Siglec-14 reduces the risk of chronic obstructive pulmonary disease exacerbation. Cell Mol Life Sci 70(17):3199–3210

Auwerx J, Francois KO, Covens K, Van Laethem K, Balzarini J (2008) Glycan deletions in the HIV-1 gp120 V1/V2 domain compromise viral infectivity, sensitize the mutant virus strains to carbohydrate-binding agents and represent a specific target for therapeutic intervention. Virology 382(1):10–19

Avril T, Wagner ER, Willison HJ, Crocker PR (2006) Sialic acid-binding immunoglobulin-like lectin 7 mediates selective recognition of sialylated glycans expressed on *Campylobacter jejuni* lipooligosaccharides. Infect Immun 74(7):4133–4141

Bax M, Kuijf ML, Heikema AP, van Rijs W, Bruijns SC, Garcia-Vallejo JJ, Crocker PR, Jacobs BC, van Vliet SJ, van Kooyk Y (2011) *Campylobacter jejuni* lipooligosaccharides modulate dendritic cell-mediated T cell polarization in a sialic acid linkage-dependent manner. Infect Immun 79(7):2681–2689

Cao H, Lakner U, de Bono B, Traherne JA, Trowsdale J, Barrow AD (2008) SIGLEC16 encodes a DAP12-associated receptor expressed in macrophages that evolved from its inhibitory counterpart SIGLEC11 and has functional and non-functional alleles in humans. Eur J Immunol 38(8):2303–2315

Carlin AF, Lewis AL, Varki A, Nizet V (2007) Group B streptococcal capsular sialic acids interact with siglecs (immunoglobulin-like lectins) on human leukocytes. J Bacteriol 189(4):1231–1237

Carlin AF, Chang YC, Areschoug T, Lindahl G, Hurtado-Ziola N, King CC, Varki A, Nizet V (2009a) Group B *Streptococcus* suppression of phagocyte functions by protein-mediated engagement of human Siglec-5. J Exp Med 206(8):1691–1699

Carlin AF, Uchiyama S, Chang YC, Lewis AL, Nizet V, Varki A (2009b) Molecular mimicry of host sialylated glycans allows a bacterial pathogen to engage neutrophil Siglec-9 and dampen the innate immune response. Blood 113(14):3333–3336

Chang YC, Olson J, Beasley FC, Tung C, Zhang J, Crocker PR, Varki A, Nizet V (2014a) Group B *Streptococcus* engages an inhibitory Siglec through sialic acid mimicry to blunt innate immune and inflammatory responses in vivo. PLoS Pathog 10(1):e1003846

Chang YC, Olson J, Louie A, Crocker PR, Varki A, Nizet V (2014b) Role of macrophage sialoadhesin in host defense against the sialylated pathogen group B *Streptococcus*. J Mol Med (Berl)

Chatterjee M, Chava AK, Kohla G, Pal S, Merling A, Hinderlich S, Unger U, Strasser P, Gerwig GJ, Kamerling JP, Vlasak R, Crocker PR, Schauer R, Schwartz-Albiez R, Mandal C (2003) Identification and characterization of adsorbed serum sialoglycans on *Leishmania donovani* promastigotes. Glycobiology 13(5):351–361

Chen W, Han C, Xie B, Hu X, Yu Q, Shi L, Wang Q, Li D, Wang J, Zheng P, Liu Y, Cao X (2013) Induction of Siglec-G by RNA viruses inhibits the innate immune response by promoting RIG-I degradation. Cell 152(3):467–478

Chen GY, Brown NK, Wu W, Khedri Z, Yu H, Chen X, van de Vlekkert D, D'Azzo A, Zheng P, Liu Y (2014a) Broad and direct interaction between TLR and Siglec families of pattern recognition receptors and its regulation by Neu1. Elife 3:e04066

Chen Y, Guo R, He S, Zhang X, Xia X, Sun H (2014b) Additive inhibition of porcine reproductive and respiratory syndrome virus infection with the soluble sialoadhesin and CD163 receptors. Virus Res 179:85–92

Cieslewicz MJ, Chaffin D, Glusman G, Kasper D, Madan A, Rodrigues S, Fahey J, Wessels MR, Rubens CE (2005) Structural and genetic diversity of group B streptococcus capsular polysaccharides. Infect Immun 73(5):3096–3103

Crocker PR, Gordon S (1989) Mouse macrophage hemagglutinin (sheep erythrocyte receptor) with specificity for sialylated glycoconjugates characterized by a monoclonal antibody. J Exp Med 169(4):1333–1346

Crocker PR, Mucklow S, Bouckson V, McWilliam A, Willis AC, Gordon S, Milon G, Kelm S, Bradfield P (1994) Sialoadhesin, a macrophage sialic acid binding receptor for haemopoietic cells with 17 immunoglobulin-like domains. EMBO J 13(19):4490–4503

Crocker PR, Paulson JC, Varki A (2007) Siglecs and their roles in the immune system. Nat Rev Immunol 7(4):255–266

Cross AS, Kelly NM (1990) Bacteria-phagocyte interactions: emerging tactics in an ancient rivalry. FEMS Microbiol Immunol 2(5–6):245–258

Dale JB, Washburn RG, Marques MB, Wessels MR (1996) Hyaluronate capsule and surface M protein in resistance to opsonization of group A streptococci. Infect Immun 64(5):1495–1501

De Baere MI, Van Gorp H, Delputte PL, Nauwynck HJ (2012) Interaction of the European genotype porcine reproductive and respiratory syndrome virus (PRRSV) with sialoadhesin (CD169/Siglec-1) inhibits alveolar macrophage phagocytosis. Vet Res 43(1):47

DeAngelis PL (2002) Microbial glycosaminoglycan glycosyltransferases. Glycobiology 12(1):9R–16R

Delputte PL, Nauwynck HJ (2004) Porcine arterivirus infection of alveolar macrophages is mediated by sialic acid on the virus. J Virol 78(15):8094–8101

Delputte PL, Van Breedam W, Delrue I, Oetke C, Crocker PR, Nauwynck HJ (2007) Porcine arterivirus attachment to the macrophage-specific receptor sialoadhesin is dependent on the sialic acid-binding activity of the N-terminal immunoglobulin domain of sialoadhesin. J Virol 81(17):9546–9550

Devi SJ, Robbins JB, Schneerson R (1991) Antibodies to poly[(2-8)-alpha-*N*-acetylneuraminic acid] and poly[(2-9)-alpha-*N*-acetylneuraminic acid] are elicited by immunization of mice with *Escherichia coli* K92 conjugates: potential vaccines for groups B and C meningococci and *E. coli* K1. Proc Natl Acad Sci U S A 88(16):7175–7179

Ding Y, Guo Z, Liu Y, Li X, Zhang Q, Xu X, Gu Y, Zhang Y, Zhao D, Cao X (2016) The lectin Siglec-G inhibits dendritic cell cross-presentation by impairing MHC class I-peptide complex formation. Nat Immunol 17(10):1167–1175

Duan X, Nauwynck HJ, Pensaert MB (1997) Effects of origin and state of differentiation and activation of monocytes/macrophages on their susceptibility to porcine reproductive and respiratory syndrome virus (PRRSV). Arch Virol 142(12):2483–2497

Duan X, Nauwynck HJ, Favoreel HW, Pensaert MB (1998) Identification of a putative receptor for porcine reproductive and respiratory syndrome virus on porcine alveolar macrophages. J Virol 72(5):4520–4523

Erdmann H, Steeg C, Koch-Nolte F, Fleischer B, Jacobs T (2009) Sialylated ligands on pathogenic *Trypanosoma cruzi* interact with Siglec-E (sialic acid-binding Ig-like lectin-E). Cell Microbiol 11(11):1600–1611

Erikson E, Wratil PR, Frank M, Ambiel I, Pahnke K, Pino M, Azadi P, Izquierdo-Useros N, Martinez-Picado J, Meier C, Schnaar RL, Crocker PR, Reutter W, Keppler OT (2015) Mouse Siglec-1 mediates trans-infection of surface-bound murine leukemia virus in a sialic acid N-Acyl side chain-dependent manner. J Biol Chem 290(45):27345–27359

Finne J (1982) Occurrence of unique polysialosyl carbohydrate units in glycoproteins of developing brain. J Biol Chem 257(20):11966–11970

Foley MJ, Wood WB Jr (1959) Studies on the pathogenicity of group A streptococci. II. The antiphagocytic effects of the M protein and the capsular gel. J Exp Med 110:617–628. https://doi.org/10.1084/jem.110.4.617

Frydas IS, Nauwynck HJ (2016) Replication characteristics of eight virulent and two attenuated genotype 1 and 2 porcine reproductive and respiratory syndrome virus (PRRSV) strains in nasal mucosa explants. Vet Microbiol 182:156–162

Hayakawa T, Angata T, Lewis AL, Mikkelsen TS, Varki NM, Varki A (2005) A human-specific gene in microglia. Science 309(5741):1693. https://doi.org/10.1126/science.1114321

Heath PT, Schuchat A (2007) Perinatal group B streptococcal disease. Best Pract Res Clin Obstet Gynaecol 21(3):411–424

Heikema AP, Bergman MP, Richards H, Crocker PR, Gilbert M, Samsom JN, van Wamel WJ, Endtz HP, van Belkum A (2010) Characterization of the specific interaction between sialoadhesin and sialylated *Campylobacter jejuni* lipooligosaccharides. Infect Immun 78(7):3237–3246

Heikema AP, Jacobs BC, Horst-Kreft D, Huizinga R, Kuijf ML, Endtz HP, Samsom JN, van Wamel WJ (2013) Siglec-7 specifically recognizes *Campylobacter jejuni* strains associated with oculomotor weakness in Guillain-Barre syndrome and Miller Fisher syndrome. Clin Microbiol Infect 19(2):E106–E112

Huizinga R, Easton AS, Donachie AM, Guthrie J, van Rijs W, Heikema A, Boon L, Samsom JN, Jacobs BC, Willison HJ, Goodyear CS (2012) Sialylation of *Campylobacter jejuni* lipo-oligosaccharides: impact on phagocytosis and cytokine production in mice. PLoS ONE 7(3):e34416

Huizinga R, van Rijs W, Bajramovic JJ, Kuijf ML, Laman JD, Samsom JN, Jacobs BC (2013) Sialylation of *Campylobacter jejuni* endotoxin promotes dendritic cell-mediated B cell responses through CD14-dependent production of IFNβ and TNFα. J Immunol 191(11):5636–5645. https://doi.org/10.4049/jimmunol.1301536

Izquierdo-Useros N, Lorizate M, Puertas MC, Rodriguez-Plata MT, Zangger N, Erikson E, Pino M, Erkizia I, Glass B, Clotet B, Keppler OT, Telenti A, Krausslich HG, Martinez-Picado J (2012) Siglec-1 is a novel dendritic cell receptor that mediates HIV-1 trans-infection through recognition of viral membrane gangliosides. PLoS Biol 10(12):e1001448

Jacobs BC, van Doorn PA, Schmitz PI, Tio-Gillen AP, Herbrink P, Visser LH, Hooijkass H, van der Meche FG (1996) *Campylobacter jejuni* infections and anti-GM1 antibodies in Guillain-Barre syndrome. Ann Neurol 40(2):181–187

Jaeger M, Pinelli M, Borghi M, Constantini C, Dindo M, van Emst L, Puccetti M, Pariano M, Ricano-Ponce I, Bull C, Gresnigt MS, Wang X, Gutierrez Achury J, Jacobs CWM, Xu N, Oosting M, Arts P, Joosten LAB, van de Veerdonk FL, Veltman JA, Ten Oever J, Kullberg BJ, Feng M, Adema GJ, Wijmenga C, Kumar V, Sobel J, Gilissen C, Romani L, Netea MG (2019) A systems genomics approach identifies SIGLEC15 as a susceptibility factor in recurrent vulvovaginal candidiasis. Sci Transl Med 11:496

Jann K, Jann B (1992) Capsules of *Escherichia coli*, expression and biological significance. Can J Microbiol 38(7):705–710

Jans J, Unger WWJ, Vissers M, Ahout IML, Schreurs I, Wickenhagen A, de Groot R, de Jonge MI, Ferwerda G (2018) Siglec-1 inhibits RSV-induced interferon gamma production by adult T cells in contrast to newborn T cells. Eur J Immunol 48(4):621–631

Jones C, Virji M, Crocker PR (2003) Recognition of sialylated meningococcal lipopolysaccharide by siglecs expressed on myeloid cells leads to enhanced bacterial uptake. Mol Microbiol 49(5):1213–1225

Kalograiaki I, Euba B, Proverbio D, Campanero-Rhodes MA, Aastrup T, Garmendia J, Solis D (2016) Combined bacteria microarray and quartz crystal microbalance approach for exploring glycosignatures of nontypeable *Haemophilus influenzae* and recognition by host lectins. Anal Chem 88(11):5950–5957

Kaper JB, Nataro JP, Mobley HL (2004) Pathogenic *Escherichia coli*. Nat Rev Microbiol 2(2):123–140

Khatua B, Ghoshal A, Bhattacharya K, Mandal C, Saha B, Crocker PR, Mandal C (2010) Sialic acids acquired by *Pseudomonas aeruginosa* are involved in reduced complement deposition and siglec mediated host-cell recognition. FEBS Lett 584(3):555–561

Khatua B, Bhattacharya K, Mandal C (2012) Sialoglycoproteins adsorbed by *Pseudomonas aeruginosa* facilitate their survival by impeding neutrophil extracellular trap through siglec-9. J Leukoc Biol 91(4):641–655

Klaas M, Crocker PR (2012) Sialoadhesin in recognition of self and non-self. Semin Immunopathol 34(3):353–364

Kuijf ML, Samsom JN, van Rijs W, Bax M, Huizinga R, Heikema AP, van Doorn PA, van Belkum A, van Kooyk Y, Burgers PC, Luider TM, Endtz HP, Nieuwenhuis EE, Jacobs BC (2010) TLR4-mediated sensing of *Campylobacter jejuni* by dendritic cells is determined by sialylation. J Immunol 185(1):748–755

Lee CH, Chang CC, Liu JW, Chen RF, Yang KD (2014) Sialic acid involved in hypermucoviscosity phenotype of *Klebsiella pneumoniae* and associated with resistance to neutrophil phagocytosis. Virulence 5(6):673–679

Linhardt RJ, Toida T (2004) Role of glycosaminoglycans in cellular communication. Acc Chem Res 37(7):431–438

Martinez-Picado J, McLaren PJ, Erkizia I, Martin MP, Benet S, Rotger M, Dalmau J, Ouchi D, Wolinsky SM, Penugonda S, Gunthard HF, Fellay J, Carrington M, Izquierdo-Useros N, Telenti A (2016) Identification of Siglec-1 null individuals infected with HIV-1. Nat Commun 7:12412

Monteiro VG, Lobato CS, Silva AR, Medina DV, de Oliveira MA, Seabra SH, de Souza W, DaMatta RA (2005) Increased association of *Trypanosoma cruzi* with sialoadhesin positive mice macrophages. Parasitol Res 97(5):380–385

Nagala M, McKenzie E, Richards H, Sharma R, Thomson S, Mastroeni P, Crocker PR (2017) Expression of Siglec-E alters the proteome of lipopolysaccharide (LPS)-activated macrophages but does not affect LPS-driven cytokine production or toll-like receptor 4 endocytosis. Front Immunol 8:1926

Pereira ME (1983) A developmentally regulated neuraminidase activity in *Trypanosoma cruzi*. Science 219(4591):1444–1446

Pereira ME, Loures MA, Villalta F, Andrade AF (1980) Lectin receptors as markers for *Trypanosoma cruzi*. Developmental stages and a study of the interaction of wheat germ agglutinin with sialic acid residues on epimastigote cells. J Exp Med 152(5):1375–1392

Pereira-Chioccola VL, Acosta-Serrano A, Correia de Almeida I, Ferguson MA, Souto-Padron T, Rodrigues MM, Travassos LR, Schenkman S (2000) Mucin-like molecules form a negatively charged coat that protects *Trypanosoma cruzi* trypomastigotes from killing by human anti-alpha-galactosyl antibodies. J Cell Sci 113(Pt 7):1299–1307

Perez-Zsolt D, Cantero-Perez J, Erkizia I, Benet S, Pino M, Serra-Peinado C, Hernandez-Gallego A, Castellvi J, Tapia G, Arnau-Saz V, Garrido J, Tarrats A, Buzon MJ, Martinez-Picado J, Izquierdo-Useros N, Genesca M (2019a) Dendritic cells from the cervical mucosa capture and transfer HIV-1 via Siglec-1. Front Immunol 10:825

Perez-Zsolt D, Erkizia I, Pino M, Garcia-Gallo M, Martin MT, Benet S, Chojnacki J, Fernandez-Figueras MT, Guerrero D, Urrea V, Muniz-Trabudua X, Kremer L, Martinez-Picado J, Izquierdo-Useros N (2019b) Anti-Siglec-1 antibodies block Ebola viral uptake and decrease cytoplasmic viral entry. Nat Microbiol. https://doi.org/10.1038/s41564-019-0453-2 (Epub ahead of print)

Prather RS, Rowland RR, Ewen C, Trible B, Kerrigan M, Bawa B, Teson JM, Mao J, Lee K, Samuel MS, Whitworth KM, Murphy CN, Egen T, Green JA (2013) An intact sialoadhesin (Sn/SIGLEC1/CD169) is not required for attachment/internalization of the porcine reproductive and respiratory syndrome virus. J Virol 87(17):9538–9546

Puryear WB, Akiyama H, Geer SD, Ramirez NP, Yu X, Reinhard BM, Gummuluru S (2013) Interferon-inducible mechanism of dendritic cell-mediated HIV-1 dissemination is dependent on Siglec-1/CD169. PLoS Pathog 9(4):e1003291

Rempel H, Calosing C, Sun B, Pulliam L (2008) Sialoadhesin expressed on IFN-induced monocytes binds HIV-1 and enhances infectivity. PLoS ONE 3(4):e1967

Robbins JB, McCracken GH Jr, Gotschlich EC, Orskov F, Orskov I, Hanson LA (1974) *Escherichia coli* K1 capsular polysaccharide associated with neonatal meningitis. N Engl J Med 290(22):1216–1220

Roy S, Mandal C (2016) *Leishmania donovani* utilize sialic acids for binding and phagocytosis in the macrophages through selective utilization of siglecs and impair the innate immune arm. PLoS Negl Trop Dis 10(8):e0004904. https://doi.org/10.1371/journal.pntd.0004904

Rutishauser U (2008) Polysialic acid in the plasticity of the developing and adult vertebrate nervous system. Nat Rev Neurosci 9(1):26–35

Saito M, Yamamoto S, Ozaki K, Tomioka Y, Suyama H, Morimatsu M, Nishijima KI, Yoshida SI, Ono E (2016) A soluble form of Siglec-9 provides a resistance against group B *Streptococcus* (GBS) infection in transgenic mice. Microb Pathog 99:106–110

Schwarz F, Landig CS, Siddiqui S, Secundino I, Olson J, Varki N, Nizet V, Varki A (2017) Paired Siglec receptors generate opposite inflammatory responses to a human-specific pathogen. EMBO J 36(6):751–760

Secundino I, Lizcano A, Roupe KM, Wang X, Cole JN, Olson J, Ali SR, Dahesh S, Amayreh LK, Henningham A, Varki A, Nizet V (2016) Host and pathogen hyaluronan signal through human siglec-9 to suppress neutrophil activation. J Mol Med (Berl) 94(2):219–233

Sewald X, Ladinsky MS, Uchil PD, Beloor J, Pi R, Herrmann C, Motamedi N, Murooka TT, Brehm MA, Greiner DL, Shultz LD, Mempel TR, Bjorkman PJ, Kumar P, Mothes W (2015) Retroviruses use CD169-mediated trans-infection of permissive lymphocytes to establish infection. Science 350(6260):563–567

Skoff TH, Farley MM, Petit S, Craig AS, Schaffner W, Gershman K, Harrison LH, Lynfield R, Mohle-Boetani J, Zansky S, Albanese BA, Stefonek K, Zell ER, Jackson D, Thompson T, Schrag SJ (2009) Increasing burden of invasive group B streptococcal disease in nonpregnant adults, 1990–2007. Clin Infect Dis 49(1):85–92

Stephenson HN, Mills DC, Jones H, Milioris E, Copland A, Dorrell N, Wren BW, Crocker PR, Escors D, Bajaj-Elliott M (2014) Pseudaminic acid on *Campylobacter jejuni* flagella modulates dendritic cell IL-10 expression via Siglec-10 receptor: a novel flagellin-host interaction. J Infect Dis 210(9):1487–1498

Stevens P, Huang SN, Welch WD, Young LS (1978) Restricted complement activation by *Escherichia coli* with the K-1 capsular serotype: a possible role in pathogenicity. J Immunol 121(6):2174–2180

Thigpen MC, Whitney CG, Messonnier NE, Zell ER, Lynfield R, Hadler JL, Harrison LH, Farley MM, Reingold A, Bennett NM, Craig AS, Schaffner W, Thomas A, Lewis MM, Scallan E, Schuchat A (2011) Bacterial meningitis in the United States, 1998–2007. N Engl J Med 364(21):2016–2025

Tomlinson S, Pontes de Carvalho LC, Vandekerckhove F, Nussenzweig V (1994) Role of sialic acid in the resistance of *Trypanosoma cruzi* trypomastigotes to complement. J Immunol 153(7):3141–3147

Uchil PD, Pi R, Haugh KA, Ladinsky MS, Ventura JD, Barrett BS, Santiago ML, Bjorkman PJ, Kassiotis G, Sewald X, Mothes W (2019) A protective role for the lectin CD169/Siglec-1 against a pathogenic murine retrovirus. Cell Host Microbe 25(1):87–100 (e110)

Uchiyama S, Sun J, Fukahori K, Ando N, Wu M, Schwarz F, Siddiqui SS, Varki A, Marth JD, Nizet V (2019) Dual actions of group B *Streptococcus* capsular sialic acid provide resistance to platelet-mediated antimicrobial killing. Proc Natl Acad Sci U S A 116(15):7465–7470

Van Breedam W, Van Gorp H, Zhang JQ, Crocker PR, Delputte PL, Nauwynck HJ (2010) The M/GP(5) glycoprotein complex of porcine reproductive and respiratory syndrome virus binds the sialoadhesin receptor in a sialic acid-dependent manner. PLoS Pathog 6(1):e1000730

van den Berg TK, Nath D, Ziltener HJ, Vestweber D, Fukuda M, van Die I, Crocker PR (2001) Cutting edge: CD43 functions as a T cell counterreceptor for the macrophage adhesion receptor sialoadhesin (Siglec-1). J Immunol 166(6):3637–3640

Van Reeth K, Labarque G, Nauwynck H, Pensaert M (1999) Differential production of proinflammatory cytokines in the pig lung during different respiratory virus infections: correlations with pathogenicity. Res Vet Sci 67(1):47–52

Vanderheijden N, Delputte PL, Favoreel HW, Vandekerckhove J, Van Damme J, van Woensel PA, Nauwynck HJ (2003) Involvement of sialoadhesin in entry of porcine reproductive and respiratory syndrome virus into porcine alveolar macrophages. J Virol 77(15):8207–8215

Varchetta S, Lusso P, Hudspeth K, Mikulak J, Mele D, Paolucci S, Cimbro R, Malnati M, Riva A, Maserati R, Mondelli MU, Mavilio D (2013) Sialic acid-binding Ig-like lectin-7 interacts with HIV-1 gp120 and facilitates infection of CD4pos T cells and macrophages. Retrovirology 10:154

Varki A (1993) Biological roles of oligosaccharides: all of the theories are correct. Glycobiology 3(2):97–130

Wang Y, Neumann H (2010) Alleviation of neurotoxicity by microglial human Siglec-11. J Neurosci 30(9):3482–3488

Weiman S, Dahesh S, Carlin AF, Varki A, Nizet V, Lewis AL (2009) Genetic and biochemical modulation of sialic acid O-acetylation on group B *Streptococcus*: phenotypic and functional impact. Glycobiology 19(11):1204–1213

Weiman S, Uchiyama S, Lin FY, Chaffin D, Varki A, Nizet V, Lewis AL (2010) O-Acetylation of sialic acid on Group B *Streptococcus* inhibits neutrophil suppression and virulence. Biochem J 428(2):163–168

Wessels MR, Rubens CE, Benedi VJ, Kasper DL (1989) Definition of a bacterial virulence factor: sialylation of the group B streptococcal capsule. Proc Natl Acad Sci U S A 86(22):8983–8987

Wessels MR, Goldberg JB, Moses AE, DiCesare TJ (1994) Effects on virulence of mutations in a locus essential for hyaluronic acid capsule expression in group A streptococci. Infect Immun 62(2):433–441

Willison HJ, Yuki N (2002) Peripheral neuropathies and anti-glycolipid antibodies. Brain 125(Pt 12):2591–2625

Willison HJ, Jacobs BC, van Doorn PA (2016) Guillain-Barre syndrome. Lancet 388(10045):717–727

Xia W, Wu Z, Guo C, Zhu S, Zhang X, Xia X, Sun H (2018) Recombinant adenovirus-delivered soluble CD163 and sialoadhesin receptors protected pigs from porcine reproductive and respiratory syndrome virus infection. Vet Microbiol 219:1–7

Xie J, Christiaens I, Yang B, Breedam WV, Cui T, Nauwynck HJ (2017) Molecular cloning of porcine Siglec-3, Siglec-5 and Siglec-10, and identification of Siglec-10 as an alternative receptor for porcine reproductive and respiratory syndrome virus (PRRSV). J Gen Virol 98(8):2030–2042

Yamanaka M, Kato Y, Angata T, Narimatsu H (2009) Deletion polymorphism of SIGLEC14 and its functional implications. Glycobiology 19(8):841–846

Yu X, Feizpour A, Ramirez NG, Wu L, Akiyama H, Xu F, Gummuluru S, Reinhard BM (2014) Glycosphingolipid-functionalized nanoparticles recapitulate CD169-dependent HIV-1 uptake and trafficking in dendritic cells. Nat Commun 5:4136

Zheng Q, Hou J, Zhou Y, Yang Y, Xie B, Cao X (2015) Siglec1 suppresses antiviral innate immune response by inducing TBK1 degradation via the ubiquitin ligase TRIM27. Cell Res 25(10):1121–1136

Zhu L, Song H, Zhang X, Xia X, Sun H (2014) Inhibition of porcine reproductive and respiratory syndrome virus infection by recombinant adenovirus- and/or exosome-delivered the artificial microRNAs targeting sialoadhesin and CD163 receptors. Virol J 11:225

Zou Z, Chastain A, Moir S, Ford J, Trandem K, Martinelli E, Cicala C, Crocker P, Arthos J, Sun PD (2011) Siglecs facilitate HIV-1 infection of macrophages through adhesion with viral sialic acids. PLoS ONE 6(9):e24559

Chapter 9
Siglecs that Associate with DAP12

Takashi Angata

Abstract Siglecs are a family of transmembrane receptor-like glycan-recognition proteins expressed primarily on leukocytes. Majority of Siglecs have an intracellular sequence motif called immunoreceptor tyrosine-based inhibitory motif (ITIM) and associate with Src homology region 2 domain-containing tyrosine phosphatase-1 (SHP-1), and negatively regulate tyrosine phosphorylation-mediated intracellular signaling events. On the other hand, some Siglecs have a positively charged amino acid residue in the transmembrane domain and associate with DNAX activation protein of 12 kDa (DAP12), which in turn recruits spleen tyrosine kinase (Syk). These DAP12-associated Siglecs play diverse functions. For example, Siglec-15 is conserved throughout vertebrate evolution and plays a role in bone homeostasis by regulating osteoclast development and function. Human Siglec-14 and -16 have inhibitory counterparts (Siglec-5 and -11, respectively), which show extremely high sequence similarity with them at the extracellular domain but interact with SHP-1. The DAP12-associated Siglec in such "paired receptor" configuration counteracts the pathogens that exploit the inhibitory counterpart. Polymorphisms (mutations) that render DAP12-associated inactive Siglecs are found in humans, and some of these appear to be associated with sensitivity or resistance of human hosts to bacterially induced conditions. Studies of mouse Siglec-H have revealed complex and intriguing functions it plays in regulating adaptive immunity. Many questions remain unanswered, and further molecular and genetic studies of DAP12-associated Siglecs will yield valuable insights with translational relevance.

Keywords Siglec · DAP12 · ITAM · Syk · Paired receptors

T. Angata (✉)
Institute of Biological Chemistry, Academia Sinica, 128, Section 2, Academia Road, Nangang District, Taipei, Taiwan
e-mail: angata@gate.sinica.edu.tw

S.-L. Hsieh (ed.), *Lectin in Host Defense Against Microbial Infections*,
Advances in Experimental Medicine and Biology 1204,
https://doi.org/10.1007/978-981-15-1580-4_9

9.1 Introduction

Glycan recognition proteins in vertebrates are often expressed by leukocytes, and are often involved in the discrimination of self versus non-self. For example, some of the C-type lectins recognize pathogen-associated molecular patterns (PAMPs) and alert immune system, as described in other chapters. On the other hand, sialic acids, a group of acidic sugars with nine-carbon backbone (Angata and Varki 2002), are rarely found in microbes. Thus, sialic acids can be a good marker of "self" (self-associated molecular patterns, i.e., SAMPs) for vertebrate immune system (Varki 2011). In fact, complement factor H, a protein that regulates the complement cascade by binding to and inhibiting complement component C3b in association with complement factor I, has been known to bind to sialic acids and protect mammalian cells from complement-mediated attack (Meri and Pangburn 1990; Blaum et al. 2015; Blaum 2017). Investigations in the past few decades have revealed that not only soluble protein (complement factor H) but also membrane-bound receptor-like proteins utilize sialic acids as SAMPs. At least some members of Siglec family appear to fall into this category.

As the properties and functions of Siglec family have been extensively reviewed elsewhere (Varki and Angata 2006; Pillai et al. 2012; Macauley et al. 2014), this chapter only reiterates some basic information here. The word "Siglec" is a composite word from "sialic acid (Sia)", "immunoglobulin (Ig) superfamily", and "lectin" (Crocker et al. 1998). As the name suggests, most of them recognize glycans that contain sialic acids. Siglecs share a basic architecture, which consists of an N-terminal V-set Ig-like domain, followed by multiple C2-set Ig-like domains (the number of which varies from 1 to 16), a single-pass transmembrane domain, and a short cytoplasmic tail. The N-terminal Ig-like domain is the most important for sialic acid recognition. A subgroup of Siglecs (i.e., sialoadhesin/Siglec-1, CD22/Siglec-2, myelin-associated glycoprotein (MAG)/Siglec-4, and Siglec-15) is conserved among multiple lineages of vertebrates, whereas others, collectively known as CD33/Siglec-3-related Siglecs (CD33rSiglecs), are less conserved (Fig. 9.1). CD33rSiglecs are encoded in a gene cluster, implying that this subfamily has expanded by repeated gene duplications; sequences of CD33rSiglec genes were further diversified through genetic recombination and exon shuffling events, in addition to the gradual accumulation of nucleotide substitutions (Angata et al. 2004).

The Siglec family can also be classified based on the partner molecule involved in the signal transduction. Majority of Siglecs have a sequence motif called immunoreceptor tyrosine-based inhibitory motif (ITIM) and another ITIM-like motif in the cytoplasmic tail, and recruit Src homology region 2 domain-containing tyrosine phosphatase-1 (SHP-1), a protein tyrosine phosphatase that plays a major role in suppressing autoimmunity (Shultz et al. 1997). This configuration implies that the majority of Siglecs play a negative regulatory role in the cells that express them. This fact fits the current view of the basic role of Siglecs, which is to sense the sialic acids on own cells as SAMPs and to prevent immune responses against them (Lajaunias et al. 2005; Duong et al. 2010; Varki 2011). In contrast, a small number

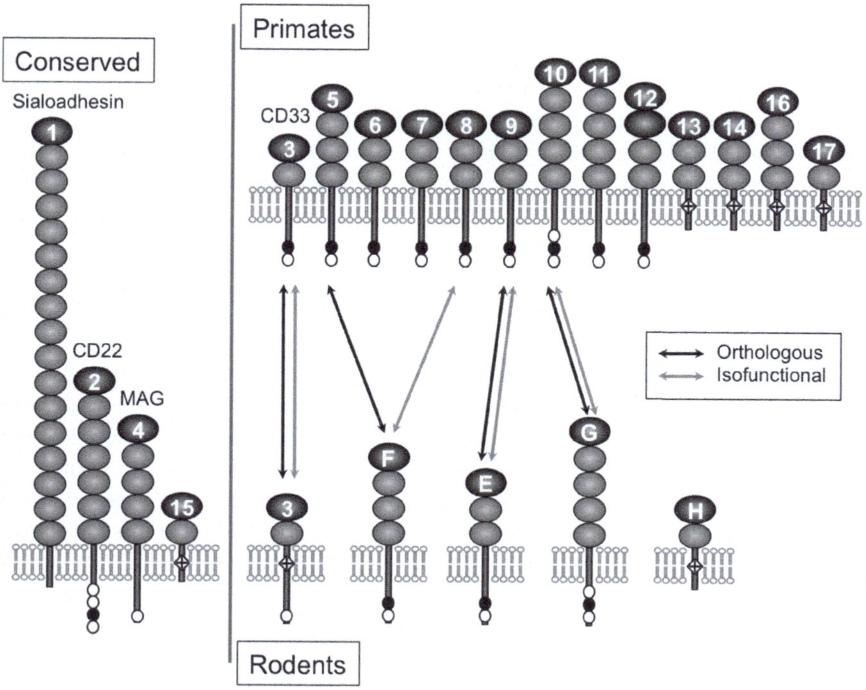

Fig. 9.1 Schematic representation of Siglecs in primates and rodents. Analyses of genomic DNA sequences of mammals so far have revealed that primate genomes encode up to 17 functional Siglecs, while the rodent genomes have nine functional Siglecs. Other lineages of mammals may have different set of Siglecs (Angata 2006). Note that not all primate Siglecs are present in all primates. For example, Siglec-13 is absent in humans, and Siglec-17 is non-functional or absent in many Old World monkeys and apes (including humans). Putative orthologous and isofunctional relationships between primate and rodent Siglecs are indicated with solid and gray lines, respectively. V-set Ig-like domains are represented by larger ovals with darker shade, whereas C2-set Ig-like domains are represented by smaller ovals with lighter shade. Other symbols used are: filled circle: ITIM; open circle: ITIM-like sequence motif; diamond with "+" symbol: positively charged amino acid residue in the transmembrane domain, which is involved in the interaction with DAP12. Modified from Angata (2017)

of Siglecs lack ITIM, and instead have a positively charged amino acid residue in the transmembrane domain, which is involved in the interaction with DNAX activation protein of 12 kDa (DAP12), a small adapter protein that has an immunoreceptor tyrosine-based activating motif (ITAM). Phosphorylated ITAM of DAP12 recruits spleen tyrosine kinase (Syk), a protein tyrosine kinase that plays a major role in the recognition and elimination of foreign agents by the immune system (Kerrigan and Brown 2011; Lowell 2011). Thus, a subset of Siglecs may recognize SAMPs and activate the immune system. This situation appears risky, as it could lead to autoimmunity. How could this risk be mitigated?

In this chapter, the Siglecs that associate with DAP12 are introduced. Siglec-15 is a DAP12-associated Siglec that appears to be conserved in vertebrates. Two human Siglecs (Siglec-14 and -16) and two primate Siglecs (Siglec-13 and -17) absent in humans have been shown to interact with DAP12. In rodents, Siglec-H is known to associate with DAP12. Functions of these Siglecs and the potential mechanisms that prevent the triggering of autoimmunity by these Siglecs will be discussed.

9.2 Primate and Rodent Siglecs that Associate with DAP12

DAP12, also known as killer-activating receptor-associated protein (KARAP), is encoded by *TYROBP* gene in humans. DAP12 is a type I transmembrane protein expressed on natural killer cells and myeloid cells, and has an aspartic acid residue in the transmembrane domain, by which it interacts with numerous membrane proteins that have a positively charged amino acid residue in the transmembrane domain (Turnbull and Colonna 2007). One of these DAP12-associated receptors is triggering receptor expressed on myeloid cells 2 (TREM2, encoded by *TREM2* gene in humans). Mutation in either *TYROBP* or *TREM2* gene causes a hereditary disease called Nasu-Hakola disease (also known as polycystic lipomembranous osteodysplasia with sclerosing leukoencephalopathy) (Paloneva et al. 2000, 2002, 2003), implying that TREM2 is one of the major receptors that use DAP12 as signaling module. The mechanism that enables various receptors that use DAP12 as a common signaling module to play diverse receptor-specific functions is not fully understood, but the specific recognition of ligands by these receptors is likely of critical importance.

None of the DAP12-associated human Siglecs has been as extensively studied as TREM2, partly because only Siglec-15 among them is conserved between human and mouse, allowing functional investigations using mouse model. Nevertheless, valuable insights have been gained by evolutionary studies, cell-based functional studies, and genetic association studies in humans.

9.2.1 Siglec-15

Siglec-15 is conserved from fish to mammals (Angata et al. 2007). In contrast to most Siglecs that have three cysteine residues in each of the two N-terminal Ig-like domains, which form not only intra-domain disulfide bonds but also an inter-domain disulfide bond that tether these two domains, Siglec-15 has even number of cysteine residues in these domains (4 and 2 cysteines in the first and second Ig-like domain, respectively), that is more typical for Ig superfamily. These facts may possibly imply that Siglec-15 was the ancestor of the Siglec family. Siglec-15 interacts with DAP12, and also with another adapter protein, DNAX activation protein of 10 kDa (DAP10), at least when these proteins are over-expressed (Angata et al. 2007).

Multiple laboratories have demonstrated that Siglec-15 is expressed on osteoclasts and is involved in osteoclast differentiation (Hiruma et al. 2011; Ishida-Kitagawa et al. 2012; Stuible et al. 2014). Siglec-15 deficient mice develop mild osteopetrosis (i.e., an increase of bone density) (Hiruma et al. 2013; Kameda et al. 2013), particularly in the trabecular bone, due to the deficiency in osteoclast differentiation. This property prompted some groups to propose Siglec-15 as a potential therapeutic target for osteoporosis (Hiruma et al. 2013; Stuible et al. 2014; Kameda et al. 2015; Shimizu et al. 2015). Importantly, antibody-mediated cross-linking of osteoclast cell-surface Siglec-15 facilitates internalization and degradation of Siglec-15 (Stuible et al. 2014), mimicking Siglec-15 deficiency.

We have recently demonstrated that CD44 is a biologically relevant ligand of Siglec-15 on an osteoclast model cell line (Chang et al. 2017). Our previous findings that Siglec-15 recognizes tumor-associated glycan structure (sialyl-Tn) and is expressed on tumor-associated macrophages and regulates the production of growth factor (Angata et al. 2007; Takamiya et al. 2013), combined with the high expression of sialyl-Tn and CD44 on various solid tumors (Ponta et al. 2003; Zoller 2011), imply that Siglec-15 may also play some roles in tumor microenvironment through interaction with the ligands displayed on tumor cells.

There may be several safeguarding mechanisms to prevent Siglec-15 from promoting autoimmunity by recognizing sialic acid-containing ligands. For example, the expression of Siglec-15 is restricted to osteoclasts (as mentioned above) and a subset of lymph node macrophages in normal tissues (Angata et al. 2007); its expression on lymph node macrophages appears to be predominantly intracellular (Angata et al. 2007); the glycan structure it preferentially recognizes (sialyl-Tn) is also expressed in a spatio-temporarily restricted manner in normal tissues (Ogata et al. 1995; Julien et al. 2012).

9.2.2 Siglec-14

The gene encoding Siglec-14 (*SIGLEC14*) was initially identified as a genetic segment that shows extreme sequence identity with *SIGLEC5*, the gene encoding Siglec-5 (Angata et al. 2004). A follow-up study revealed that this segment encodes a functional protein, which was designated Siglec-14 (Angata et al. 2006). Siglec-14 is conserved among primates and some other lineages of mammals, but is absent in rodents (Angata et al. 2004, 2006; Angata 2006). Siglec-14 is expressed on myeloid cells and enhances pro-inflammatory responses by interacting with DAP12 (Yamanaka et al. 2009). Given that the glycan binding preference of Siglec-14 is somewhat promiscuous (Angata et al. 2006), it is possible that Siglec-14 may recognize endogenous sialylated glycans and induce pro-inflammatory responses.

One possible mechanism by which the immune system circumvents this problem is to couple it with an inhibitory counterpart. The gene encoding Siglec-14 (*SIGLEC14*) is located right next to *SIGLEC5* encoding Siglec-5, a SHP-1-associated inhibitory Siglec (Cornish et al. 1998), in a tandem (head-to-tail) orientation. The expression patterns of Siglec-14 and Siglec-5 overlap at least partially (Yamanaka et al. 2009). [A caution is required for the interpretation of the earlier reports on the expression pattern of Siglec-5, as most antibodies against Siglec-5 cross-react with Siglec-14 (Angata et al. 2006).] As mentioned above, the segment of *SIGLEC14* gene that encodes the N terminus of the protein (signal peptide and first two Ig-like domains) shows extreme sequence similarity with the corresponding segment in *SIGLEC5* gene, and thus Siglec-14 and Siglec-5 show similar glycan binding preferences (Angata et al. 2006). This extreme sequence similarity between *SIGLEC14* and *SIGLEC5* is observed in multiple species, likely implying that the sequence similarities between these two genes have been maintained through repeated gene conversion events in multiple lineages and selected for under selective pressure (Angata et al. 2006). Based on the findings that Siglec-5 is exploited by some bacterial pathogens (Carlin et al. 2009; Angata et al. 2013; Ali et al. 2014) and that Siglec-14 appears to counteract this exploitation by binding to the bacteria and triggering cell-activating signals (Angata et al. 2013; Ali et al. 2014), it was proposed that Siglec-14 may have arisen to counter the pathogens that exploit Siglec-5.

This hypothesis gained support from human genetic studies. An allele in which *SIGLEC14* and *SIGLEC5* are fused (resulting in *SIGLEC14–SIGLEC5* fusion gene) is common (allele frequency > 0.05) in all human populations tested, and is even a dominant allele in some parts of East Asia (Yamanaka et al. 2009). This fusion gene lacks the segment that is unique to Siglec-14, and encodes a protein identical to Siglec-5 in amino acid sequence; thus, it is functionally a *SIGLEC14*-null allele. By focused genetic association studies, this *SIGLEC14*-null allele was found to be associated with the increased risk of pre-term delivery in the presence of group B *Streptococcus* infection (Ali et al. 2014), whereas the same allele is associated with the reduced risk of exacerbation among the patients of chronic obstructive pulmonary disease (Angata et al. 2013) and the development of meningitis among the individuals infected by *Mycobacterium tuberculosis* (Graustein et al. 2017). While the first case implies that Siglec-14 counters the pathogens that exploit Siglec-5 (thus protect the host), the latter two imply that strong pro-inflammatory responses caused by bacterial engagement of Siglec-14 may be detrimental to the host under some circumstances. This fact may explain why *SIGLEC14*-null allele is dominant in some geographical areas; in the area where some environmental condition that causes chronic lung inflammation is widespread and the inflammation can be exacerbated by some bacteria (e.g., non-typeable *Haemophilus influenzae*, a commensal bacteria present in the airways of a majority of healthy individuals), carriers of *SIGLEC14*-null allele may have had some selective advantage. However, definitive proof to support this hypothesis is still lacking.

9.2.3 Primate Siglec-16

Siglec-16 is another Siglec that is coupled with inhibitory counterpart (in this case Siglec-11). The gene encoding human Siglec-16 (*SIGLEC16*) was initially considered a pseudogene (designated *SIGLECP16* or *SIGLEC16P*) because of a 4-nucleotide deletion in exon 2 (Angata et al. 2002), but an independent study correctly pointed out that this deletion in human *SIGLEC16* is polymorphic and a functional allele also exists (Cao et al. 2008). The latter study also demonstrated that Siglec-16 interacts with DAP12 and transduces cell-activating signal. Thus, Siglec-16 associates with DAP12, whereas Siglec-11 associates with SHP-1 and transduces inhibitory signal (Angata et al. 2002). These proteins show high sequence similarity, which is particularly high at the N-terminal three Ig-like domains (Hayakawa et al. 2005). A putative ortholog of *SIGLEC11* gene is found in multiple mammalian lineages, whereas *SIGLEC16* appears to be confined to primate lineage (Cao et al. 2008).

Similar to *SIGLEC14* and *SIGLEC5*, *SIGLEC16* and *SIGLEC11* genes are also located next to each other, but in this case the two genes are situated in an inverted (head-to-head) orientation (Angata et al. 2002). In the lineage leading to modern humans, these two genes have undergone several rounds of gene conversion events (Wang et al. 2012a). In addition, our recent study has revealed that *SIGLEC16* and *SIGLEC11* have also undergone concerted evolution in other lineages of primates (Hayakawa et al. 2017). The frequency of gene conversion events appears to be lower than that of *SIGLEC14/SIGLEC5* pair, likely because of the inverted orientation of the genes (Hayakawa et al. 2017). Siglec-11 was reported to preferentially recognize α2–8-linked oligosialic acids (Angata et al. 2002), and the gene conversions appear to have maintained this binding preference of Siglec-11 and Siglec-16 proteins (Hayakawa et al. 2017). These findings imply that the Siglec-11 and Siglec-16 have also been maintained as paired receptors under some selective pressure.

Siglec-11 and Siglec-16 are both expressed on myeloid cells, including brain microglia (Hayakawa et al. 2005; Cao et al. 2008). The expression of Siglec-11 (and/or Siglec-16) in brain is considered to be a human-specific event (Hayakawa et al. 2005), which was likely caused by a gene conversion by the promoter region of the non-functional allele of *SIGLEC16* gene (i.e., *SIGLEC16P*) (Wang et al. 2012a). Whereas earlier reports on the expression patterns of Siglec-11 (Hayakawa et al. 2005; Wang et al. 2011) using a monoclonal antibody against Siglec-11 may require re-evaluation due to the cross-reactivity of the antibody with Siglec-16, the expression of Siglec-11 and Siglec-16 on myeloid cells was confirmed with a set of mono-specific antibodies that recognize either Siglec-11 or Siglec-16 alone (Schwarz et al. 2017).

Curiously, the non-functional allele of *SIGLEC16* appears to be dominant in modern humans (Cao et al. 2008; Wang et al. 2012a; Hayakawa et al. 2017). A study showing genetic association of this polymorphism with any human disease is yet to be seen. Regardless, a functional study using mouse model, in which Siglec-E (an inhibitory Siglec widely expressed on myeloid cells) was converted to an activating-type "Siglec-E16" by replacing its native transmembrane-intracellular segment with that of human Siglec-16, revealed that the mice expressing Siglec-E16 is protected against K1 strain of *Escherichia coli*, a bacterial strain covered with capsular polysaccharide consisting of polysialic acids and causes meningitis in humans (Schwarz et al. 2017). Thus, an overall picture that was proposed for Siglec-14–Siglec-5 pair, that is, activating-type Siglec counters the pathogen that exploits inhibitory counterpart, appears to be applicable to Siglec-16–Siglec-11 pair as well. The risk of Siglec-16-mediated autoimmune reaction by recognition of SAMP may be mitigated by the presence of Siglec-11. In addition, low prevalence of oligo/polysialic acids in non-neuronal tissues compared with those of other linkages (i.e., α2-3 and α2-6-linked sialic acids) may also contribute to the avoidance of SAMP-mediated stimulation of innate immune system by Siglec-16 engagement [although the distribution of oligo/polysialic acids in mammalian tissues is much broader than it was once assumed in the past (Sato and Kitajima 2013; Colley et al. 2014)].

Another study using transgenic mice that express Siglec-11 in mononuclear phagocytes demonstrated that Siglec-11 has anti-inflammatory function by recognizing polysialic acids (Karlstetter et al. 2017). As inflammation in nervous system can lead to detrimental consequences (e.g., neurodegenerative disorders), the prevalence of non-functional *SIGLEC16* allele in modern humans may imply that the carrier of the allele had some selective advantage (Hayakawa et al. 2017). However, again, definitive proof to support this hypothesis is still lacking.

9.2.4 Primate Siglec-13 and Siglec-17

The gene encoding Siglec-13 (*SIGLEC13*) was eliminated from human genome, while it is present in the genomes of other primates (Angata et al. 2004). The deletion of this gene is uniform in all humans examined (Wang et al. 2012b). The same study demonstrated that chimpanzee Siglec-13 is expressed in myeloid cells, recognizes sialylated glycans, interacts with DAP12, and enhances inflammatory cytokine responses upon lipopolysaccharide (LPS; a ligand for Toll-like receptor (TLR) 4) stimulation of the myeloid cells that express it (Wang et al. 2012b). Paradoxically, stimulation of the same myeloid cells (RAW264.7 mouse macrophage cell line) that express chimpanzee Siglec-13 with sialylated bacterial pathogen resulted in the reduced production of pro-inflammatory cytokine (tumor necrosis factor α; TNFα), which is opposite to what was observed by the stimulation with LPS.

The same study revealed that the gene for primate Siglec-17 (gene symbol *SIGLEC17P*, originally designated *SIGLECP3*, in humans) is a pseudogene in human (due to 1 bp deletion in coding sequence) and many other Old World monkeys (due to independent events that rendered it non-functional or non-existent), but it is a functional gene in marmoset, a New World monkey (Wang et al. 2012b). The *SIGLEC17P* transcript is selectively expressed in natural killer cells in humans. The authors "resurrected" human Siglec-17 (i.e., re-created functional human Siglec-17 cDNA), which recognizes sialylated glycans, interacts with DAP12, and enhances inflammatory cytokine responses upon LPS stimulation of the myeloid cells that express it (Wang et al. 2012b).

Given that the glycan binding patterns of these Siglecs appear to be rather promiscuous (Wang et al. 2012b), the mechanisms that prevented these activating-type Siglecs from causing inflammatory responses against own cells is not clear. As mentioned above, chimpanzee Siglec-13 expression on myeloid cells enhanced TNFα production in response to LPS stimulation but suppressed the TNFα response to sialylated bacteria. A similar suppressive signal via DAP12-associated Siglec was reported for Siglec-H (as described below), implying that Siglec-13 may have acquired a new signaling modality to transduce suppressive signaling in a sialylated ligand-dependent manner. In either case, the pseudogenization of *SIGLEC13* and *SIGLEC17* appears to be selected for, possibly suggesting that the loss of these genes were favorable at some stage of primate evolution (Wang et al. 2012b).

9.2.5 Rodent Siglec-H

Mouse Siglec-H was the first Siglec to be reported as a DAP12-associated Siglec (Blasius et al. 2006). Siglec-H is a rodent-specific Siglec (Zhang et al. 2006) and shows unique expression on plasmacytoid dendritic cells (pDCs) (Blasius et al. 2004, 2006; Zhang et al. 2006), a subset of macrophages in spleen and in lymph nodes (Zhang et al. 2006), and microglia (Konishi et al. 2017). DAP12 is required for the cell-surface expression of Siglec-H (Blasius et al. 2006), which is similar to many other DAP12-associated proteins (Turnbull and Colonna 2007).

Functional studies using antibody have shown that Siglec-H is involved in the regulation of type I interferon (IFN) production in response to TLR9 and TLR7 engagement. However, unexpectedly, Siglec-H does not enhance the production of type I IFN but rather suppresses it (Blasius et al. 2004, 2006). This antibody-mediated suppression of type I IFN response appears selective, in that an independent study revealed that the pDC treatment with TLR9 ligand in the presence of anti-Siglec-H antibody did not alter the production of several other cytokines (TNF-α, interleukin (IL)-10, and IL-6) (Zhang et al. 2006).

Type I IFN is a key coordinator of the host defense against viral infection, and pDC is the primary source of type I IFN. Therefore, some studies have addressed whether Siglec-H plays any significant role in the immune reaction against viral infection. A study, with intention to investigate the role of pDCs in immunity, developed a mouse line that turned out to be deficient in Siglec-H (Takagi et al. 2011). This study revealed that Siglec-H deficiency leads to exaggerated activation of NF-κB pathway in pDCs, enhanced production of type I IFN and IL-12p40 in response to TLR9 stimulation in vivo, and stronger antigen-specific $CD4^+$ T cell responses and weaker antigen-specific $CD8^+$ T cell responses in vivo, as compared with control mice. The Siglec-H-deficient mice were less efficient in clearing herpes simplex virus (Takagi et al. 2011). Another study, using an independent line of Siglec-H deficient mice, revealed that the type I IFN production in response to mouse cytomegalovirus (mCMV) infection was indeed enhanced, but it did not result in enhanced clearance of the virus (Puttur et al. 2013). Interestingly, the Siglec-H-deficient mice develop autoimmune disease similar to lupus several weeks after the mCMV infection, in a type I IFN-dependent manner (Schmitt et al. 2016). Aging Siglec-H-deficient mice are also reported to develop mild autoimmunity (Schmitt et al. 2016).

Siglec-H has endocytic function (Zhang et al. 2006) that is dependent on DAP12 (Kopatz et al. 2013). This property was exploited to efficiently deliver antigen to pDCs for cross-presentation, facilitating the development of antigen-specific $CD8^+$ (cytotoxic) T cells (Zhang et al. 2006). On the other hand, another study found that the delivery of antigen to pDC via Siglec-H inhibits the development of $CD4^+$ (helper) T cell-dependent immunity (Loschko et al. 2011). Thus, the delivery of antigen to pDCs via Siglec-H appears to have opposite effects on $CD4^+$ T versus $CD8^+$ T cell-mediated immunity. These results appear to be in line with the phenotype of Siglec-H-deficient mice, which showed stronger antigen-specific $CD4^+$ T cell responses and weaker antigen-specific $CD8^+$ T cell responses compared with control mice (Takagi et al. 2011). The molecular mechanism behind this biased support for T cell subsets by Siglec-H is not fully understood. It appears possible that the antibody-mediated internalization of Siglec-H (mimicking Siglec-H deficiency), rather than the pDC-targeted antigen delivery, induced the phenotype observed.

Although mouse Siglec-H has canonical sequence features of Siglecs, including conserved arginine residue on the β-strand F of the first Ig-like domain that makes contact with the carboxyl group of sialic acid, glycan ligand for Siglec-H has not been found so far (Zhang et al. 2006). Given that rat Siglec-H lacks the conserved arginine residue essential for sialic acid recognition, it appears possible that rodent Siglec-H does not interact with glycans that contain sialic acid, but rather with some other ligand (Zhang et al. 2006). Whereas a study found that Siglec-H binds to glioma cell lines but not to normal astrocytes or other normal cells, the molecular entity of the ligand was not identified (Kopatz et al. 2013). If Siglec-H does not recognize common sialylated ligands (i.e., SAMPs), it is not likely to trigger autoimmunity against own tissues. Siglec-H may be also sequestered in an intracellular compartment in certain cell types (Zhang et al. 2006), which may prevent the contact between Siglec-H and its ligand.

The signal transduction mechanism by which DAP12-mediated signal from Siglec-H leads to suppression of type I IFN production is not fully understood. Based on previous findings that DAP12-deficient myeloid cells show enhanced responses to TLR agonists (Hamerman and Lanier 2006; Hamerman et al. 2005), it was proposed that low-degree Siglec-H engagement by antibody or endogenous ligand leads to sequestration of Siglec-H or recruitment of inhibitory mediator, whereas high-degree engagement by pathogen may lead to cell activation and enhanced type I IFN production (Blasius and Colonna 2006). In an alternative model proposed by the same authors, co-engagement of TLR and Siglec-H by pathogen may be required for strong type I IFN response (Blasius and Colonna 2006). Definitive proof for either of these models is yet to be published.

9.2.6 *Other Siglecs that Potentially Associate with DAP12*

Some other primate and rodent Siglecs have been implied to interact with DAP12. For example, mouse CD33/Siglec-3 was predicted to interact with DAP12, based on the sequence similarity in the transmembrane domain with that of mouse Siglec-H (Blasius et al. 2006). Mouse CD33/Siglec-3 is a putative ortholog of primate CD33/Siglec-3, but unlike primate CD33/Siglec-3 that has canonical ITIM and associates with SHP-1 (Paul et al. 2000; Taylor et al. 1999; Ulyanova et al. 1999), it lacks canonical ITIM (although it retains a membrane-distal ITIM-like motif). Several other genes/pseudogenes for CD33-related Siglecs in mammals were also predicted to have a conserved lysine residue in the transmembrane domain, which may be engaged in the interaction with DAP12 (Cao et al. 2008). Experimental confirmation of these predictions is awaited.

Sialoadhesin/Siglec-1 was also reported to be associated with DAP12 (Wu et al. 2016; Zheng et al. 2015), although the transmembrane domain of sialoadhesin/Siglec-1 lacks positively charged amino acid residue, and thus it is unlikely to directly interact with DAP12. Further investigation would be required to find out how sialoadhesin/Siglec-1 interacts with DAP12 (e.g., by way of another adapter protein that bridges between sialoadhesin/Siglec-1 and DAP12).

9.3 Summary

Siglecs that associate with DAP12 have diverse functions, ranging from bone homeostasis, pathogen recognition to the regulation of adaptive immune responses, as described in the previous section. Evolutionary processes appear to have employed several mechanisms to prevent self-inflicted injury by the immune system (i.e., autoimmunity) triggered by the engagement of DAP12-associated Siglecs by sialylated glycans (SAMP), such as the sequestration of Siglec into intracellular compartment (Siglec-15 and Siglec-H), restricted distribution of sialylated ligands (Siglec-15), or loss of binding to common sialylated ligands (Siglec-H), coupling with inhibitory counterpart (Siglec-14 and Siglec-16), acquisition of alternative signaling modality (Siglec-13 and Siglec-H), and ultimately, pseudogenization or loss of the gene encoding DAP12-associated Siglec (Siglec-13, -14, -16, -17) (Fig. 9.2).

Despite the efforts by several groups in the past decade, many unanswered questions remain regarding the functions of DAP12-associated Siglecs and the molecular mechanisms they use for immune regulations. For example, biologically relevant ligands for these Siglecs are not clearly identified in many cases. Such information may shed new light on the function of these Siglecs. Further studies on the genetic association of polymorphisms of human Siglecs would be necessary to further reveal appropriate biological contexts in which these Siglecs play significant roles. Identification of intracellular molecular partner(s) in addition to DAP12 may also allow us to better understand how these Siglecs operate (e.g., the mechanism by which

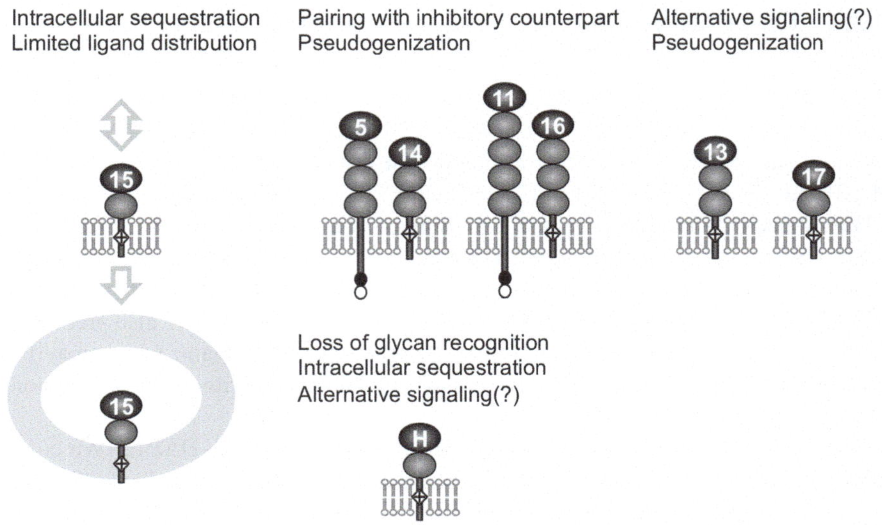

Fig. 9.2 Possible mechanisms that prevent the development of autoimmunity through the recognition of sialylated glycans by DAP12-associated Siglecs. Some potential mechanisms that may prevent self-oriented attacks (autoimmunity) by the immune cells triggered by the interaction between sialylated glycans (self-associated molecular patterns) and DAP12-associated Siglecs are illustrated. Such mechanisms include sequestration into intracellular compartments (Siglec-15 and Siglec-H), restricted distribution of sialylated ligands (Siglec-15) or loss of binding toward sialylated ligands (Siglec-H), coupling with inhibitory counterpart (Siglec-14 and Siglec-16), altered signaling modality (Siglec-13 and Siglec-H), and pseudogenization or gene deletion (Siglec-13, -14, -16, -17)

primate Siglec-13 and rodent Siglec-H transduce suppressive regulatory signaling). Further investigations of DAP12-associated Siglecs will lead to deeper understanding of Siglec family as a whole, and potentially lead to clinical translations.

Acknowledgements The work in the author's laboratory has been supported by intramural funding from Academia Sinica and extramural funding from the Ministry of Science and Technology, Taiwan [MOST 104-2311-B-001-017-MY3, 105-2627-M-007-001, and 106-2321-B-001-032].

References

Ali S, Fong J, Carlin A, Busch T, Linden R, Angata T, Areschoug T, Parast M, Varki N, Murray J, Nizet V, Varki A (2014) Siglec-5 and Siglec-14 are polymorphic paired receptors that modulate neutrophil and amnion signaling responses to group B Streptococcus. J Exp Med 211:1231–1242

Angata T (2006) Molecular diversity and evolution of the Siglec family of cell-surface lectins. Mol Divers 10:555–566

Angata T (2017) Polymorphisms and mutations in SIGLEC genes and their associations with diseases. J Jpn Biochem Soc 89:652–659

Angata T, Varki A (2002) Chemical diversity in the sialic acids and related alpha-keto acids: an evolutionary perspective. Chem Rev 102:439–470

Angata T, Kerr S, Greaves D, Varki N, Crocker P, Varki A (2002) Cloning and characterization of human Siglec-11. A recently evolved signaling molecule that can interact with SHP-1 and SHP-2 and is expressed by tissue macrophages, including brain microglia. J Biol Chem 277:24466–24474

Angata T, Margulies E, Green E, Varki A (2004) Large-scale sequencing of the CD33-related Siglec gene cluster in five mammalian species reveals rapid evolution by multiple mechanisms. Proc Natl Acad Sci U S A 101:13251–13256

Angata T, Hayakawa T, Yamanaka M, Varki A, Nakamura M (2006) Discovery of Siglec-14, a novel sialic acid receptor undergoing concerted evolution with Siglec-5 in primates. FASEB J 20:1964–1973

Angata T, Tabuchi Y, Nakamura K, Nakamura M (2007) Siglec-15: an immune system Siglec conserved throughout vertebrate evolution. Glycobiology 17:838–846

Angata T, Ishii T, Motegi T, Oka R, Taylor R, Soto P, Chang Y, Secundino I, Gao C, Ohtsubo K, Kitazume S, Nizet V, Varki A, Gemma A, Kida K, Taniguchi N (2013) Loss of Siglec-14 reduces the risk of chronic obstructive pulmonary disease exacerbation. Cell Mol Life Sci 70:3199–3210

Blasius A, Colonna M (2006) Sampling and signaling in plasmacytoid dendritic cells: the potential roles of Siglec-H. Trends Immunol 27:255–260

Blasius A, Vermi W, Krug A, Facchetti F, Cella M, Colonna M (2004) A cell-surface molecule selectively expressed on murine natural interferon-producing cells that blocks secretion of interferon-alpha. Blood 103:4201–4206

Blasius A, Cella M, Maldonado J, Takai T, Colonna M (2006) Siglec-H is an IPC-specific receptor that modulates type I IFN secretion through DAP12. Blood 107:2474–2476

Blaum BS (2017) The lectin self of complement factor H. Curr Opin Struct Biol 44:111–118

Blaum BS, Hannan JP, Herbert AP, Kavanagh D, Uhrin D, Stehle T (2015) Structural basis for sialic acid-mediated self-recognition by complement factor H. Nat Chem Biol 11:77–82

Cao H, Lakner U, de Bono B, Traherne J, Trowsdale J, Barrow A (2008) SIGLEC16 encodes a DAP12-associated receptor expressed in macrophages that evolved from its inhibitory counterpart SIGLEC11 and has functional and non-functional alleles in humans. Eur J Immunol 38:2303–2315

Carlin A, Chang Y, Areschoug T, Lindahl G, Hurtado-Ziola N, King C, Varki A, Nizet V (2009) Group B Streptococcus suppression of phagocyte functions by protein-mediated engagement of human Siglec-5. J Exp Med 206:1691–1699

Chang L, Chen YJ, Fan CY, Tang CJ, Chen YH, Low PY, Ventura A, Lin CC, Chen YJ, Angata T (2017) Identification of Siglec ligands using a proximity labeling method. J Proteome Res 16:3929–3941

Colley KJ, Kitajima K, Sato C (2014) Polysialic acid: biosynthesis, novel functions and applications. Crit Rev Biochem Mol Biol 49:498–532

Cornish A, Freeman S, Forbes G, Ni J, Zhang M, Cepeda M, Gentz R, Augustus M, Carter K, Crocker P (1998) Characterization of siglec-5, a novel glycoprotein expressed on myeloid cells related to CD33. Blood 92:2123–2132

Crocker P, Clark E, Filbin M, Gordon S, Jones Y, Kehrl J, Kelm S, le Douarin N, Powell L, Roder J, Schnaar R, Sgroi D, Stamenkovic K, Schauer R, Schachner M, van den Berg T, van der Merwe P, Watt S, Varki A (1998) Siglecs: a family of sialic-acid binding lectins [letter]. Glycobiology 8:v

Duong BH, Tian H, Ota T, Completo G, Han S, Vela JL, Ota M, Kubitz M, Bovin N, Paulson JC, Nemazee D (2010) Decoration of T-independent antigen with ligands for CD22 and Siglec-G can suppress immunity and induce B cell tolerance in vivo. J Exp Med 207:173–187

Graustein AD, Horne DJ, Fong JJ, Schwarz F, Mefford HC, Peterson GJ, Wells RD, Musvosvi M, Shey M, Hanekom WA, Hatherill M, Scriba TJ, Thuong NTT, Mai NTH, Caws M, Bang ND, Dunstan SJ, Thwaites GE, Varki A, Angata T, Hawn TR (2017) The SIGLEC14 null allele is associated with *Mycobacterium tuberculosis*- and BCG-induced clinical and immunologic outcomes. Tuberculosis (Edinb) 104:38–45

Hamerman JA, Lanier LL (2006) Inhibition of immune responses by ITAM-bearing receptors. Sci STKE re1

Hamerman JA, Tchao NK, Lowell CA, Lanier LL (2005) Enhanced Toll-like receptor responses in the absence of signaling adaptor DAP12. Nat Immunol 6:579–586

Hayakawa T, Angata T, Lewis A, Mikkelsen T, Varki N, Varki A (2005) A human-specific gene in microglia. Science 309:1693

Hayakawa T, Khedri Z, Schwarz F, Landig C, Liang SY, Yu H, Chen X, Fujito NT, Satta Y, Varki A, Angata T (2017) Coevolution of Siglec-11 and Siglec-16 via gene conversion in primates. BMC Evol Biol 17:228

Hiruma Y, Hirai T, Tsuda E (2011) Siglec-15, a member of the sialic acid-binding lectin, is a novel regulator for osteoclast differentiation. Biochem Biophys Res Commun 409:424–429

Hiruma Y, Tsuda E, Maeda N, Okada A, Kabasawa N, Miyamoto M, Hattori H, Fukuda C (2013) Impaired osteoclast differentiation and function and mild osteopetrosis development in Siglec-15-deficient mice. Bone 53:87–93

Ishida-Kitagawa N, Tanaka K, Bao X, Kimura T, Miura T, Kitaoka Y, Hayashi K, Sato M, Maruoka M, Ogawa T, Miyoshi J, Takeya T (2012) Siglec-15 protein regulates formation of functional osteoclasts in concert with DNAX-activating protein of 12 kDa (DAP12). J Biol Chem 287:17493–17502

Julien S, Videira PA, Delannoy P (2012) Sialyl-tn in cancer: (how) did we miss the target? Biomolecules 2:435–466

Kameda Y, Takahata M, Komatsu M, Mikuni S, Hatakeyama S, Shimizu T, Angata T, Kinjo M, Minami A, Iwasaki N (2013) Siglec-15 regulates osteoclast differentiation by modulating RANKL-induced phosphatidylinositol 3-kinase/Akt and Erk pathways in association with signaling Adaptor DAP12. J Bone Miner Res 28:2463–2475

Kameda Y, Takahata M, Mikuni S, Shimizu T, Hamano H, Angata T, Hatakeyama S, Kinjo M, Iwasaki N (2015) Siglec-15 is a potential therapeutic target for postmenopausal osteoporosis. Bone 71:217–226

Karlstetter M, Kopatz J, Aslanidis A, Shahraz A, Caramoy A, Linnartz-Gerlach B, Lin Y, Luckoff A, Fauser S, Duker K, Claude J, Wang Y, Ackermann J, Schmidt T, Hornung V, Skerka C, Langmann T, Neumann H (2017) Polysialic acid blocks mononuclear phagocyte reactivity, inhibits complement activation, and protects from vascular damage in the retina. EMBO Mol Med 9:154–166

Kerrigan AM, Brown GD (2011) Syk-coupled C-type lectins in immunity. Trends Immunol 32:151–156

Konishi H, Kobayashi M, Kunisawa T, Imai K, Sayo A, Malissen B, Crocker PR, Sato K, Kiyama H (2017) Siglec-H is a microglia-specific marker that discriminates microglia from CNS-associated macrophages and CNS-infiltrating monocytes. Glia 65:1927–1943

Kopatz J, Beutner C, Welle K, Bodea LG, Reinhardt J, Claude J, Linnartz-Gerlach B, Neumann H (2013) Siglec-h on activated microglia for recognition and engulfment of glioma cells. Glia 61:1122–1133

Lajaunias F, Dayer J, Chizzolini C (2005) Constitutive repressor activity of CD33 on human monocytes requires sialic acid recognition and phosphoinositide 3-kinase-mediated intracellular signaling. Eur J Immunol 35:243–251

Loschko J, Heink S, Hackl D, Dudziak D, Reindl W, Korn T, Krug AB (2011) Antigen targeting to plasmacytoid dendritic cells via Siglec-H inhibits Th cell-dependent autoimmunity. J Immunol 187:6346–6356

Lowell CA (2011) Src-family and Syk kinases in activating and inhibitory pathways in innate immune cells: signaling cross talk. Cold Spring Harb Perspect Biol 3

Macauley M, Crocker P, Paulson J (2014) Siglec-mediated regulation of immune cell function in disease. Nat Rev Immunol 14:653–666

Meri S, Pangburn MK (1990) Discrimination between activators and nonactivators of the alternative pathway of complement: regulation via a sialic acid/polyanion binding site on factor H. Proc Natl Acad Sci U S A 87:3982–3986

Ogata S, Ho I, Chen A, Dubois D, Maklansky J, Singhal A, Hakomori S, Itzkowitz SH (1995) Tumor-associated sialylated antigens are constitutively expressed in normal human colonic mucosa. Cancer Res 55:1869–1874

Paloneva J, Kestila M, Wu J, Salminen A, Bohling T, Ruotsalainen V, Hakola P, Bakker AB, Phillips JH, Pekkarinen P, Lanier LL, Timonen T, Peltonen L (2000) Loss-of-function mutations in TYROBP (DAP12) result in a presenile dementia with bone cysts. Nat Genet 25:357–361

Paloneva J, Manninen T, Christman G, Hovanes K, Mandelin J, Adolfsson R, Bianchin M, Bird T, Miranda R, Salmaggi A, Tranebjaerg L, Konttinen Y, Peltonen L (2002) Mutations in two genes encoding different subunits of a receptor signaling complex result in an identical disease phenotype. Am J Hum Genet 71:656–662

Paloneva J, Mandelin J, Kiialainen A, Bohling T, Prudlo J, Hakola P, Haltia M, Konttinen YT, Peltonen L (2003) DAP12/TREM2 deficiency results in impaired osteoclast differentiation and osteoporotic features. J Exp Med 198:669–675

Paul S, Taylor L, Stansbury E, McVicar D (2000) Myeloid specific human CD33 is an inhibitory receptor with differential ITIM function in recruiting the phosphatases SHP-1 and SHP-2. Blood 96:483–490

Pillai S, Netravali I, Cariappa A, Mattoo H (2012) Siglecs and immune regulation. Annu Rev Immunol 30:357–392

Ponta H, Sherman L, Herrlich PA (2003) CD44: from adhesion molecules to signalling regulators. Nat Rev Mol Cell Biol 4:33–45

Puttur F, Arnold-Schrauf C, Lahl K, Solmaz G, Lindenberg M, Mayer CT, Gohmert M, Swallow M, van Helt C, Schmitt H, Nitschke L, Lambrecht BN, Lang R, Messerle M, Sparwasser T (2013) Absence of Siglec-H in MCMV infection elevates interferon alpha production but does not enhance viral clearance. PLoS Pathog 9:e1003648

Sato C, Kitajima K (2013) Disialic, oligosialic and polysialic acids: distribution, functions and related disease. J Biochem 154:115–136

Schmitt H, Sell S, Koch J, Seefried M, Sonnewald S, Daniel C, Winkler TH, Nitschke L (2016) Siglec-H protects from virus-triggered severe systemic autoimmunity. J Exp Med 213:1627–1644

Schwarz F, Landig CS, Siddiqui S, Secundino I, Olson J, Varki N, Nizet V, Varki A (2017) Paired Siglec receptors generate opposite inflammatory responses to a human-specific pathogen. EMBO J 36:751–760

Shimizu T, Takahata M, Kameda Y, Endo T, Hamano H, Hiratsuka S, Ota M, Iwasaki N (2015) Sialic acid-binding immunoglobulin-like lectin 15 (Siglec-15) mediates periarticular bone loss, but not joint destruction, in murine antigen-induced arthritis. Bone 79:65–70

Shultz LD, Rajan TV, Greiner DL (1997) Severe defects in immunity and hematopoiesis caused by SHP-1 protein-tyrosine-phosphatase deficiency. Trends Biotechnol 15:302–307

Stuible M, Moraitis A, Fortin A, Saragosa S, Kalbakji A, Filion M, Tremblay G (2014) Mechanism and function of monoclonal antibodies targeting siglec-15 for therapeutic inhibition of osteoclastic bone resorption. J Biol Chem 289:6498–6512

Takagi H, Fukaya T, Eizumi K, Sato Y, Sato K, Shibazaki A, Otsuka H, Hijikata A, Watanabe T, Ohara O, Kaisho T, Malissen B, Sato K (2011) Plasmacytoid dendritic cells are crucial for the initiation of inflammation and T cell immunity in vivo. Immunity 35:958–971

Takamiya R, Ohtsubo K, Takamatsu S, Taniguchi N, Angata T (2013) The interaction between Siglec-15 and tumor-associated sialyl-Tn antigen enhances TGF-beta secretion from monocytes/macrophages through the DAP12-Syk pathway. Glycobiology 23:178–187

Taylor V, Buckley C, Douglas M, Cody A, Simmons D, Freeman S (1999) The myeloid-specific sialic acid-binding receptor, CD33, associates with the protein-tyrosine phosphatases, SHP-1 and SHP-2. J Biol Chem 274:11505–11512

Turnbull I, Colonna M (2007) Activating and inhibitory functions of DAP12. Nat Rev Immunol 7:155–161

Ulyanova T, Blasioli J, Woodford-Thomas T, Thomas M (1999) The sialoadhesin CD33 is a myeloid-specific inhibitory receptor. Eur J Immunol 29:3440–3449

Varki A (2011) Since there are PAMPs and DAMPs, there must be SAMPs? Glycan "self-associated molecular patterns" dampen innate immunity, but pathogens can mimic them. Glycobiology 21:1121–1124

Varki A, Angata T (2006) Siglecs—the major subfamily of I-type lectins. Glycobiology 16:1R–27R

Wang X, Chow R, Deng L, Anderson D, Weidner N, Godwin A, Bewtra C, Zlotnik A, Bui J, Varki A, Varki N (2011) Expression of Siglec-11 by human and chimpanzee ovarian stromal cells, with uniquely human ligands: implications for human ovarian physiology and pathology. Glycobiology 21:1038–1048

Wang X, Mitra N, Cruz P, Deng L, Varki N, Angata T, Green E, Mullikin J, Hayakawa T, Varki A (2012a) Evolution of siglec-11 and siglec-16 genes in hominins. Mol Biol Evol 29:2073–2086

Wang X, Mitra N, Secundino I, Banda K, Cruz P, Padler-Karavani V, Verhagen A, Reid C, Lari M, Rizzi E, Balsamo C, Corti G, de Bellis G, Longo L, Beggs W, Caramelli D, Tishkoff S, Hayakawa T, Green E, Mullikin J, Nizet V, Bui J, Varki A (2012b) Specific inactivation of two immunomodulatory SIGLEC genes during human evolution. Proc Natl Acad Sci U S A 109:9935–9940

Wu Y, Lan C, Ren D, Chen GY (2016) Induction of Siglec-1 by endotoxin tolerance suppresses the innate immune response by promoting TGF-beta1 production. J Biol Chem 291:12370–12382

Yamanaka M, Kato Y, Angata T, Narimatsu H (2009) Deletion polymorphism of SIGLEC14 and its functional implications. Glycobiology 19:841–846

Zhang J, Raper A, Sugita N, Hingorani R, Salio M, Palmowski M, Cerundolo V, Crocker P (2006) Characterization of Siglec-H as a novel endocytic receptor expressed on murine plasmacytoid dendritic cell precursors. Blood 107:3600–3608

Zheng Q, Hou J, Zhou Y, Yang Y, Xie B, Cao X (2015) Siglec1 suppresses antiviral innate immune response by inducing TBK1 degradation via the ubiquitin ligase TRIM27. Cell Res 25:1121–1136

Zoller M (2011) CD44: can a cancer-initiating cell profit from an abundantly expressed molecule? Nat Rev Cancer 11:254–267

The manufacturer's authorised representative in the EU is Springer Nature Customer Service Centre GmbH, Europaplatz 3, 69115 Heidelberg, Germany. If you have any concerns regarding our products, please contact ProductSafety@springernature.com

Printed and bound by CPI Group (UK) Ltd, Croydon, CR0 4YY
30/09/2026
02225458-0008